STRIVE FOR A 5:

PREPARING FOR THE AP* BIOLOGY EXAM

to accompany

PRINCIPLES OF LIFE
Hillis, Sadava, Heller, Price

John Lepri

Franklin Bell

W. H. Freeman, New York City

ISBN-13: 978-1-4292-9849-0
ISBN-10: 1-4292-9849-9

Printed in the United States of America

Second Printing

W. H. Freeman and Company
41 Madison Avenue
New York, NY 10010
Houndmills, Basingstoke RG21 6XS, England

www.whfreeman.com

CONTENTS

Test Preparation Section

Preface

This book, *Strive for a 5: Preparing for the AP* Biology Examination*, is designed for use with Hillis, Sadava, Heller, and Price's *Principles of Life.* It is intended to help you evaluate your understanding of the material covered in the textbook, reinforce the key concepts you need to learn, and prepare you to take the AP Biology exam.

This book is divided into two sections: a study guide section and a test preparation section.

The Study Guide Section

The study guide section is designed for you to use throughout your AP course. As each chapter is covered in your class, use the study guide to help you identify and reinforce the important concepts in biology.

For each chapter, the study guide contains a general overview of the chapter, an overview of each numbered concept in the chapter, and a set of study questions for you to complete as you review each concept. These study questions are intended to make you apply what you have learned and to use your knowledge of biology. Very few of them are simple recall knowledge questions. We suggest that you read over the questions in the study guide before you read the chapter. If you find that the questions are difficult and require you to think, then we have succeeded! Many students find it helpful to form study groups to work out the answers to these questions. The more practice you do for each chapter, the better you will be prepared for success on the AP Biology exam.

Each chapter concludes with a question that ties material in the chapter with the Science Practices. These questions run the gamut from short and long free response questions to grid-in calculation items.

Answers to all of the study questions are found under the Adopted Resources tab at the *Principles of Life* Book Companion Site: bcs.whfreeman.com/POL1e

Give yourself some quality "Biology" time every single day, even if it's only 30 minutes per day, to work on reading the text according to your teacher's directions and completing the problems. You can work the *Strive* questions after completing your studies of each concept, or wait until you reach the end of the chapter, and do the whole chapter at once.

The Test Preparation Section

It is a good idea to read through the exam prep section early in the course so that you have a solid understanding of what you are preparing for from the start. To help you prepare to take a full, timed exam, this guide features a full-length practice exam. After completing the exam, you can check the answers in the back of the book. Be sure to look closely at the questions that you answered incorrectly and spend extra time with the concepts covered in those questions.

Best of luck on the AP exam!

John Lepri
Franklin Bell

Chapter 1: Principles of Life

Chapter Outline

1.1 - Living organisms share common aspects of structure, function, and energy flow.
1.2 - Genetic systems control the flow, exchange, storage, and use of information.
1.3 - Organisms interact with and affect their environments.
1.4 - Evolution explains both the unity and diversity of life.
1.5 - Science is based on quantifiable observations and experiments.

Living organisms share many common aspects as a result of having evolved from a common ancestor. This chapter provides you with an overview of life, genetics, and evolution that will help you to understand life and how scientists look at the world. In the laboratory portion of this course, you will be doing many activities and laboratories using inquiry that will help you to think like a scientist. The science practices at the end of each chapter will assist you to establish lines of evidence and use them to develop and refine testable explanations and predictions of natural phenomena.

Chapter One ties principally with **Big Idea 1: The process of evolution drives the diversity and unity of life** in the AP Biology Curriculum Framework**.** Below are the essential knowledge areas found in Chapter One:

1.a.1 Natural selection is a major mechanism of evolution.
1.b.1 Organisms share many conserved core processes and features that evolved and are widely distributed among organisms today.
1.b.2 Phylogenetic trees and cladograms are graphical representations (models) of evolutionary history that can be tested.
1.d.1 There are several hypotheses about the natural origin of life on Earth, each with supporting scientific evidence.
1.d.2 Scientific evidence from many different disciplines supports models of the origin of life.

Chapter Review

Section 1.1 is an overview of living organisms and how they are connected by their shared traits.

1. Organisms share many conserved processes that are widely distributed among organisms today. Briefly outline the distinctive characteristics of life shared by all living organisms.

a. ____

b. ____

c. ____

d. ____

e. ____

f. ____

g. ____

h. ____

2. How do the above shared characteristics (among widely distributed organisms) provide evidence for evolution?

____
____
____
____
____
____
____
____

3. There are several hypotheses about the evolution of early life on Earth. Once the first cells evolved, they needed energy and raw materials for metabolism. Briefly explain how the first cells found on the earth obtained these nutrients and raw materials.

____
____
____
____
____
____
____
____

4. Briefly describe and discuss how living organisms have altered the oxygen content of the atmosphere over the past three billion years.

____
____
____
____
____
____
____
____

5. Phylogenetic trees and cladograms are graphical representations (models) of evolutionary history that can be tested. In the figure below, label the two vertical boxes as "mitochondria" or "chloroplasts" at the arrows representing "endosymbiotic events." In each of the 3 horizontal boxes, write the name of the domain for that group of organisms.

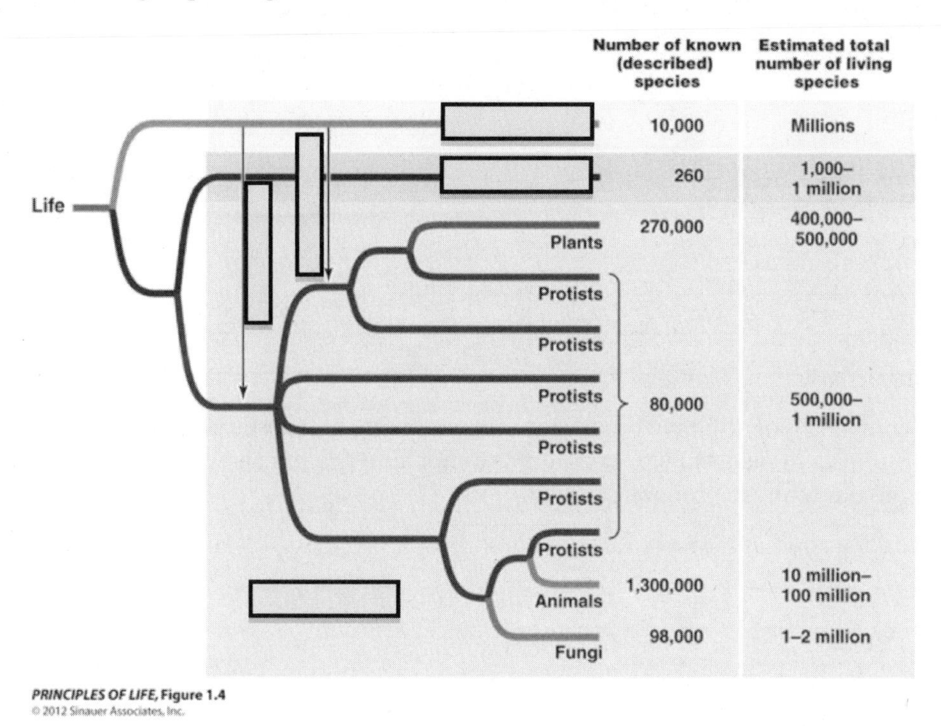

PRINCIPLES OF LIFE, Figure 1.4

Section 1.2 outlines some basic genetics to help you know how information is transmitted from one generation to the next.

6. In the figure to the right, label: DNA, nucleotide, gene, protein.

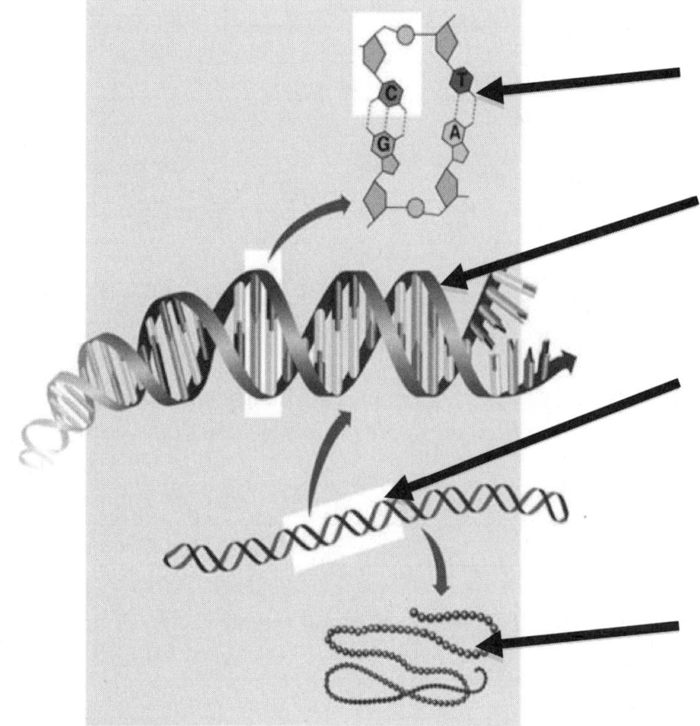

PRINCIPLES OF LIFE, Figure 1.5

7. Each of the cells in an individual mouse contain the same genes, but the mouse has many different types of cells: for example, muscle cells, nerve cells, skin cells, and many more. Describe how cells that have identical DNA and genes can become different types of cells.

____

____

____

Section 1.3 explains how all living organisms interact with the environment and with other organisms.

8. Define homeostasis:

____

____

____

9. Regulation is a key component of all living organisms in maintaining homeostasis. Explain how the opening and closing of stomata in plants (a feedback loop) resembles a heating and cooling system that keeps the temperature of your room comfortable.

____

____

____

10. The interactions of organisms with others is a topic of primary interest in the field of ecology. List 3 specific examples of how organisms interact with one another, and describe the interaction as competition, predation or cooperation.

a. ____

____

b. ____

____

c. ____

____

Section 1.4 discusses how evolution is the central unifying theme of biology and provides a framework for organizing how we think about living systems.

11. The term "theory" is an important vocabulary word in science and is often used differently in common usage and in the scientific community. How do scientists define a theory?

____

____

____

12. Explain how evolution is both fact and theory.

____

____

____

Section 1.5 focuses on how science is based on experimentation involving data collection and observation. Scientists are guided in their work by the principles of experimental design as they work to uncover the aspects of our natural world.

13. Biologist Tyrone Hayes and his co-workers investigated the effects of the herbicide atrazine on sexual development in frogs. In their experiments, they exposed each group of tadpoles to a specific amount of atrazine, and they repeated each experiment multiple times for each treatment. Their observations suggested that frogs exposed to atrazine early in life developed multiple, mixed gonads or became de-masculinized as a result.

Below is an excerpt from the design of his experiment from the original paper found at http://www.pnas.org/content/99/8/5476.full.pdf+html

> In experiment 1, we exposed larvae to atrazine at nominal concentrations of 0.01, 0.1, 1.0, 10.0, and 25 parts per billion (ppb). Concentrations were confirmed by two independent laboratories (PTRL West, Richmond, CA, and the Iowa Hygienic Laboratory, Univ. of Iowa, Iowa City, IO). All stock solutions were made in ethanol (10 ml), mixed in 15-gallon containers, and dispensed into treatment tanks. Controls were treated with ethanol such that all tanks contained 0.004% ethanol. Water was changed and treatments were renewed once every 72 h. Each treatment was replicated 3 times with 30 animals per replicate (total of 90 animals per treatment) in both experiments. All treatments were systematically rotated around the shelf every 3 days to ensure that no one treatment or no one tank experienced position effects. Experiments were carried out at 22°C with animals under a 12-h-12-h light-dark cycle (lights on at 6 a.m.).

Every experiment has several well-defined elements. Identify the elements below found in the atrazine experiment.

a. Independent Variable: ______________________________

b. Range of the Independent Variable: ______________________________

c. Dependent Variable: ______________________________

d. Control: ______________________________

e. Constants: ______________________________

f. Repeated Trials: ______________________________

Science Practices & Inquiry

A key component of any biology class is the laboratory experience. Science students must be able to design and plan experiments using inquiry. In the AP Biology Curriculum Framework, there is a set of 7 Science Practices. In this chapter, we will focus on **Science Practice 4: The student can plan and implement data collection strategies appropriate to a particular scientific question.**

Specifically, question #14 asks you to *justify the selection of the kind of data* needed to answer a particular scientific question, *design a plan* for collecting data to answer a particular scientific question, *collect data* to answer a particular scientific question, and to *evaluate sources of data* to answer a particular scientific question. (LO's 4.1, 4.2, 4.3, and 4.4)

14. After looking at the many types of fish food available at a local store, a friend asked you if flake fish food or shrimp pellets would make his goldfish grow faster. Design a controlled experiment to test this question.

> **Hint:** You are not simply writing a procedure. You should discuss all of the elements above in question #10 in paragraph form in sentences. Begin by identifying your independent and dependent variables and writing a hypothesis. This experiment will utilize many goldfish in separate containers.

Chapter 2: Life Chemistry and Energy

Chapter Outline

2.1 - Atomic Structure Is the Basis for Life's Chemistry
2.2 - Atoms Interact and Form Molecules
2.3 - Carbohydrates Consist of Sugar Molecules
2.4 - Lipids Are Hydrophobic Molecules
2.5 - Biochemical Changes Involve Energy

Living organisms such as birds and fish are made up of cells, and these cells are collections of molecules that work together. Interacting atoms make up the molecules, and it is necessary for you to understand a few details about atoms and molecules if you are going to be able to understand life. All life exists at the expense of its surrounding environment and is dependent on biochemical transformations of matter. These transformations occur within the laws of thermodynamics, specifying that energy is neither created nor destroyed and that disorder (entropy) increases during transformations.

Chapter 2 continues consideration of **Big Idea 1**, evolution, and Chapter 2 also begins your exploration of **Big Idea 2**, wherein you examine energy use by cells as you begin to catalogue the molecular building blocks of life processes. The following list includes these specific parts (essential knowledge) of the AP-Biology curriculum that are covered in Chapter 2.

1.d.2: Scientific evidence from many different disciplines supports models of the origin of life.
2.a.1: All living systems require constant input of free energy.
2.a.3: Organisms must exchange matter with the environment to grow, reproduce, and maintain organization.
4.a.1: The subcomponents of biological molecules and their sequence determine the properties of that molecule.
4.b.1: Interactions between molecules affect their structure and function.
4.c.1: Variation in molecular units provides cells with a wider range of functions.

Chapter Review

Concept 2.1 reviews some details about atomic structure in order to understand how molecules function in living organisms.

1. For each of the following, provide the number of electrons, protons, neutrons and atomic number in its elemental form.

	electrons	protons	neutrons	atomic number
a. hydrogen	________	________	________	________
b. carbon	________	________	________	________
c. oxygen	________	________	________	________
d. phosphorus	________	________	________	________

Concept 2.2 explains how molecules result from interactions between atoms.

2. Place each of these types of atomic interactions on the list below, based on the strength of the atomic interaction: van der Waals forces, covalent bonds, hydrogen bonds, ionic bonds.

strongest → weakest

________________ > ________________ > ________________ > ________________

3. Define "cation" and "anion."

cation: ____
____

anion: ____
____

4. Use the example of sodium chloride to explain how the electrons of these two atoms are redistributed when these two atoms interact with each other.

____
____
____
____

5. Name the molecule shown by the two models:

____

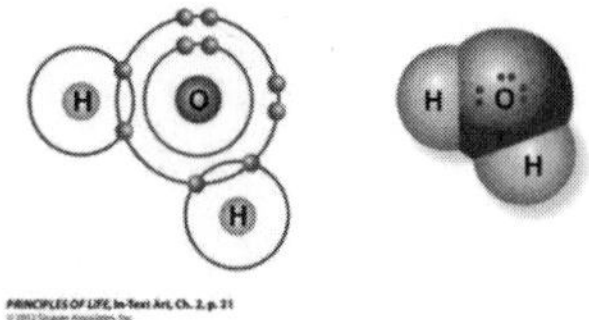

Explain how the electrons of these atoms are affected by their atomic interaction, and describe what this does to the distribution of charge around the molecule.

____
____
____
____

6. Drawings (A) and (B) are shown at different magnifications, and represent three molecules, two of which are interacting with each other, and the other which is interacting with itself. Explain the interactions in (A) and then in (B) and then explain why you think (A) and (B) represent the same or different numbers of atoms.

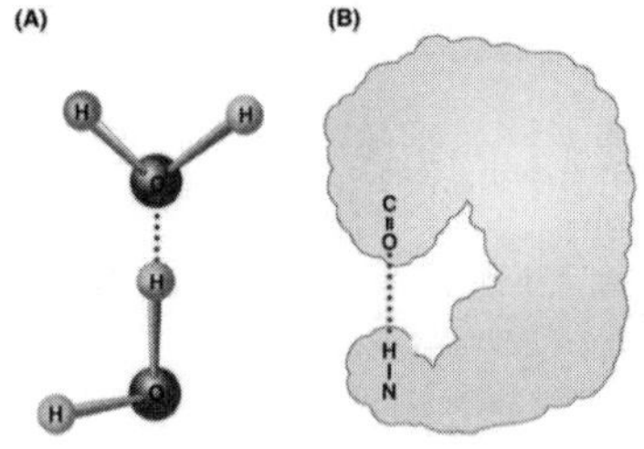

(A)
____
____

(B) ____
____

More atoms are represented in drawing_______ because____________________
____
____

7. These two chemicals found in the body differ in their solubility in water: one is quite soluble in water, and the other is much less soluble. Explain using the prompts below.

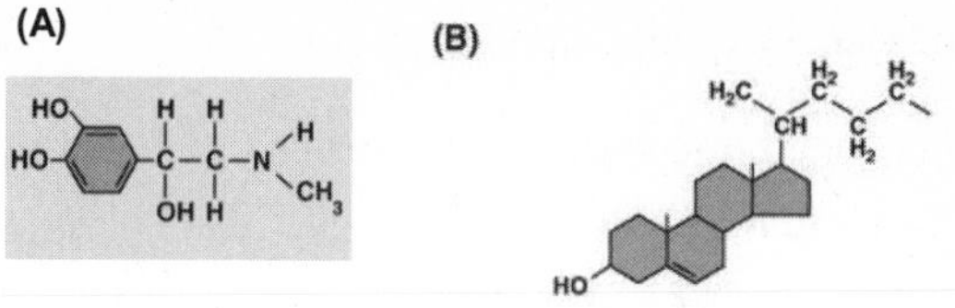

(A) (B)

Choice________is more water-soluble because______________________________________

____

____

Choice________is less water-soluble because_______________________________________

____

____

Concept 2.3 explains how carbohydrates, or sugar molecules, yield chemical energy when taken apart. Many organisms, including plants, catabolize (take apart) glucose and other sugars to liberate energy for their own use. Plants, of course, also synthesize sugars, by using solar energy and environmental sources of carbon dioxide and water.

8. Fill in the blanks: Solar energy drives _____________ in green plants, resulting in the synthesis of ___________, a monosaccharide. Sucrose is a disaccharide resulting from the formation of a __________ linkage between two monosaccharides. The starch molecule, also known as __________, is an even larger polymer of these products of these synthetic processes, and the most abundant member of this group on earth is ___________.

9. Number the un-numbered carbons and provide the names of each of these two monosaccharides.

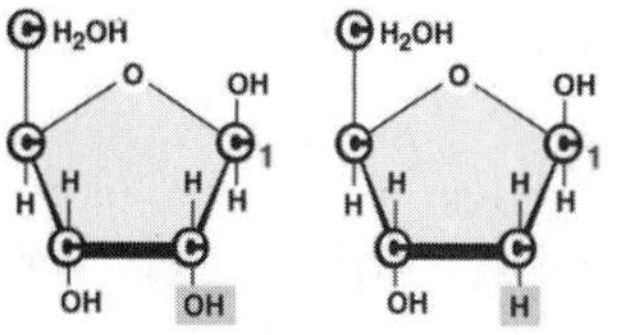

________________ ________________

Concept 2.4 discusses lipids (fats), which are energy-rich storage molecules.

10. Provide labels for the 4 different areas of the molecule, indicated by the 4 different shades on both representations (two models are shown).

The hydrophobic tail includes

_________________________________.

The hydrophilic head includes

_________________________________.

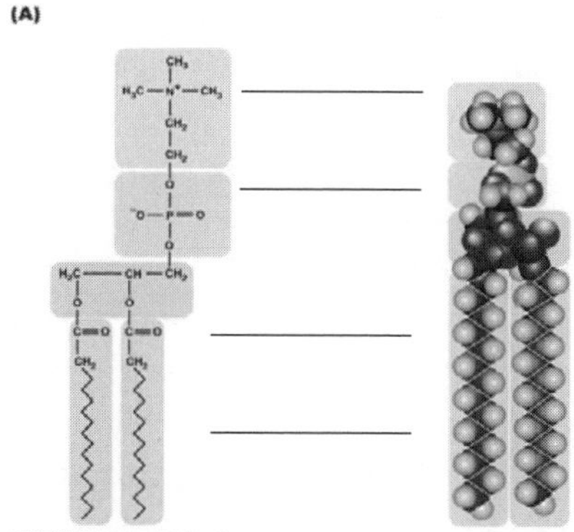

PRINCIPLES OF LIFE, Figure 2.13 (Part 1)
© 2012 Sinauer Associates, Inc.

11. Steroids and other fatty substances pass readily through most cellular membranes because

____

____

____

Concept 2.5 explains how energy for life comes from biochemical changes in molecules.

12. The "anabolic steroids" are drugs that are sometimes misused by people who want to increase their athletic prowess. Describe what is meant by "anabolic" as a label in this term.

____

____

____

Science Practices & Inquiry

In the AP Biology Curriculum Framework, there is a set of 7 Science Practices. In this chapter, we will focus on **Science Practice 6: The student can work with scientific explanations and theories.** More specifically, practice 6.2: The student can construct explanations of phenomena based on evidence produced through scientific practices.

Questions 13 – 16 ask you to construct explanations based on evidence of how variation in molecular units provides cells with a wider range of functions. (LO 4.22)

In 1953, Stanley Miller and Harold Urey set up an apparatus, depicted here, to simulate the Earth's early atmosphere. The gases they added in their original set-up were methane (CH_4), ammonia (NH_3), hydrogen (H_2), water (H_2O), carbon dioxide (CO_2), and nitrogen gas (N_2). Energy was added by passing a spark across two electrodes and by boiling the reactants. After one week of continuous sparking and boiling of this "primordial soup," several amino acids, including aspartic acid, glycine, and alanine, were found in the condensed fluid from the apparatus.

H_2O

N_2 CH_4

NH_3 H_2 CO_2

The final lines from the original paper state:

"In this apparatus an attempt was made to duplicate a primitive atmosphere of the earth, and not to obtain the optimum conditions for the formation of amino acids. Although in this case the total yield was small for the energy expended, it is possible that, with more efficient apparatus ... this type of process would be a way of commercially producing amino acids.

A more complete analysis of the amino acids and other products of the discharge is now being performed and will be reported in detail shortly."

13. Define abiogenesis:

____

____

14. Define biogenesis:

____

____

15. Explain whether or not abiogenesis and biogenesis were demonstrated in the Miller-Urey experiment.

____

____

____

16. Discuss this claim:

"The Miller-Urey apparatus proves that life originated in a primordial sea."

____

____

____

Chapter 3: Nucleic Acids, Proteins, and Enzymes

Chapter Outline

3.1 - Nucleic Acids Are Informational Macromolecules
3.2 - Proteins Are Polymers with Important Structural and Metabolic Roles
3.3 - Some Proteins Act as Enzymes to Speed up Biochemical Reactions
3.4 - Regulation of Metabolism Occurs by Regulation of Enzymes

As you saw in Chapter 2, smaller molecules (monomers) can bond together to form larger molecules (macromolecules). Chapter 3 focuses on the relationships between two related groups of monomers: the nucleic acids and the proteins. In later chapters, you will see that proteins form the structural components of many cells and serve as enzymes that speed up many metabolic processes in the cells.

Chapter 3 continues consideration of **Big Idea 1**, evolution, and Chapter 3 also begins your exploration of **Big Ideas 3 & 4**, in which you study the structure of RNA and DNA and examine proteins and enzymes. The following list includes the specific parts (essential knowledge) of the AP-Biology curriculum that are covered in Chapter 3.

1.d.2: Scientific evidence from many different disciplines supports models of the origin of life.
3.a.1: DNA, and in some cases RNA, is the primary source of heritable information.
4.a.1: The subcomponents of biological molecules and their sequence determine the properties of that molecule.
4.b.1: Interactions between molecules affect their structure and function.

Chapter Review

Concept 3.1 explains how the nucleic acids DNA and RNA are polymers of nucleotides connected each to the next by a bond called a phosphodiester linkage. These macromolecules are used by cells to code genetic information because the sequence of nucleotides determines the sequence of amino acids in proteins. The structures of proteins, in turn, determine their structural, enzymatic, and other functions.

1. What are the three major differences between RNA and DNA?

a. ____
b. ____
c. ____

2. What is the difference between a polynucleotide and an oligonucleotide? Give an example of each.

____
____
____
____
____
____
____
____
____

3. In the diagram to the right, identify **all** of the:

3' ends

5' ends

purines

pyrimidines

hydrogen bonds

phosphodiester bonds

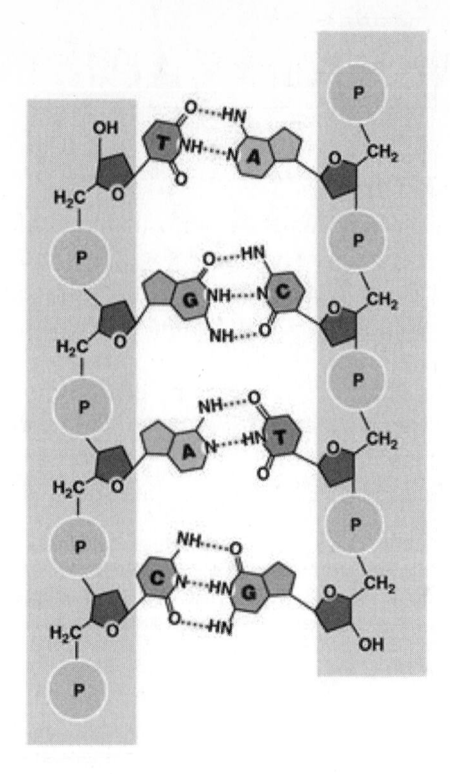

This drawing is an example of (circle one):

DNA or RNA

4. DNA is frequently examined to determine how closely two groups of organisms are related in terms of their evolutionary past. For example, the Asian and African elephants were believed to be the only two living species of elephants. However, recent DNA testing showed enough DNA differences between Africa's forest and savanna elephants to identify them as two separate species. Explain why DNA sequences, but not carbohydrates or lipids, are used for this type of taxonomic analysis.

___

___

___

Concept 3.2 explains how proteins are macromolecules comprised of amino acids linked together by peptide linkages. Proteins serve many diverse roles, including structural, metabolic, enzymatic, regulatory, and many more functions. There are at least 4 levels of structure that can be used to describe a protein.

5. There are 20 different amino acids in humans. You do not need to memorize all 20 but you should know the structure of a generalized amino acid. Draw a generic amino acid, using the letter "R" to designate a generic side chain, and be sure to include and label the carboxyl group and amino group.

6. Explain how the diversity of different proteins is created.

____
____
____

7. Most proteins have at least four levels of structure. For each level, briefly explain it and identify where it is found in a protein molecule.

Primary____
____

Secondary____
____

Tertiary____
____

Quaternary____
____

8. Draw the two amino acids, glutamic acid and lysine.

a. Which has two carboxyl groups? ____________ Circle and label it.
b. Which has two amino groups? ______________Circle and label it.
c. In the space below, draw a dipeptide made up of lysine and glutamic acid bound together by a peptide bond. Identify the peptide bond with an arrow.

9. Different sequences of amino acids result in different structures of proteins. A change in the amino acid sequence of a protein is called a mutation. Explain what might happen to the structure of a protein if the mutations below happened.

Leucine substituted for Cysteine:

____

____

____

Arginine substituted for Phenylalanine:

____

____

____

Alanine substituted for Aspartic Acid:

____

____

____

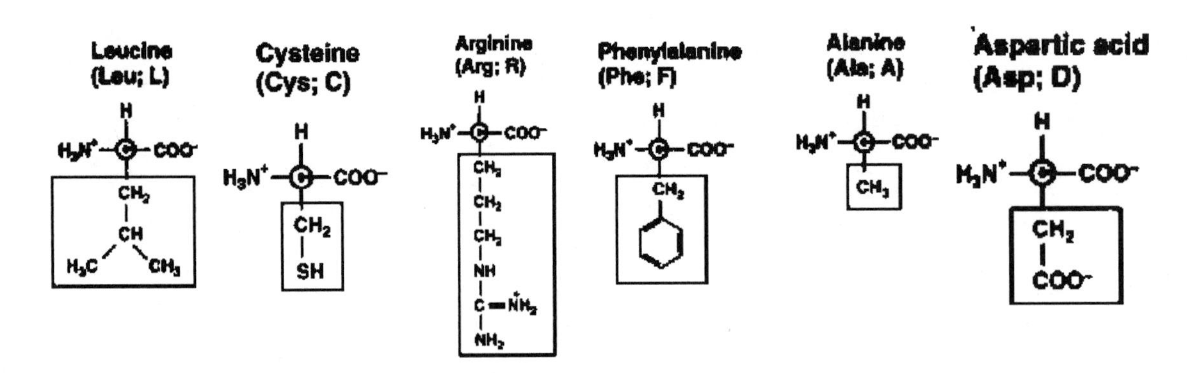

10. Which protein below shows a protein with quaternary structure? Explain why.

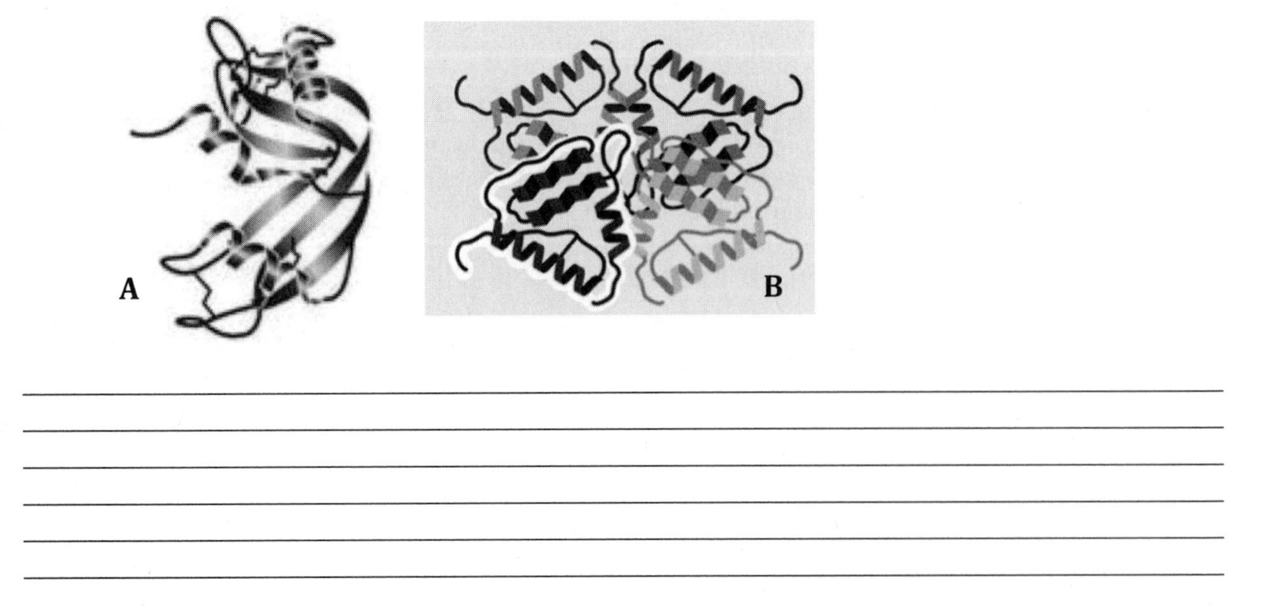

____

____

____

____

____

____

12. A student added a strong acid to a beaker containing a solution with a functional protein. After adding the acid, the protein no longer functioned. Explain how adding the acid altered the protein's structure and function. Be sure you include the following terms in your answer: protons, three dimensional structure, carboxyl groups, polarity, tertiary structure, denaturation.

____
____
____
____
____
____

Concept 3.3 focuses on one of the major roles of proteins, that of enzymes or biological catalysts. Enzymes speed up some of the biochemical reactions of life by lowering the activation rate of such reactions.

13. Enzymes can speed up synthetic reactions in living organisms. How does an enzyme speed up a reaction between two substrate molecules?

____
____
____

14. In the diagram below, label: the energy state of the products, the energy state of the reactants, the amount of energy of activation for the catalyzed reaction, and the amount of energy of activation for the uncatalyzed reaction.

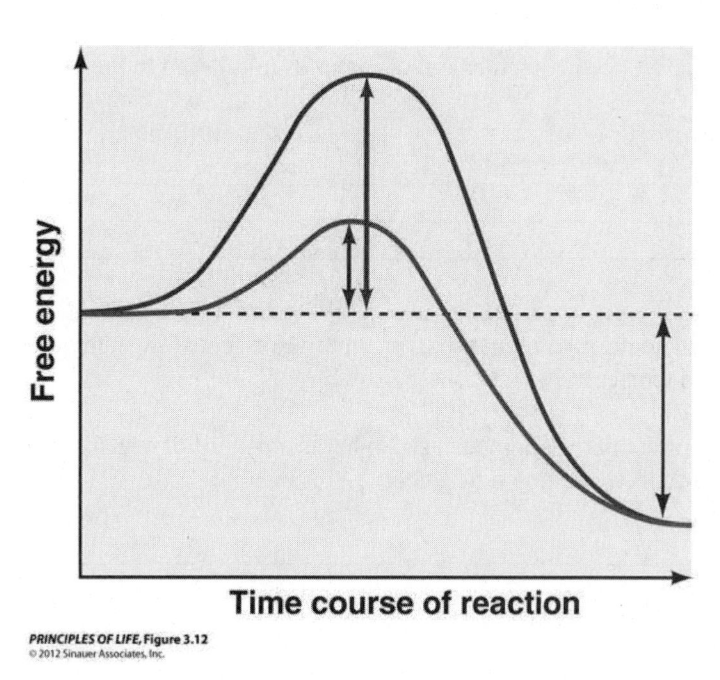

PRINCIPLES OF LIFE, Figure 3.12

15. The diagram to the right is the enzyme sucrase. The specificity of most enzymes is such that each only "recognizes" very particular substrates. Using your knowledge of protein structure, explain how the enzyme sucrase is specific to the disaccharide sucrose but does not bind to other disaccharides.

PRINCIPLES OF LIFE, Figure 3.9

16. An "induced fit" occurs when enzymes change their shape in response to interacting with a substrate. Explain how binding to a substrate can cause an enzyme to change its shape.

17. Which of these three molecules—DNA, RNA, or proteins—most likely operated in the ancient past prior to the appearance of the other two molecules? Explain how this earliest "proto-life" chemical serves more than one function.

18. Explain the difference between cofactors and coenzymes in relation to the different functions of different proteins.

Concept 3.4 shows how all organisms maintain a constant internal state called homeostasis through regulation by enzymes. In addition to regulation, enzymes are affected by many different environmental factors, including pH and temperature.

19. Regulation is an important part of homeostasis. What is a benefit of organisms being able to regulate enzymatic activity, such as the breakdown of glucose?

20. Which picture in the diagram to the right is an example of allosteric regulation? Explain why.

(A) Competitive inhibition

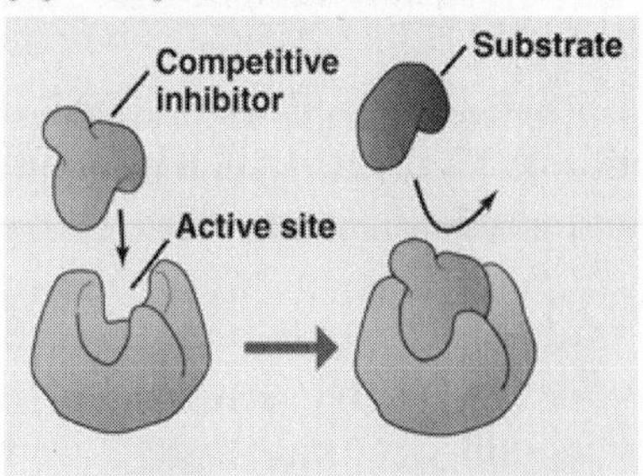

(B) Noncompetitive inhibition

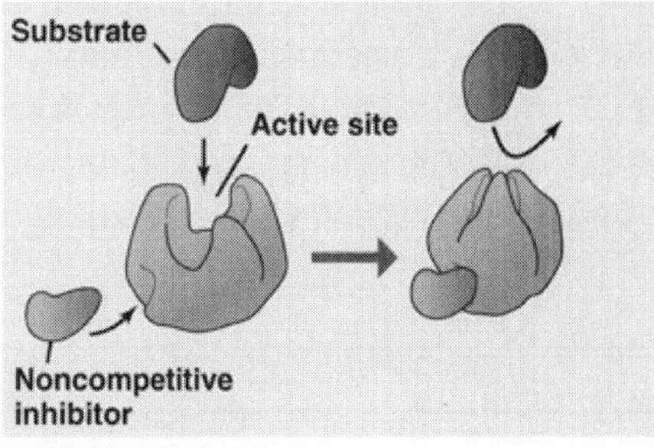

PRINCIPLES OF LIFE, Figure 3.17

Science Practices & Inquiry

Of the 7 Science Practices in the AP Biology Curriculum Framework, Chapter 3 addresses **Science Practice 4: The student can plan and implement data collection strategies appropriate to a particular scientific question.**

Here is an exercise based more specifically on **Practice 4.1:** The student can justify the selection of the kind of data needed to answer a particular scientific question; and on **Practice 4.2:** The student can design a plan for collecting data to answer a particular scientific question.

21. Edward Stone's letter to the Royal Society, dated 1763, detailed his ideas on aspirin as a pain killer. Stone tested 50 people to formulate his ideas. Design a sample that Edward Stone would use today in modern day research in the drug industry. You may wish to research double-blind studies and include this in your design.

Chapter 4: Cells – The Working Units of Life

Chapter Outline

4.1 - Cells provide compartments for biochemical reactions.
4.2 - Prokaryotic cells do not have a nucleus.
4.3 - Eukaryotic cells have a nucleus and other membrane-bound organelles.
4.4 - The cytoskeleton provides strength and movement.
4.5 - Extracellular structures allow cells to communicate with the external environment.

You already know that living organisms are made up of cells. Think of cells as small water-filled balloons holding a mixture of molecules and ions that work together. In this chapter you learn about intracellular organelles that compartmentalize and maximize biochemical activities necessary for cellular life. Not all cells have these organelles. In fact, the most numerous cells on the planet, those of the *Archaea* and *Bacteria*, lack organelles, and this absence of organelles is a defining characteristic of simple, ancient cells called prokaryotic cells. Eukaryotic cells, such as those in our bodies, are larger in size and have numerous membranous internal structures—these are the organelles. By concentrating certain types of processes to certain types of organelles, eukaryotic cells are extremely efficient, making larger cellular size sustainable.

The content of Chapter Four spans all four of the **Big Ideas** found in the AP Biology Curriculum Framework. The Big Ideas are a means for organizing the vast amount of information in biology. Try to develop your understanding across these **Big Ideas** and their corresponding **Enduring Understandings**.

Big Idea 1 recognizes that evolution ties together all parts of biology. In Chapter 4 we look at a theory for the development of cell complexity, thus noting:

1.d.2: Scientific evidence from many different disciplines supports models of the origin of life.

Big Idea 2 states that the utilization of free energy and use of molecular building blocks are characteristic fundamental of life processes. Specifically, Chapter 4 includes:

2.a.3: Organisms must exchange matter with the environment to grow, reproduce and maintain organization.

2.b.3: Eukaryotic cells maintain internal membranes that partition the cell into specialized regions, including the rough endoplasmic reticulum, mitochondria, chloroplasts, Golgi apparatus, nucleus, and smooth endoplasmic reticulum.

Big Idea 3 states that living systems store, retrieve and transmit information essential to life processes. Specifically, Chapter 4 lays this groundwork:

3.d.2: Cells communicate with each other through direct contact with other cells or interact from a distance via chemical signaling; examples include immune cells and plasmodesmata between plant cells.

Big Idea 4 states that biological systems interact in complex ways. Included in Chapter 4:

4.a.2: The structure and function of sub-cellular components, and their interactions, provide essential cellular processes.

Chapter Review

Concept 4.1 introduces the idea that multicellular life forms like humans contain billions of cells. These cells come from pre-existing cells, and they are the basic unit of most life forms in biology. What we learn from studying the activities of single cells applies to whole organisms. Just like whole organisms, cells can stay alive and persist only when nutrients are available and waste materials do not reach dangerous or toxic levels.

1. Most cells are quite small. Limits on cell size are related to limits on the rate of movement of "good stuff in" and "bad stuff out" across cell membranes. Movement rates are greatly influenced by the surface-area-to-volume ratio of the cells.

Imagine three cube-shaped cells, similar to what you saw in FIGURE 4.2 in the text. Given the dimensions shown for each cube-shaped cell here, calculate that cell's surface area, its volume and its surface-area-to-volume ratio.

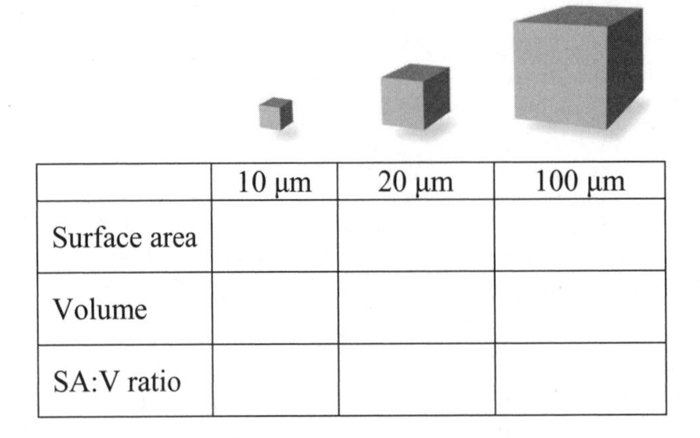

	10 μm	20 μm	100 μm
Surface area			
Volume			
SA:V ratio			

EXTRA: Now you are ready to solve the problem of circular cells on p. 59 of the text.

2. As the amount of a toxin increases around the outside of the three cube-shaped cells in the above diagram, which size of "cell" would be the first to have an enriched concentration of the toxin in its center (core) region? Explain your answer using the surface-area-to-volume ratio.

____

____

____

____

3. Use the provided logarithmic scale to determine how many 100 μm cells would you have to stack on top of each other to make the stack as tall as an athlete of 2 m height (hint: 2m = ? μm)?

0.1 nm 1 nm 10 nm 100 nm 1 μm 10 μm 100 μm 1 mm 1 cm 0.1 m 1 m 10 m 100 m 1 km

***PRINCIPLES OF LIFE*, Figure 4.1**

____

____

____

____

EXTRA: How many 10 μm prokaryotic cells would be needed for the same objective?

____

Concept 4.2 shows how the nucleoid region in a prokaryotic cell serves the same hereditary functions served by the nucleus in a eukaryotic cell.

4. Explain how prokaryotes carry out enzymatically-catalyzed biochemical conversions without the use of cellular organelles.

5. Describe, in general terms, the structural components of ribosomes, including a brief explanation of their function. Explain whether ribosomes are present only in eukaryotes, only in prokaryotes, or in both eukaryotes and prokaryotes.

6. Humans, perhaps unjustly, claim credit for inventing "the wheel." Discuss the argument that prokaryotes with flagella long preceded the human "invention" of the wheel.

7. Some models of the cell show it as a plastic bag full of alphabet soup with a golf ball thrown to represent the nucleus. Discuss how this model is not a good representation of a cell, and be sure you discuss the cytoskeleton in your answer.

Concept 4.3 describes how the nucleus in a eukaryotic cell serves the same hereditary functions served by the nucleoid region in a prokaryotic cell, and Eukarya have many other organelles.

8. Many hormonal signals, for example, insulin, are proteins secreted by cells. Describe the structure and function of as many cellular organelles as you can in regard to the synthesis and secretion of protein signals.

9. Identify the two primary groups of molecules that interact to become ribosomes, and include a description of where ribosomes are synthesized in eukaryotic cells.

10. It's your turn to be the teacher. One of your fellow classmates tells you that: "The mitochondrion is the site where ingested glucose molecules are made into into ATP molecules." Kindly point out the error of your classmate's statement by offering a statement that is more correct. Hint: is glucose biochemically converted to ATP?

Concept 4.4 explains how the cytoskeleton of eukaryotic cells provides strength and coordinates movement.

11. Describe how the movement of a paramecium using cilia is similar to and different from that of an amoeba.

12. Compare eukaryotic flagella and cilia in terms of structural size and in number present on a flagellar (Euglena) and a ciliated (Paramecium) cell.

____

____

____

____

____

____

13. The longest cells of eukaryotes, as you might guess, are found as neurons in giraffes. Such cells can be 2 or more meters in length. The length of such cells results in an impressive mechanism for moving proteins from one end of the cell to the other. Describe the intracellular transport system, including vesicles, microtubules, and motor-proteins, for such long and thin cells.

____

____

____

____

____

____

____

____

____

____

Concept 4.5 discusses how cells can interact with other cells and send and receive chemical signals at specialized regions on the surface of the cell.

14. Correct and expand upon this statement: "Adjacent plant cells are joined together by walls made up of only phospholipid molecules and proteins."

____

____

____

____

____

____

____

____

15. Correct and expand on this statement: "Sugar molecules hold adjacent animal cells together."

____

____

____

____

____

____

____

____

16. Specialized connections between adjacent cells in your heart hold them together closely so that blood does not leak out between the cells as the heart pumps. The pressure of pumping would blow apart adjacent cells were they not held tightly together by a second specialized connection. Furthermore, coordinated pumping activity of these cells relies on a third specialization between these cells. Describe how these three types of intercellular connections work together for the functioning of the heart.

____
____
____
____
____
____
____
____
____
____
____
____

17. What do the characteristics of modern cells tell us about how the first cells originated? Using your knowledge of cellular organelles, create a model (flow chart) showing the steps involved with eukaryotic cells evolving from a chemical rich environment.

Science Practices & Inquiry

In the AP Biology Curriculum Framework, there is a set of 7 Science Practices. In this chapter, we will focus on **Science Practice 2: The student can use mathematics appropriately.** More specifically, practice 2.2: The student can *apply mathematical routines* to quantities that describe natural phenomena.

Question #18 ties with Science Practice 2 relating your knowledge of math and biology together. This question asks you to use calculated surface area-to-volume ratios to predict which cell(s) might eliminate wastes or procure nutrients faster by diffusion (LO 2.6). The ability to use math and relate it to your knowledge of biology is important. Why are cells the size they are?

18. Calculate the SA:V ratio for a cube whose dimensions are typical of an eukaryotic cell, one that is 0.1 mm on a side. Explain why bacteria need to divide well before they get this large.

Chapter 5: Cell Membranes and Signaling

Chapter Outline

5.1 - Biological Membranes Have a Common Structure and Are Fluid
5.2 - Some Substances Can Cross the Membrane by Diffusion
5.3 - Some Substances Require Energy to Cross the Membrane
5.4 - Large Molecules Cross the Membrane via Vesicles
5.5 - The Membrane Plays a Key Role in a Cell's Response to Environmental Signals
5.6 - Signal Transduction Allows the Cell to Respond to Its Environment

Living organisms such as birds and fish are made up of cells, and these cells are collections of molecules that work together. Surrounding each cell is a plasma membrane that serves as a boundary between the cell and the environment. The cell membrane is much more than a wall; it regulates what goes into or out of the cell. Membranes are also the location where many communication messages from other cells are received. Intercellular communication is essential for multicellular forms of life, provides precision in homeostasis, is the site of self-recognition and cell defense (immune system), and must respond to changes in the environment.

As you will soon realize, most chapters have ideas that cover more than one Big Idea. The Big Ideas are a means for organizing the vast amount of information in biology. Try to develop your understanding across these **Big Ideas** and the corresponding **Enduring Understandings**. Chapter 5's emphasis is primarily in Big Idea 3, but also includes some of Big Idea 2 and Big Idea 4 as well.

Big Idea 2 states that the utilization of free energy and use of molecular building blocks are characteristic fundamental of life processes. Specifically, Chapter 5 includes:

2.b.1: Cell membranes are selectively permeable due to their structure.
2.b.2: Growth and dynamic homeostasis are maintained by the constant movement of molecules across membranes.

Big Idea 3 states that living systems store, retrieve and transmit information essential to life processes. Specifically, Chapter 5 lays the groundwork of cell communication:

3.b.2: A variety of intercellular and intracellular signal transmissions mediate gene expression.
3.d.1: Cell communication processes share common features that reflect a shared evolutionary history.
3.d.3: Signal transduction pathways link signal reception with cellular response.
3.d.4: Changes in signal transduction pathways can alter cellular response.

Big Idea 4 states that biological systems interact in complex ways. Included in Chapter 5:

4.c.1: Variation in molecular units provides cells with a wider range of functions.

Chapter Review

Concept 5.1 shows how the membranes surrounding cells have a common structure through all forms of life: that of a phospholipid bilayer with proteins and carbohydrates adding functionality. This bilayer has a hydrophobic middle while the membrane towards the outside and inside of the cell has hydrophilic properties.

1. For the diagram below, explain what information you would use to determine which side of the membrane faces the inside of the cell and which side faces the extracellular environment. Label these items: phospholipid, cholesterol, cytoskeleton, cell interior (cytoplasm), integral protein, peripheral protein, and carbohydrate. Write your explanation below the figure.

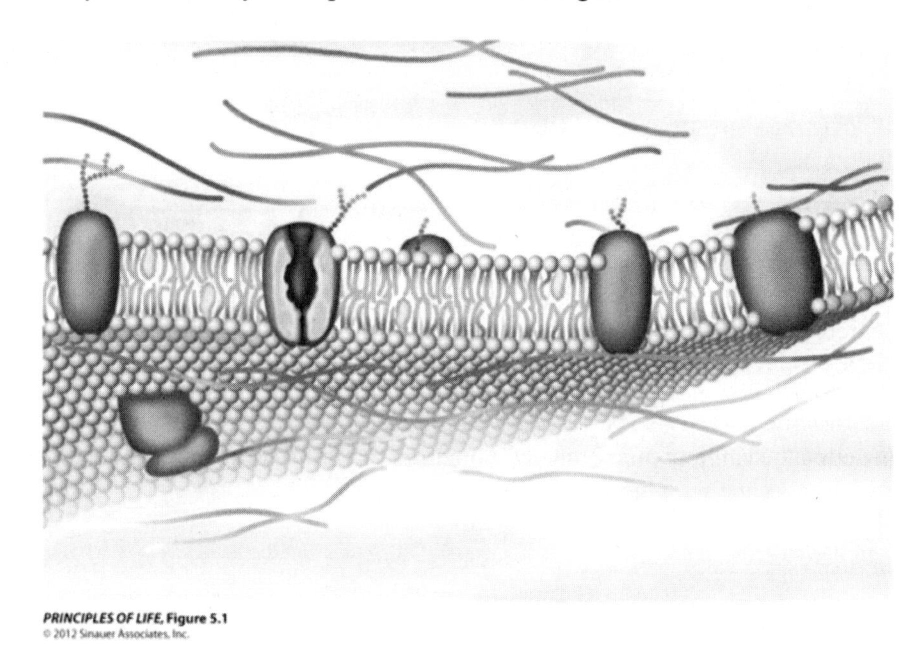

PRINCIPLES OF LIFE, **Figure 5.1**

Evidence for inside versus outside:

2. The current model of the plasma membrane is referred to as the fluid mosaic model. Provide evidence that the membrane is "fluid" and describe the "mosaic" of this model.

3. Explain how the structure of a phospholipid molecule is amphipathic and can form a membrane layer that is nonpolar in the middle and polar on the outsides.

4. What are the two primary factors that influence membrane fluidity?

5. Molecules that are amphipathic have both polar and nonpolar regions. For a large, amphipathic protein embedded in the phospholipid membrane, describe how this characteristic facilitates its placement in membranes. Draw a diagram of such an amphipathic protein embedded in the membrane below and label the polar and nonpolar regions.

PRINCIPLES OF LIFE, In-Text Art, Ch. 5, p. 80

6. Describe a biochemical change in membrane composition that helps organisms that endure hot summers and cold winters cope with their temperature extremes.

7. Describe the two major structural components of glycoproteins and describe one function of glycoproteins.

Concept 5.2 explains how diffusion is the movement of substances from an area of high concentration to an area of low concentration, and is driven by kinetic energy. For a substance that might be expected to cross cell membranes, the molecular size and polarity of the molecules determine whether or not they will diffuse across the plasma membrane. A special form of diffusion, osmosis, refers to the diffusion of water across membranes.

8. In this example, a drop of ink was placed into a bowl of gelatin. Explain how the ink diffused throughout the gel even though there were no currents to help move it around.

9. Describe two differences between active and passive transport.

10. Briefly explain how each of the three factors below can impact the diffusion of solutes across membranes.

size of the diffusing solute: ____

____

____

temperature: ____

____

____

concentration gradient: ____

____

____

11. Some topical anesthetics dissolve into the membranes of sensory neurons. Describe two structural properties of an anesthesia-inducing molecule that would make it a likely candidate for this route of anesthetic effect.

____

____

____

12. Even though water can readily move across many natural membranes, explain why it might be expected to move slowly or not at all through artificial membranes constructed without proteins.

____

____

____

13. The three terms below are used when comparing solute concentration on either side of a cell membrane. Define each term and provide a description how that condition might affect a cell's shape.

isotonic: ____

____

____

____

hypotonic: ____

____

____

____

hypertonic: ____

____

____

____

14. Facilitated diffusion refers to a special type of transport: for example, the entry of glucose in to the muscles in your body. Is this type of trans-membrane movement considered to be an example of active or passive transport? Explain why.

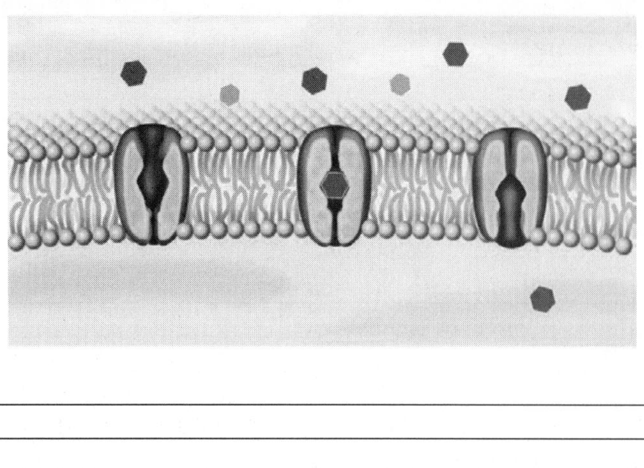

15. After several days without watering, plants tend to wilt. When the plant is watered, it will often return to its normal shape. Explain how cells are involved in the transition from wilted to normal.

16. Explain how the carrier protein in the diagram below is facilitating the diffusion of a molecule. Include in your answer an explanation why the protein is needed.

Concept 5.3 considers active transport and describes the movement of substances against their chemical concentration gradient thus requiring energy to accomplish this task.

17. Describe the primary chemical process that drives active transport.

18. Complete the table below:

	Simple Diffusion	Osmosis	Facilitated Diffusion	Active Transport
Cellular energy required?				
Driving force				
Membrane protein required?				
Directional?				
Specificity?				

19. The Na+-K+ ATPase is the most active and wide-spread active-transport system in the human body. Label five items on the diagram below detailing how this pump functions.

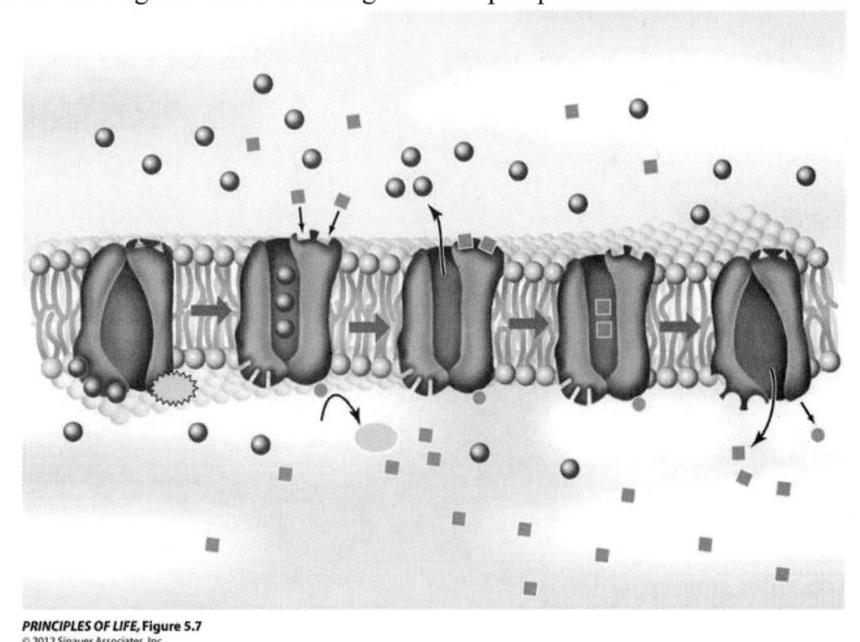

PRINCIPLES OF LIFE, Figure 5.7

Concept 5.4 examines how many large molecules cannot cross membranes via transporters embedded in membranes; rather, such compounds enter or leave cells via vesicles, in a process called endocytosis or exocytosis, respectively.

20. Explain how phagocytosis and pinocytosis are similar and different.

____

____

____

21. Describe receptor-mediated endocytosis with enough detail to explain whether or not this process meets the criteria for active transport or passive transport.

____

____

____

Concept 5.5 explains how organisms and cells respond to stimuli (signals) from their environment. The signal may be a physical stimulus such as light or heat, or a chemical such as a hormone. In order to respond, the cell must have a specific receptor that becomes modified by the stimulus. Once a receptor in the membrane is activated by the signal, it sets off a series of biochemical changes within the cell. These pathways are sequences of events and chemical reactions that lead to a cell's response to a signal. This ability to respond to the environment is critical to the organism's or cell's ability to maintain precision in its homeostatic mechanisms.

22. Describe each of the three major steps in cell signaling.

a. ____

____

____

b. ____

____

____

c. ____

____

____

23. Different receptor proteins for different signals are found in the cytoplasm or on the membrane of the cell. Give an example of each and discuss the properties of the ligand (signal molecule) that activates this receptor.

intracellular receptor protein:

____

____

____

membrane-bound receptor protein:

____

____

24. If a cell had no proteins in its membrane, will it be able to respond to any environmental stimuli? Explain your answer.

____

____

____

Concept 5.6 shows how physical or chemical signals initiate responses from cells that have a signal transduction pathway for that signal. For example, signals that modify receptor proteins in membranes rapidly initiate a series of biochemical changes in the cell. The pathways affected by these biochemical changes are typically components in a signaling cascade that amplifies and distributes responses by effector proteins in the cell.

25. Complete the diagram to the right showing an example of a signal cascade.

26. Explain how the signal cascade in the diagram to the right achieves amplification.

PRINCIPLES OF LIFE, Figure 5.17

27. Describe how the signal cascade above is terminated after the necessary response has been obtained.

Science Practices & Inquiry

In the AP Biology Curriculum Framework, there is a set of 7 Science Practices. In this chapter, we will focus on **Science Practices 1, 3, and 7.**

Science Practice 1: The student can use representations and models to communicate scientific phenomena and solve scientific problems.

Science Practice 3: The student can engage in scientific questioning to extend thinking or to guide investigations within the context of the AP course.

Science Practice 7: The student is able to connect and relate knowledge across various scales, concepts and representations in and across domains.

Question #28 ties Science Practices 1, 3, and 7 together asking you to use representations and models to pose scientific questions about the properties of cell membranes and selective permeability based on molecular structure. You need to construct models that connect the movement of molecules across membranes with membrane structure and function. (LO 2.10)

Question #29 focuses on Science Practices 3 and 7. This question asks you to generate scientific questions involving cell communication as it relates to the process of evolution; and to describe how organisms exchange information in response to internal changes or environmental cues. (LO 3.32)

28. To the right is a diagram of caffeine. Caffeine acts by binding with a receptor on the cell surface. Why doesn't it enter the cell? Label the parts of the molecule that make it difficult for caffeine to enter the cell.

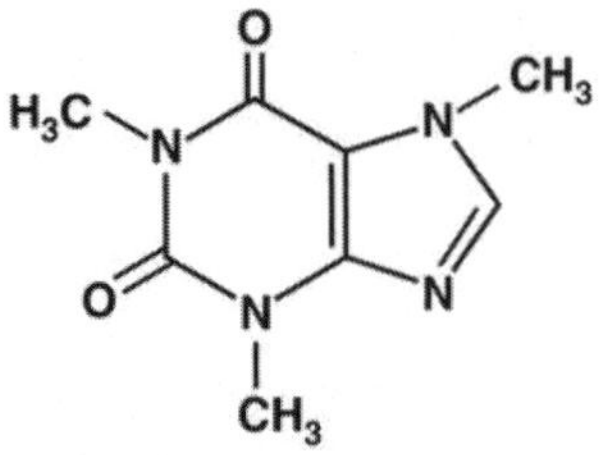

O
H_3C
N
N
CH_3
O
N
N
CH_3

____

____

____

____

____

____

____

____

29. The drug ouabain inhibits the activity of the Na+–K+ATPase. A nerve cell is incubated in ouabain. Make a table in which you predict what would happen to the concentrations of Na+ and K+ inside the cell, as a result of the action of ouabain. Explain why.

____

____

____

____

____

____

____

____

____

Chapter 6: Pathways that Harvest and Store Chemical Energy

Chapter Outline

- 6.1 - ATP, Reduced Coenzymes, and Chemiosmosis Play Important Roles in Biological Energy Metabolism.
- 6.2 - Carbohydrate Catabolism in the Presence of Oxygen Release a Large Amount of Energy.
- 6.3 - Carbohydrate Catbolism in the absence of Oxygen Releases a Small Amount of Energy.
- 6.4 - Catabolic and Anabolic Pathways are Integrated.
- 6.5 - During Photosynthesis, Light Energy is Converted to Chemical Energy.
- 6.6 - Photosynthetic Organisms Use Chemical Energy to convert CO_2 to Carbohydrates.

Plants use sunlight-driven photosynthesis to synthesize carbohydrates. The energy released from taking apart (catabolizing) these carbohydrates is a primary source of chemical energy for both plants and animals. A sequence of controlled enzyme-catalyzed pathways allows these catabolic reactions to operate without causing too much damage along the way.

Some of the energy made available from the catabolism of carbohydrates and other fuel molecules drives the synthesis of adenosine triphosphate (ATP). Inside the cells, ATP molecules function as a form of energy currency; the hydrolysis of ATP, an exergonic reaction that yields adenosine diphosphate (ADP) and phosphate ion, liberates a small amount of chemical energy, some of which activates the biochemical reactions needed in cells.

The two primary means of maintaining ATP supplies are substrate-level phosphorylation and oxidative phosphorylation. In substrate phosphorylation, a phosphate group on an organic molecule is transferred to ADP, restoring it to ATP. This pathway makes ATP quickly, but phosphorylated organic molecules are present in only limited quantity in cells, so it is only a short-term solution to increased ATP demand. Oxidative phosphorylation, which bonds ADP to free phosphate ions in a process linked to the activity of the respiratory chain in mitochondria, makes lots of ATP, but its continued action requires continual access to oxygen and reduced coenzymes. The reduced coenzymes are NADH and $FADH_2$. The reduced coenzymes are continuously supplied by the continuous catabolism of glucose via glycolysis and the citric-acid cycle, while oxygen molecules come from the environment.

Big Idea 1 recognizes that evolution ties together all parts of biology. Chapter 6 reviews energy transfers that are conserved across all categories of animals and plants. It also affords brief consideration of the ways that agriculture and fermented beverages developed among early humans, supporting:

- 1.d.1: There are several hypotheses about the natural origin of life on Earth, each with supporting scientific evidence.

Big Idea 2 focuses on free energy and the use of molecular building blocks that are fundamental to life processes. Chapter 6 will allow you to tie together many pieces of essential knowledge, including:

- 2.a.1: All living systems require constant input of free energy. Examples included in Chapter 6 are the Calvin cycle, glycolysis, the Krebs cycle and fermentation.
- 2.a.2: Organisms capture and store free energy for use in biological processes. Examples in Chapter 6 include NADP in the reactions of photosynthesis and the importance of oxygen in cellular respiration.

Big Idea 4 states that biological systems interact in complex ways. Included in Chapter 6 are:

- 4.a.2: The structure and function of sub-cellular components, and their interactions, provide essential cellular processes.
- 4.c.1: Variation in molecular units provides cells with a wider range of functions. Chapter 6 describes the role chlorophyll in photosynthesis.

Chapter Review

Concept 6.1 introduces how ATP, reduced coenzymes, and chemiosmosis play important roles in biological energy metabolism. The energy needed for many biochemical reactions in cells is provided by the hydrolysis of ATP, yielding ADP and either phosphorylated proteins or free phosphate ions (HPO_4^-). As ATP is "used" in this way, it is also being continuously produced by two processes, substrate phosphorylation and oxidative phosphorylation.

In substrate phosphorylation, phosphate groups on proteins and other molecules are transferred to ADP to quickly restore it to ATP. Although substrate phosphorylation rapidly delivers ATP, there is a limited supply of phosphorylated substrates that can "give up" phosphate groups in this manner.

In contrast, ATP production resulting from oxidative phosphorylation yields much more ATP, although oxidative phosphorylation requires more ingredients: oxygen, reduced coenzymes (NADH and $FADH_2$), and, of course, ADP and HPO_4^-. The mitochondrion is the intracellular organelle where most of the components of oxidative phosphorylation are found.

The catabolism of fuel molecules such as glucose supports both pathways of ATP production, yielding energy transfers resulting in substrate phosphorylation directly, and yielding the reduced coenzymes needed for oxidative phosphorylation. As NADH and $FADH_2$ are oxidized, this energy transfer develops a gradient of hydrogen ions (H^+) inside the mitochondrion. The gradient provides energy transfers to an enzyme, ATP synthase, accelerating its role in binding ADP and HPO_4^- to make ATP.

1. The hydrolysis of ATP to support an anabolic process includes both endergonic and exergonic reactions, depending on which perspective one takes: the hydrolysis of ATP versus the formation of anabolic products. Discuss this statement.

____

____

____

____

____

____

2. The diagram below shows the conversion of compound AH to compound A and the conversion of compound BH to compound B, with interconversions of NAD+ and NADH. To each of the four boxes, add either "oxidation" or "reduction." Explain your label choices.

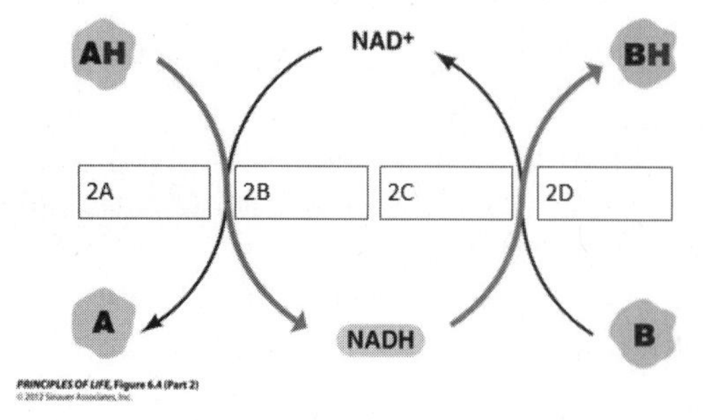

Box 2A: ____

____

Box 2B: ______________________________

____

Box 2C: ______________________________

____

Box 2D: ______________________________

____

3. Use the figure to the right to complete the following questions based on the molecules:

propanoic acid	propanol	propane
C_2H_6COOH	C_3H_7OH	C_3H_8

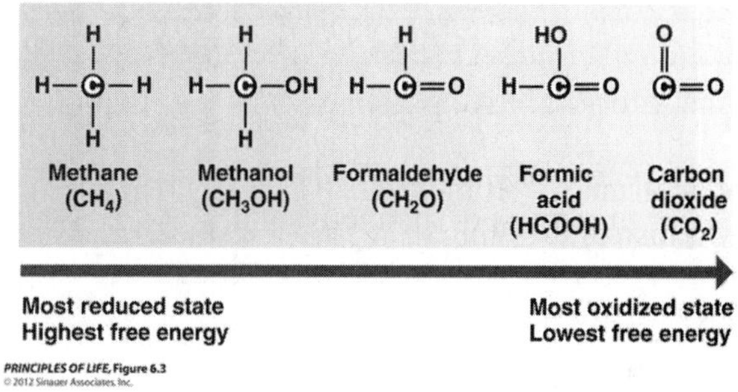

Methane (CH_4) Methanol (CH_3OH) Formaldehyde (CH_2O) Formic acid ($HCOOH$) Carbon dioxide (CO_2)

Most reduced state
Highest free energy

Most oxidized state
Lowest free energy

PRINCIPLES OF LIFE, Figure 6.3

Which compound is in the most reduced state? ______________

Which compound has the lowest free energy? ______________

Which compound is in the most oxidized state? ______________
Which compound has the highest free energy? ______________

4. Explain how two different membrane-embedded proteins in mitochondria simultaneously influence the gradient of hydrogen ions and ATP synthesis.

Proton Pump ______________________________

____

____

____

ATP Synthase ______________________________

____

____

____

Concept 6.2 describes how carbohydrate catabolism in the presence of oxygen releases a large amount of energy. In the presence of oxygen, catabolizing fuel molecules releases a large amount of energy. In cellular respiration, the fuel molecule under consideration is usually glucose, a monosaccharide carbohydrate. The catabolic pathways are well studied:

Glycolysis → Pyruvate oxidation → Citric acid (Krebs) cycle.

Along the catabolic pathways, ATP is made directly by substrate phosphorylation and reduced coenzymes are produced, thus supporting oxidative phosphorylation, a process that generates considerably more ATP synthesis than does substrate phosphorylation.

5. The complete catabolism of glucose can yield 686 kcal/mol energy transfer. For each of the following statements, indicate whether the statement is true or false and then explain your answer.

A. All 686 kcal/mol is directly transferred to ATP synthesis.
TRUE FALSE [choose one, then explain]

____

____

____

____

B. Less than half of the 686 kcal/mol is directly transferred to ATP synthesis.
TRUE FALSE [choose one, then explain]

____

____

____

____

C. Only 10% of the 686 kcal/mol is directly transferred to ATP synthesis, in accordance with the principles of thermodynamics. TRUE FALSE [choose one, then explain]

____

____

____

____

____

____

____

6. Assume that the diagram on the right refers to the catabolism of one glucose molecule.

→ Add arrows in the appropriate places to show where NADH and $FADH_2$.are generated. Include the number of each produced.

→ Then, show with arrows in the diagram where the reduced coenzymes participate in energy-transfer reactions.

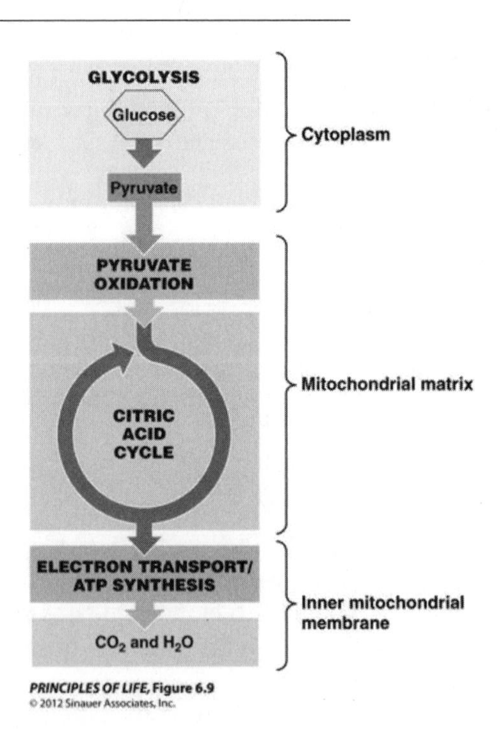

PRINCIPLES OF LIFE, Figure 6.9

7. Starting with THREE molecules of glucose, insert the appropriate numbers in the blanks below, assuming complete catabolism, with oxygen available.

_________ molecules of ATP must be hydrolyzed to start the process.
_________ molecules of NADH are produced.
_________ molecules of $FADH_2$ are produced.
_________ molecules of ATP are produced via substrate phosphorylation.
_________ molecules of water are produced in the electron transport chain.
_________ molecules of carbon dioxide are released from the process.

Concept 6.3 examines how carbohydrate catbolism in the absence of oxygen releases a small amount of energy. Organisms that live in conditions in which molecular oxygen is periodically unavailable or is never available use the process called fermentation to reoxidize NADH to NAD^+. Without NAD^+, glycolysis and all subsequent steps of catabolism come to a halt and ATP production stops. Complex organisms like ourselves, as well as some microbes, produce lactic acid or lactate as a byproduct of fermentation, whereas yeasts and some plants produce an alcohol known as ethanol as a byproduct.

8. Lactic acid fermentation and alcoholic fermentation both result in alterations to three-carbon pyruvate molecules. Which type of fermentation converts pyruvate to the smaller catabolites? Provide specific details.

____
____
____
____
____

Concept 6.4 shows how catabolic and anabolic pathways are integrated**.** Changes in cellular activity and in the availability of fuel molecules occur frequently in nature. During lean times or under high metabolic activity, energy transfers occur by taking apart (catabolizing) a variety of fuel molecules, adding catabolism of proteins and fats to whatever catabolism of carbohydrates is taking place. By contrast, in times of plenty, cellular reserves are restored. For example, many types of smaller molecules can be converted to lipids. In addition, the storage-polymer of glucose, called glycogen, is synthesized in muscle and liver.

9. Carbohydrates are just one source of fuel molecules. Identify two additional pools of fuel molecules whose catabolism can yield energy transfers that result in ATP synthesis. For each of the two categories of fuel molecules you've identified, briefly describe how the molecules are utilized and comment on their similarity to glucose catabolism.

Fuel-molecule pool#1: _______________________

____
____
____
____
____
____

Fuel-molecule pool#2: _______________________

____
____
____
____
____

Briefly define each term in each of the following pairs, and explain whether that process is more active during times of plenty access to food or during times of limited access to food.

glycogenolysis *vs*. glycogenesis

____
____
____
____
____

lipolysis *vs*. lipogenesis

____
____
____
____
____

proteolysis *vs*. proteins synthesis

____
____
____
____
____

10. Cells of the brain and the heart are highly specialized to carry out specific functions in the body, and their metabolic needs must always be met or death will soon follow. Part of this specialization includes reliance on glucose as a key fuel molecule. Explain how glucose is made available to the tissues during times when no carbohydrates are available as food.

____
____
____
____
____
____
____

Concept 6.5 studies photosynthesis, and shows how light energy is converted to chemical energy. Most plants do not appear to eat, yet they grow. Until it became clear to scientists that plants take up carbon dioxide from the environment, and use solar energy to fuel sugar synthesis, it was thought that most of their growth was fueled by things they take from the soil. Now we understand the process of photosynthesis much better and it works in this way. In the "light reactions," sunlight's energy activates chlorophyll molecules to transfer energy that is captured in ATP and NADPH synthesis. As these compounds accumulate, water molecules are catabolized, liberating hydrogen ions sufficient to build a chemiosmotic gradient, similar to that in mitochondria. Molecular oxygen is a byproduct of photosynthesis. Energy released during ATP hydrolysis and the oxidation of NADPH drive the "Calvin cycle," in which carbohydrates such as glucose are made (see Concept 6.6).

11. There are two photosystems, I and II, directly activated by different wavelengths of photon energy. Describe each and discuss the interdependence between them.

Photosystem II

____
____
____
____
____
____

Photosystem I

____
____
____
____
____
____

Concept 6.6 shows how photosynthetic organisms use chemical energy to convert CO_2 to carbohydrates. Carbon dioxide, ATP, and NADPH are the key rquirements for the synthesis of sugars by plants. The ATP and the NADPH are generated by the light reactions of photosynthesis (Concept 6.5) and CO_2 is taken up from the environment, especially the atmosphere, via the leaves of plants.

12. Briefly describe each of the three major segments of the Calvin cycle, noting the key ingredients needed in each segment.

____
____
____
____

____
____
____
____

____
____
____
____

13. Explain the claim, "rubisco is the most abundant protein on the planet" by describing its key role in the Calvin cycle.

____
____
____
____
____
____
____
____
____

14. Help Nathan and Elijah settle an argument. Elijah says that since plants can carry out photosynthesis, they do not need cellular respiration. Nathan says that photosynthesis without respiration is a wasted effort. Who is correct? Explain your answer.

Science Practices & Inquiry

In the AP Biology Curriculum Framework, there is a set of 7 Science Practices. In this chapter, we will focus on **Science Practice 6:** The student can work with scientific explanations and theories. More specifically, **practice 6.2:** The student can construct explanations of phenomena based on evidence produced through scientific practices.

Question 15 asks you to to construct explanations of the mechanisms and structural features of cells that allow organisms to capture, store or use free energy (LO 2.5) and to construct explanations based on scientific evidence as to how interactions of subcellular structures provide essential functions (LO 4.5).

15. In the absence of electron transport, an artificial H+ gradient is sufficient for ATP synthesis in cellular organelles. In an experiment, chloroplasts were isolated from plant cells and incubated at pH 7. The chloroplasts were then subjected to six different conditions. They were incubated with ADP, phosphate (P_i), and magnesium ions (Mg^{2+}) at pH 7 and at pH 3.8. The chloroplasts were then incubated at pH 3.8 with one of the four elements (ADP, P_i , Mg^{2+}, chloroplasts) missing.

ATP formation was measured using luciferase, which catalyzes the formation of a luminescent (light-emitting) molecule if ATP is present. Here are the data from the original paper:

	Luciferase activity (light emission)	
Reaction mixture	Raw data	Corrected data
Complete, pH 3.8	141	
Complete, pH 7.0	12	
Complete, pH 3.8 – P_i	12	
" " – ADP	4	
" " – Mg^{2+}	60	
" " – chloroplasts	7	

a) Identify which reaction mixture is the control by circling it in the table above.
b) Use the control data to correct the raw data for the other, experimental reaction mixtures and fill in the table.
c) Summarize the results of the experiment.
d) Why did ATP production go down in the absence of P_i?
e) Explain why ATP production could be negative in this experiment.
f) Discuss where the free energy comes from to drive the production of ATP.

Chapter 7: The Cell Cycle and Cell Division

Chapter Outline

7.1 - Different Life Cycles Use Different Modes of Cell Reproduction
7.2 - Both Binary Fission and Mitosis Produce Genetically Identical Cells
7.3 - Cell Reproduction Is Under Precise Control
7.4 - Meiosis Halves the Nuclear Chromosome Content and Generates Diversity
7.5 - Programmed Cell Death Is a Necessary Process in Living Organisms

A cell, like any other measure of a living thing, does not live forever, so the continuity of life in a single-celled organism requires that it produce copies. Multicellular organisms like ourselves start life as a single cell, and then produce many, many copies of that cell during growth and development. To understand cell replication, you must learn how DNA is carefully duplicated in a "parent" cell and then evenly shared between two "offspring" cells. After the duplication and segregation of DNA are complete, the "parent" cell splits into two "offspring" cells, each enclosed by its own membranes via a process called cytokinesis.

Binary fission and mitosis are the two primary mechanisms by which cells produce copies of themselves. Prokaryotes are unicellular organisms that typically contain only one chromosome, and they achieve cell duplication via binary fission. The DNA in the chromosome is copied end-to-end, with one copy ending on one side of the cell and the other copy on the other side of the cell. As this separation occurs, the cell membranes grow toward the middle of the "parent" cell to effectively pinch it in half, with each half receiving one complete copy of the chromosome. In contrast, eukaryotic cells undergo mitosis. Because they have multiple chromosomes, the copying of DNA must be coordinated, and the full set of chromosomes has to be moved to opposite ends of the "parent" cell before it separates into two "offspring" cells.

The phases of the cycle of eukaryotic cell duplication make up the cell cycle, which proceeds as follows. We will begin with the cell carrying out its normal function, e.g., a cell in the liver being involved in metabolism. The cell grows and carries out its function during the G1 phase of the cell cycle. As some point, a growth regulator protein called cyclin activates the cellular machinery that replicates DNA, and this is known as the S phase of the cycle. The next phase of the cycle, during which the components needed for mitosis are made is called the G2 phase. At the end of the G2 phase, the stage is set for mitosis. Mitosis proceeds, and the cell cycle moves from interphase to prophase, during which distinct duplicated chromosomes are microscopically visible. Next, in metaphase, the duplicated chromosomes line up across the middle of the cell, and they are pulled to opposite poles during anaphase. Finally, during telophase, each set of chromosomes becomes surrounded by nuclear membranes, and cytokinesis can proceed.

Eukaryotes that reproduce sexually gain the evolutionary benefits of mixing together the genes of two different parents, resulting in offspring that are not identical to either parent. In this scenario, the parents' reproductive systems have cells that undergo two nuclear divisions, producing haploid cells called gametes via the process of meiosis. Gametes are cells with only one copy of the otherwise paired chromosomes typical of diploid cells. The gametes become fused into a single diploid cell called a zygote, as a result of mating.

As in previous chapters, you have seen many new vocabulary words. Two words—haploid and diploid—are very important to understand the nature of life cycles of living organisms. Additionally, you should now realize that cell reproduction is highly regulated; otherwise, living organisms would soon use up all of their resources and perish. When regulation breaks down, unregulated growth and cancer can result.

Finally, Chapter 7 concludes with description of cell death, which takes place either by extensive damage to the cell, killing it outright, or by the initiation of genetic programming that results in hydrolytic destruction. This type of "regulated" cell death is called apoptosis, and it can protect an individual from a spreading infection and can facilitate recycling of materials.

In the AP Biology Curriculum Framework, Chapter 7 develops Big Idea 2, invoking regulatory mechanisms of living organisms, but it is more directed at understanding Big Idea 3, information transfer.

Big Idea 2 states that utilization of free energy and use of molecular building blocks are characteristic of life processes. Specifically, by revealing that signals initiate cell duplication and cell death, you will note that Chapter 7 includes:

2.c.1: Organisms use negative feedback mechanisms to maintain their internal environments and respond to external environmental changes.

Big Idea 3 states that living systems store, retrieve and transmit information essential to life processes. Specifically, Chapter 7 lays the groundwork for cellular reproduction:

3.a.2: In eukaryotes, heritable information is passed to the next generation via processes that include the cell cycle and mitosis, or meiosis plus fertilization.
3.c.1: Changes in genotype can result in changes in phenotype.
3.c.2: Biological systems have multiple processes that increase genetic variation.

Chapter Review

Concept 7.1 examines how different life cycles use different modes of cell reproduction. The reproduction of cells is the basis for the continuity of life. Cells reproduce through both sexual and asexual means. Diploid cells carry two sets of homologous chromosomes, one from each parent. Mitosis replicates these chromosomes identically to create a new cell via asexual reproduction. In sexual reproduction, meiosis halves the diploid number of chromosomes creating haploid gametes. When two gametes unite to create a new generation, a zygote results.

1. Identify the three major reasons for the reproduction of cells.

a. ____
b. ____
c. ____

2. Clonal production of cells result from the process called __________ __________.

3. Discuss the possible consequences of an error that results in the inexact duplication of DNA.

____
____
____

4. Describe the differences between haploid and diploid cells, and identify the processes by which these two types of cells are produced.

____
____
____
____
____
____
____

5. Discuss the formation of the type of cell called the zygote.

____
____
____
____

6. Assuming success in its first week, briefly describe what happens to a zygote after it is formed.

____
____
____

7. Complete the three life-cycle diagrams below, including the following words: haplontic, diplontic, meiosis, fertilization, zygote, spore, gamete, mature organism, haploid, alternation of generations, and diploid. Add the following as examples to the titles: mold, plant, animal.

8. During a prolonged period of constant environmental conditions with abundant resources, discuss the relative advantage that asexual reproduction has over sexual reproduction.

____
____

9. During a prolonged period of varying environmental conditions with fluctuations in resource availability, discuss the relative advantage that sexual reproduction has over asexual reproduction.

____
____

Concept 7.2 explores how both binary fission and mitosis produce genetically identical cells. Asexual reproduction by either binary fission or mitosis produces two genetically identical cells. In single-celled organisms such as the amoeba, this asexual reproduction is where new amoeba arise. For multicellular organisms such as birds or insects, mitosis is the mechanism of growth and development, or repair, or replacement of dead, dying, or worn out tissues.

10. Before a cell divides by mitosis, 4 key events must occur. Briefly describe each.

a. ____
____

b. ____
____

c. ____
____

d. ____

11. Discuss two ways that binary fission differs from mitosis.

____
____
____

12. Compare cytokinesis in plant and animal cells.

____
____
____

13. The diploid cell shown below is, before DNA replication, 2n = 6. Draw the five steps of mitosis this cell would undergo during replication.

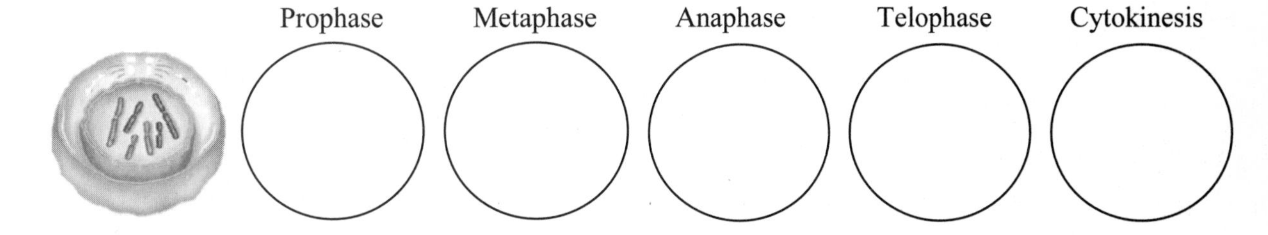

13. The "interphase" of the cell cycle is divided into 3 distinct phases, G1, S, and G2. Briefly describe what happens during each part of interphase.

G1 ____
____

S ____
____

G2 ____
____

14. Draw a chromosome that has 2 chromatids and label the centromere and sister chromatids.

15. Draw the nuclear contents of a cell with 2n=8 at the beginning of and end of the S phase of interphase.

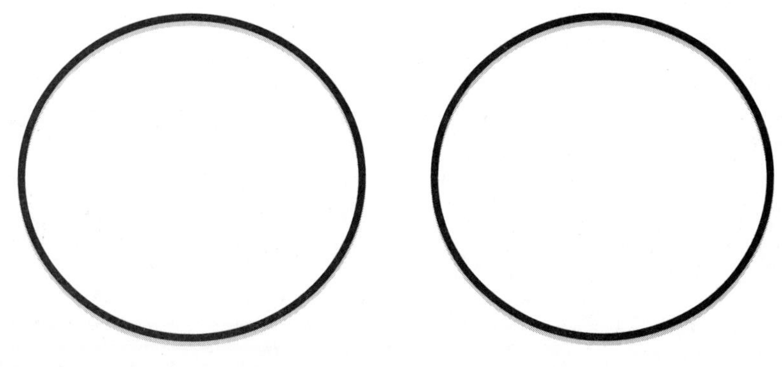

16. Assume that a mutation in a cell results in nonfunctional proteins in the kinetochore. Describe what would happen to that cell during attempted duplication.

Concept 7.3 shows how cell reproduction is under precise control. Cell reproduction requires access to nutrients and space, and is closely regulated by signaling mechanisms. A single-celled species with unregulated reproduction would soon exceed the environmental carrying capacity and would starve to death. In a multicellular organism, cell reproduction is closely regulated to maintain the forms and functions of different parts of the body. Cancer is largely a disease of deregulated mitosis, and it deprives normal body parts of access to nutrients and waste removal.

17. Identify and describe the key restriction point of the cell cycle. Identify the phase where this regulation takes place.

18. Explain the role of protein kinases during regulation of the cell cycle.

19. The CDK-cyclins regulate the cell cycle by many of the mechanism explained in previous chapters. For each idea below, briefly explain how it plays a role in the regulation of the R checkpoint of interphase.

Allosteric regulation

Gene expression

Protein synthesis

Signal transduction

Cell division

Concept 7.4 describes how meiosis halves the nuclear chromosome content and generates diversity. Meiosis forms haploid gametes that unite to initiate the next generation. Meiosis also results in offspring that are inexact copies of any one parent, providing fodder for natural selection. Without meiosis, each new generation would be a clone of the previous generation and would be at a disadvantage to react to new changes in the environment. Crossing over and independent assortment rearrange existing genetic variation between gametes, while other meiotic errors can create new variation often leading to cancer and other diseases.

20. Meiosis serves several important purposes for sexual reproduction. What would happen if each of these did not occur prior to sexual reproduction?

a. ____

____

b. ____

____

c. ____

____

21. Complete the chart below comparing mitosis and meiosis:

	Mitosis	Meiosis
Number of daughter cells		
Chromosome number of parent cell	____n	____n
Chromosome number of daughter cell	____n	____n
Number of nuclear divisions		
Pairing of homologous chromosomes? (Yes/No)		
Daughter nuclei are genetically identical? (Y/N)		

22. Identify and describe two processes that ensure that daughter nuclei formed during meiosis are genetically different.

a. ____

____

b. ____

____

23. For a cell that develops a mutation resulting in nonfunctional dynein proteins in the kinetochore, describe the consequences for mitosis and meiosis.

____

____

____

24. Complete the below for an organism that has a diploid number of 2n=12.

Number of chromatids at prophase of mitosis: _____
Number of chromosomes at metaphase of mitosis: _____
Number of centromeres at prophase of meiosis I: _____
Number of chromosomes in a gamete: _____
Number of chromosomes in a skin cell: _____
Number of daughter nuclei after mitosis: _____
Number of chromosomes after meiosis I: _____

Concept 7.5 explains why programmed cell death is a necessary process in living organisms. Cells die in two primary ways. One major type of cell death, necrosis, occurs when cells die of starvation, are cut open (wounded) or poisoned. This frequently causes inflammation: the redness and swelling we associate with an injury. But more often, cell death is due to apoptosis, which is a genetically programmed series of events that result in cell death.

25. Explain the difference between apoptosis and necrosis. Give an example of each.

____
____
____

26. Why is apoptosis important to the normal, healthy development of an organism? Give an example of when and where it occurs in humans.

____
____
____

27. Explain how a caterpillar can form a cocoon and then develop into a moth.

____
____
____
____
____

28. Two types of proteins regulate the cell cycle: oncogene proteins and tumor suppressors. Briefly describe the role that each plays in cancer.

Oncogene proteins: ____
____

Tumor suppressors: ____
____

Science Practices & Inquiry

In the AP Biology Curriculum Framework, there is a set of 7 Science Practices. In this chapter, we will focus on **Science Practices 6 and 7.**

Science Practice 6: The student can work with scientific explanations and theories.
Science Practice 7: The student is able to connect and relate knowledge across various scales, concepts and representations in and across domains.

Question #29 looks at Science Practice 7, asking you to represent the connection between meiosis and increased genetic diversity necessary for evolution. (LO 3.10)

Question #30 focuses on Science Practice 6. This question asks you to construct an explanation using a description of an experiment of cell regulation with mitosis. (LO 3.9)

29. An organism has a diploid number of 2n=6. Draw a series of diagrams of chromosomes in the nuclei for this organism as it produces gametes through meiosis, and explain how these gametes are genetically unique from each other.

30. In an elegant set of experiments, Rao and Johnson (published in Nature, 1970) determined some of the important elements of cell cycle regulation. Rao and Johnson fused together mammalian cells at different times of the cell cycle (G1, S, and G2). After fusion, the nuclei were monitored and the time measured for mitosis to occur. Below are three of their experiments with the results.

For each, write a 2-3 sentence conclusion about what the experiment shows.

a. Fusion of S-phase cells with G2-phase cells.
Result: Chromosome replication continued in the S nucleus, while the G2 nucleus was unable to synthesize DNA.

Conclusion:

____

____

____

b. Fusion of S-phase cells with G1-phase cells.
Result: The G1 nuclei rapidly moved into S phase.

Conclusion:

____

____

____

c. Fusion of G1-phase with G2-phase cells,
Result: The entry of the G2 nucleus into mitosis was delayed.

Conclusion:

____

____

____

Chapter 8: Inheritance, Genes and Chromosomes

Chapter Outline

8.1 - Genes are particulate and are inherited according to Mendel's laws
8.2 - Alleles and genes interact to produce phenotypes.
8.3 - Genes are carried on chromosomes.
8.4 - Prokaryotes can exchange genetic material.

Genetic inheritance is explainable and predictable, as explained in Chapter 8. You must gain familiarity with the specialized vocabulary needed to describe this knowledge. With that, you will be able to accurately describe how the inherited characteristics of organisms, such as the spherical versus wrinkled appearances of the garden peas that Gregor Mendel studied, are based on what the parental generation provided to the offspring generation.

In the simple case of the seed appearance in peas, there are two alleles, or alternate forms of the gene. One of the seed-appearance alleles codes for a spherical and smooth appearance on the seed's surface, and another allele codes for a wrinkled and irregular appearance on the surface of the seed. By interbreeding "true-breeding" spherical-seed-producing plants with "true-breeding" wrinkled-seed-producing plants to produce F1 offspring, Mendel found that all of the seeds in the F1 offspring were spherical and smooth. "True-breeding" means that after many generations, the offspring of members of the strain always have the same appearance as the parent, and are assumed to be genetically identical.

An important detail from Mendel's studies is that the appearance (phenotype) of the pea seeds in his crosses was **either** spherical or wrinkled, not something in-between. This shows that even if an individual pea plant inherited the allele for wrinkled seeds from one parent, receiving an allele for spherical and smooth seeds from the other parent results in all of the offspring's seeds being smooth and spherical. Thus, we say that the allele for spherical seed appearance is dominant over the recessive allele for wrinkled seed appearance. Such knowledge gives us predictability for what seeds will look like in offspring, provided we know about their parents.

Geneticists utilize several different systems for symbols of genes; we will utilize a simple system in this chapter by describing the alleles for seed appearance in peas in the following way. The upper-case "S" represents the dominant allele for spherical and the lower-case "s" represents the recessive allele for wrinkled. By custom, most alleles are abbreviated by the first letter(s) of the dominant allele.

The greatest breakthrough Mendel made was not in the demonstration of simple dominant and recessive traits of garden pea phenotypes. Rather, it was what happened when he bred together the spherical F1 offspring (discussed in the previous paragraph). These F1 seeds and the plants that grew from the seeds must have received one allele for spherical and the other allele must have been of the wrinkled trait. We know that this is true because when producing gametes (pollen and ova in plants), only one of the two alleles for a gene will be present in any given gamete. Thus, there will be some pollen that have a spherical allele for the gene of coat appearance and other pollen that will have the wrinkled allele, with the same either-or situation being true for the ova. When breeding the F1 plants together to make the F2 generation, Mendel kept track of how many smooth seeds were produced by the F2 plants, and produced that famous result, that 75% of the F2 were smooth and spherical while the other 25% were wrinkled. Mendel's results grew from his having examined tens of thousands of peas.

The possible genotypes for seed appearance in the F2 offspring described in the previous paragraph is limited to this set: SS, Ss, sS, and ss, with the reminder that each parent delivers one of the alleles in its offspring and the other parent is the source of the other. Any F2 plant with a single "S" will produce spherical seeds, and we see that there are three such combinations (SS, Ss, and sS). By the same rule, there is only one combination of alleles (ss) that will result in wrinkled seeds. Therefore, there are three times as many spherical seeds (75%) as there are wrinkled seeds (25%) in the F2 generation.

In addition to understanding straightforward genetics, there are some tricks of nature presented in this chapter. For example, mitochondria and chloroplasts are organelles that almost exclusively come from the maternal parent because their gametes, the ova, are much larger cells than sperm. The larger cells are the ones that provide mitochondria and chloroplasts, as pollen and spermatozoa are generally too small to carry any cargo other than the primary genetic information. Since mitochondria and chloroplasts have some genes (and, therefore, DNA) of their own, the "mother" is typically the sole source of mitochondria and chloroplast genes.

Big Idea 3, stating that living systems store, retrieve, and transmit information essential to life processes, is the dominant theme in Chapter 8. In more detail, Chapter 8 includes:

3.a.3: The chromosomal basis of inheritance provides an understanding of the pattern of passage transmission) of genes from parent to offspring.
3.a.4: The inheritance pattern of many traits cannot be explained by simple Mendelian genetics.
3.c.2: Biological systems have multiple processes that increase genetic variation.
3.c.3: Viral replication results in genetic variation, and viral infection can introduce genetic variation into the hosts.

Big Idea 4 states that biological systems interact in complex ways. Included in Chapter 8:

4.c.2: Environmental factors influence the expression of the genotype in an organism.

Chapter Review

Concept 8.1 shows how genes are particulate and are inherited according to Mendel's laws. Here are some key vocabulary terms in genetics that you need to be familiar with: character, trait, parental generation (P), first filial generation (F1), second filial generation (F2), test cross, monohybrid cross, dominant, recessive, allele, homozygous, heterozygous, dihybrid cross, phenotype, genotype, law of segregation, Punnett square, law of independent assortment, and pedigree analysis.

1. What can be deduced about the observation that all offspring have the same phenotype of only one of the parents when two true-breeding animals with different traits are bred?

____

____

____

____

2. Predict and explain the expected result on the F1 of breeding one type of wheat that is homozygous for a dominant trait with another strain that is homozygous with the recessive allele.

____

____

____

____

3. Why was it important for Mendel to make observations of the F2 generation after he completed his careful observations of the F1 generation?

____

____

____

____

Concept 8.2 shows how alleles and genes interact to produce phenotypes. Over the long term, it is clear that there have been many changes in the appearance of plants or animals. For example, the phenotype of today's human beings is likely quite different than what was 200,000 years ago. However, we don't need to wait for 10,000 generations of breeding before we see changes in phenotypes. A single gene with two or more alleles can result in two or more phenotypes. An allele present in 99% or more of the phenotypes seen in nature is labeled as the "wild" type allele of the gene, with alternate, uncommon alleles becoming labeled as "mutants." In addition, the phenotype might be under the control of several genes that act in different patterns of expression when present in different combinations of alleles.

Pair each of the four following terms to one of the four descriptions below, and then explain why that pairing demonstrates that term.

a. epistasis
b. dominance
c. co-dominance
d. incomplete dominance

4. True-breeding spherical peas bred with true-breeding wrinkled peas produced offspring that were all spherical.

Matching term: ________ Explanation: ____

5. True-breeding white-flowered snapdragons bred with true-breeding red-flowered snapdragons produced offspring that were all pink-flowered.

Matching term: ________ Explanation: ____

6. A man with blood type A and a woman with blood type B produce a daughter of blood type O.

Matching term: ________ Explanation: ____

7. A male Labrador dog with black coat-color and a female Labrador dog with chocolate coat-color produce a puppy that has a yellow coat color.

Matching term: ________ Explanation: ____

Concept 8.3 examines how genes are carried on chromosomes. The seven phenotypic traits selected by Mendel for detailed study are all examples of the simple dominant/recessive pattern of trait characterization. His idea that there are two "determinants" of each trait meshed well with later observations that chromosomes also exist as pairs in cells (except in sperm and ova). Thus, it was recognized, and later demonstrated, that one "determinant" can be on one chromosome, while the other member of the chromosome pair might have the same or another "determinant."

We now know that DNA, the genetic code, is arranged in the form of the chromosomes that are found in the nucleus of eukaryotic cells. The "law of independent assortment" most directly applies to genes that reside on different chromosomes, as genes that are on different chromosomes are inherited independently of each other.

Not all genes are on different chromosomes, of course, and, in general, two different genes that are located on the same chromosome are more likely to be inherited together than are two different genes that are located on different chromosomes. Thus, genes on the same (autosomal) chromosome are said to be "linked" in their pattern of inheritance. Linkage by virtue of sharing space on the same chromosome does not forever bind two genes to be inherited together. In the production of gametes, it is possible for a type of chromosomal exchange called "crossing over" in which genes undergo recombination when exchanged between homologous chromosomes, as long as these are not sex chromosomes (review Figures 7.13 and 8.14 in the text).

For the special case of genes located on the X and Y (sex) chromosomes, it is impossible for exchange between X and Y chromosomes, and among flies and humans, each male offspring (XY sex chromosomes) will have a Y chromosome identical to its paternal parent's Y. Each male offspring must also receive an X chromosome from its mother. Although the Y chromosome has only a very limited number of genes related to the differentiation of the testes, the X chromosome contains many, many more genes, and therefore the maternal parent determines many more of its male offspring's traits than does its paternal parent, especially if the allele is recessive. Note that a male with a mutation or rare allele on the X chromosome can indeed pass along that mutation or rare allele, but only to his female offspring.

8. Plant scientists studying inheritance in sweet peas developed two pure-breeding strains of peas, one with purple flowers and the other pure-breeding strain with red flowers. When these two strains were crossed, all of the F1 were purple. Another trait of interest was pollen grain appearance, and two pure-breeding strains were produced, one strain yielding long (tube-like) pollen grains and the other round pollen grains. When these two strains were crossed all of the offspring had long pollen grains.

For a mating between pure-breeding purple-flowered, long-grained pollen peas and pure-breeding red-flowered, round-grained pollen peas, predict the flower color and pollen shape of 1,000 of the F1 generation.

Now predict the flower color and pollen shape of 1,000 of the F2 back-cross generation, assuming Mendel's Law of Independent Assortment applies to the flower and pollen traits.

Develop the alternative prediction for what 1,000 of the F1 generation from the cross above would have if Mendel's Law of Independent Assortment did not apply to the flower and pollen traits.

9. Steroid hormones such as testosterone (gonadal androgens) lead to differentiation of the male reproductive system. However, if the receptors for the hormones are non-functional due to a mutation in the gene for the androgen-receptor protein, a condition called "complete androgen insensitivity" is likely to develop and result in a female-like external phenotype. Note that all affected individuals have a Y chromosome that is typically normal. The syndrome is not seen in genetic (XX) females, however. Speculate on the chromosomal location of the mutation that causes this developmental abnormality and explain your answer fully.

10. Thomas Hunt Morgan reported on many different genetic crosses he made with fruit flies, including an example of eye color inheritance. Red eye-color is dominant (R) over white eye-color (r). For the cross of a white-eyed female (X^rX^r) with a red-eyed male (X^RY), show all possible genotypes, including sex, that can be formed and give an explanation of the eye color in each of these offspring.

11. The karyotype figure* to the right shows the chromosomes of a person afflicted with Down syndrome, a trisomy in which there are more than the usual number of chromosome copies.

Find the trisomy in the karyotype, and explain its likely origin in the karyotype's source. As a clue, consider whether or not the trisomy might more likely result from an error in mitosis or an error in meisosis.

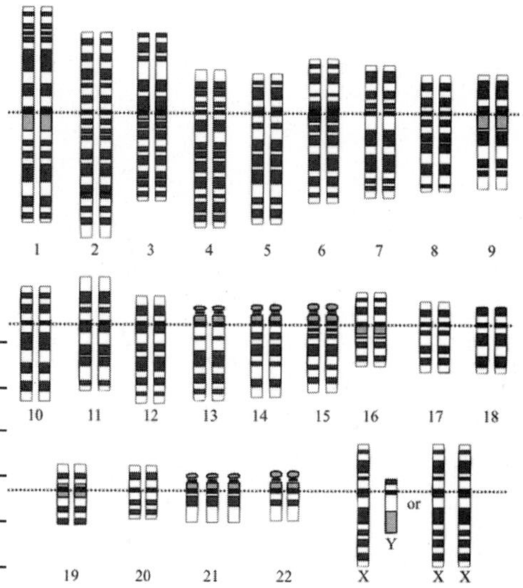

*Courtesy: National Human Genome Research Institute

Concept 8.4 examines how prokaryotes can exchange genetic material. Most prokaryotes reproduce asexually by cloning (binary fission) and have only a single chromosome that is found inside the cytosol of their cells. Evolution of prokaryotes is strongly driven by mutational changes in DNA, but there exist a limited number of ways that gene exchange between individuals can occur, even without sex. Bacteria can form a connection called a sex pilus between two organisms, and after DNA is moved through pili or conjugation tubes, segments of the DNA can be interchanged between the two genomes, via a process called bacterial conjugation and genetic recombination. In a second variation of gene exchange, plasmid DNA of bacteria, a small circle of DNA independent of the larger segment of DNA in the chromosome, can move between bacteria and result in DNA transfer between individuals.

12. A patient was admitted to a hospital with a new pathogenic strain of E. coli that shows resistance to antibacterial soap. Identify through which process the E coli was able to acquire this new trait.

____
____
____
____
____
____
____
____

13. As you go through the next week of school, identify every product that you use that is marketed as "anti-bacterial." Do you think it is wise to make so many products so widely available?

____
____
____
____
____
____
____

Science Practices and Inquiry

Chapter 8 provides many opportunities to extract knowledge from numbers. Mendel not only came up with ideas to explain inheritance in pea plants, he also made the genetic crossings and spent long hours counting phenotypes in progeny. Chapter 8 includes opportunities to use these practices:

Science Practice 1: The student can use representations and models to communicate scientific phenomena and solve scientific problems.
Science Practice 2: The student can use mathematics appropriately.
Science Practice 3: The student can engage in scientific questioning to extend thinking or to guide investigations within the context of the AP course.
Science Practice 6: The student can work with scientific explanations and theories.
Science Practice 7: The student is able to connect and relate knowledge across various scales, concepts and representations in and across domains.

Questions #14 and #15 ask you to analyze the empirical outcomes from genetic crosses in order to extract knowledge about the pattern of inheritance of genes. Question #16 provides data you will use to estimate the proximity of two genes.

In a completely hypothetical case, assume that "yogurt flies" were recently discovered in the space station, and that these flies complete their life cycle in only 90 hours, making them ideal for genetic experiments. The astronauts observed that there was variation in fly phenotypes. With lots of time on their hands, the astronauts decided to run some genetic crosses to see if the inheritance patterns in space match those back on the earth.

14. First, they crossed a true-breeding female with a purple abdomen to a true-breeding male with a pink abdomen. Here are the results of that cross:

	Females	Males
Purple abdomen	787	774
Pink abdomen	0	0

Being careful scientists, they arranged another cross, a true-breeding female with a pink abdomen to a true-breeding male with a purple abdomen, yielding:

	Females	Males
Purple abdomen	646	702
Pink abdomen	0	0

Discuss the inheritance pattern shown above.

____

____

____

____

____

15. Continuing on with hypothetical results, another trait was tested by crossing strains. First, they crossed a true-breeding female with four antennae to a true-breeding male with two antennae. Here are the results of that cross:

	Females	Males
Four antennae	827	904
Two antennae	0	0

Seeking balanced, they did the reciprocal cross: mating a true-breeding female with two antennae to a true-breeding male with four antennae. They carefully checked their results and these are the numbers:

	Females	Males
Four antennae	757	0
Two antennae	0	690

Discuss this second example of an inheritance pattern, and compare it with the results on abdomen color.

____

____

____

____

____

16. A fruit scientist conducted genetic experiments to breed a pomegranate fruit that is more juicy and that stays fresh longer on the grocery shelf. She isolated two genes, F and J. She developed and then crossed these two pure-breeding lines of pomegranate shrubs: FfJj x ffjj

Predict what the cross would yield in percentages of offspring type if the gene are not linked, i.e., subject to Mendel's Law of Independent Assortment.

%	offspring of the FJ type
%	offspring of the FJ type
%	offspring of the FJ type
%	offspring of the FJ type

Explain the percentages you placed in the table above.

The actual results from her cross follow:

518 offspring of the FJ type
175 offspring of the Fj type
168 offspring of the fJ type, and
492 offspring of the fj type
1,530 total offspring

Discuss what these data suggest about the linkage and the distance between genes F and J.

____

____

____

____

____

____

Chapter 9: DNA and Its Role in Heredity

Chapter Outline

9.1 - DNA Structure Reflects Its Role as the Genetic Material
9.2 - DNA Replicates Semiconservatively
9.3 - Mutations Are Heritable Changes in DNA

In this chapter we examine the structure and function of another macromolecule, DNA. DNA is a nucleic acid that is responsible for transmitting heredity from one generation to another. DNA occurs as a double-stranded helix, typically packaged in the form of a chromosome. DNA replication occurs in a semiconservative fashion, meaning that each of the two "new" DNA copies made in this fashion have one strand of the original DNA and one strand that has been newly synthesized. Although the vast majority of DNA copying is exact and precise, changes in DNA called mutations frequently occur. Some mutations are small and inconsequential while others change the organism and can result in the whole-scale rearrangement of a chromosome.

As you read this chapter, pay particular attention to the experimental evidence presented. This evidence reveals how we know what we know about DNA. Many of the DNA research experiments were "wet lab" procedures, like those of Rosalind Franklin, Maurice Wilkins, Hershey and Chase, and Messelson and Stahl. Others, including Watson and Crick, did little to no experimentation, but rather built models based on the work of multiple experiments. Both experimentation and modeling are important scientific practices that are highlighted in this chapter.

Chapter 9's coverage of the AP Biology Curriculum Framework is centered primarily in Big Idea 3, but also includes the idea of emergent properties found in Big Idea 4.

Big Idea 3 states that living systems store, retrieve, and transmit information essential to life processes. Specifically, Chapter 9 lays the groundwork for heredity with its discussion of the structure and function of DNA.

3.a.1: DNA, and in some cases RNA, is the primary source of heritable information.
3.a.4: The inheritance pattern of many traits cannot be explained by simple Mendelian genetics.
3.c.1: Changes in genotype can result in changes in phenotype.
3.c.2: Biological systems have multiple processes that increase genetic variation.
3.c.3: Viral replication results in genetic variation, and viral infection can introduce genetic variation into the hosts.
3.d.1: Cell communication processes share common features that reflect a shared evolutionary history.

Big Idea 4 states that biological systems interact in complex ways. Included in Chapter 9 is
4.a.1: The subcomponents of biological molecules and their sequence determine the properties of that molecule.

Chapter Review

Concept 9.1 shows how DNA structure reflects its role as the genetic material. DNA is found in chromosomes as a double-stranded helix. It is a nucleic acid, containing the nucleotides adenine (A) paired with thymine (T), and cytosine (C) paired with guanine (G). The two strands of DNA are antiparallel, meaning that the code of each strand is "read" in the direction opposite to that of the other strand.

1. Describe the major contributions of the following scientists regarding the discovery of the structure of DNA.

Hershey and Chase ____

____

____

Erwin Chargaff ____

____

____

Rosalind Franklin ____

____

____

Watson and Crick ____

____

____

2. Define bacterial transformation and discuss how studies of this phenomenon influenced DNA research.

____

____

3. In living organisms, the amount of adenine is equal to the amount of thymine, and the amount of cytosine is equal to that of guanine. A researcher measured the amount of adenine in a cell, and found it to be 15% of the DNA. Calculate the percent amount of the remaining nucleotides.

____

____

____

4. Explain why the ratio of A+T:C+G is always the same within a single species, yet differs across more than one species.

____

____

____

5. Explain how the double helical structure of DNA allows:

a. Storage of genetic information.

____

____

____

b. Precise replication during the cell division cycle.

____

____

____

c. Susceptibility to mutations.

____

____

____

d. Expression of the coded information as phenotypes.

____

____

____

Concept 9.2 shows how DNA replicates semiconservatively**.** The two strands of DNA are held together by weak hydrogen bonds. These bonds are easily broken as the two strands are pulled apart and the complementary base pairs are laid down to form a new strand by an enzyme called DNA polymerase.

6. The diagram below shows a strand of DNA being replicated.
Label the following: a phosphate, sugar, nitrogenous base, DNA polymerase, growing strand, and template strand. For each strand, label the 5' and 3' ends.

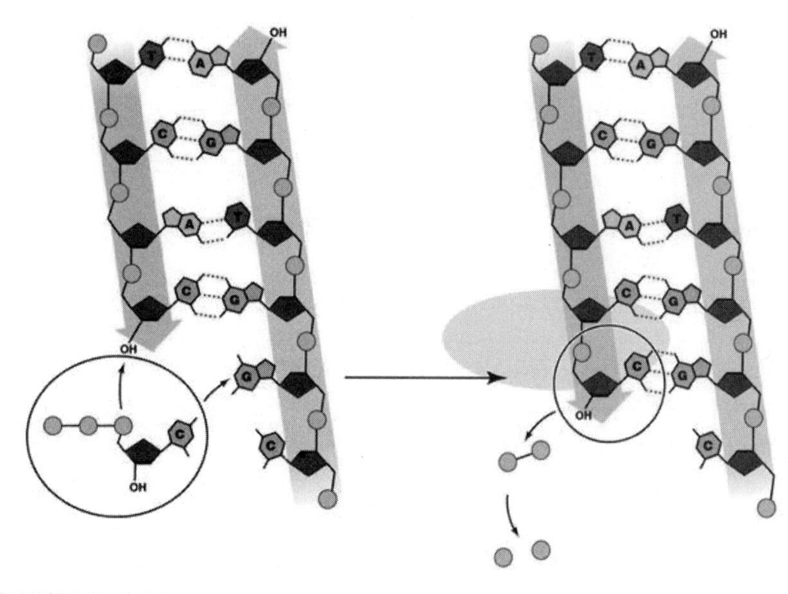

PRINCIPLES OF LIFE, Figure 9.7

7. In the diagram above, the strand on the left shows the addition of a cytosine with three phosphates attached to it. Two of the phosphates will ultimately become detached. What result is achieved by the departure of the two phosphate groups?

____

____

____

8. Briefly describe the function of each of these three enzymes.

Primase: ____

____

____

DNA Polymerase: ____

____

____

Ligase: ____

____

____

9. Discuss continuous and discontinuous replication, using the terms, "leading strand," and "lagging strand."

10. Compare the point of origin of DNA replication between eukaryotes and prokaryotes, and explain how this difference serves an important function.

11. Explain how adjacent Okazaki fragments become linked together to form a continuous strand of DNA.

12. Explain how the ends of a chromosome are shortened each time a chromosome replicates and describe how telomeres help prevent the loss of genetic material.

13. Define the function of telomerase and describe what types of cells are particularly dependent on its continual function.

14. Describe the mechanism by which PCR proceeds, and discuss PCR's use.

15. Assume you need to amplify (copy) a single gene from a eukaryotic organism with 8 chromosomes. Describe the "ingredients" you would need and state the function of each ingredient.

16. The diagram below shows two strands of DNA being replicated.
   a. Draw in the DNA on the continuous side being formed by DNA polymerase.
   b. Draw in two Okazaki fragments on the discontinuous side, one formed, and the second still being formed by DNA polymerase. Label the spot to be filled in by ligase.

Concept 9.3 describes how mutations are heritable changes in DNA. Mutations can occur by the substitution of single nucleotides or by rearrangement of large segments of chromosomes.

17. When a person develops skin cancer as an adult, is this caused by a somatic mutation or germ line mutation? Explain your answer.

____

____

____

18. Explain the difference between silent and loss-of-function mutations, and explain which type is more commonly seen.

____

____

____

19. Describe and discuss the differences between point mutations and chromosomal mutations.

____

____

____

20. Identify 5 different mutagens that are in your environment and indicate how you might avoid each.

a. ____

b. ____

c. ____

d. ____

e. ____

21. Describe the type of mutations that provide the raw material for natural selection. Explain your answer.

____

____

____

22. Discuss how is PCR used to examine the DNA of Neanderthals.

____

____

____

23. If you search the Internet for images of Neanderthals, you will find many older images that depict Neanderthals as dumb, ape-like, and inferior to modern humans. Yet figure 9.20 in your text (shown to the right) depicts a Neanderthal looking very much human-like. Explain why.

PRINCIPLES OF LIFE, Figure 9.20
© 2012 Sinauer Associates, Inc.

Science Practices & Inquiry

In the AP Biology Curriculum Framework, there is a set of 7 Science Practices. In this chapter, we will focus on **Science Practices 2 and 6.**

Science Practice 2: The student can use mathematics appropriately.

Science Practice 6: The student can work with scientific explanations and theories.

Question #24 asks you to use a mutation rate to perform a calculation and then generate a scientific explanation on this topic using your knowledge of genetics. Scientists frequently must draw on their knowledge to explain phenomena in our world. In this case, you are asked to explain a mutation rate, based on your knowledge of the human genome. (LO 3.6)

24. In the 30 April 2010 issue of **Science**, Roach, J. C., *et al.*, reported that the mutation rate for **humans** is approximately 1.1×10^{-8} mutations per base pair in the haploid genome. Humans have a diploid genome of 6×10^{9} base pairs.

a. Calculate the number of mutations in each new child. Show your work.

b. These are spontaneous mutations. Explain why the majority of these mutations have no effect on a new organism.

Chapter 10: From DNA to Protein: Gene Expression

Chapter Outline

10.1 - Genetics Shows That Genes Code for Proteins
10.2 - DNA Expression Begins with Its Transcription to RNA
10.3 - The Genetic Code in RNA Is Translated into the Amino Acid Sequences of Proteins
10.4 - Translation of the Genetic Code is Mediated by tRNA and Ribosomes
10.5 - Proteins Are Modified after Translation

In Chapter 10, we make the transition from stored information, usually DNA, to the retrieval of that information, usually the synthesis of "effector" proteins that change what is happening in and around the cells of living organisms. The overall sequence is:

DNA --transcription--> RNA --splicing--> mRNA --translation--> Polypeptide

Once the transcription of the DNA code to the RNA code is complete, RNA modification occurs, and the exon sequences of RNA get spliced back together, producing "messenger RNA (mRNA)." The mRNA departs from the nucleus and moves to the ribosomes, where it guides protein synthesis. Polypeptides are often modified within cells to become specific effector proteins. Such modifications occur after translation has been completed.

The regulation of which DNA sequences, i.e., genes, are transcribed to make RNA sequences is most directly controlled in the cells by signals called transcription factors (Chapter 11). Each of us, whether male or female, has enough genetic information to make most of the male and female anatomical parts of the reproductive system. That said, it is apparent that most individuals do not have both sets of parts. Why? We only make the reproductive parts we need because the expression of genetic information is closely regulated. In the case of reproductive development, hormones and other signals make sure the correct genes are expressed. Understanding the points of regulation is an important focus of Chapter 11.

Chapter 10 expands on your understanding of Big Idea 3, concerning the nature of information transfer in living organisms. In particular, Chapter 10 examines the underpinnings of genetics, and includes this essential knowledge:

3.a.1: DNA, and in some cases RNA, is the primary source of heritable information.

Chapter Review

Concept 10.1 examines how genetics show that genes code for proteins. Enzymes were an obvious and measurable category of proteins to early researchers, and their observations led to the hypothesis that genetically-determined diseases are often based on mutations in the genes that code for enzymes. Subsequent work led to the one gene-one protein hypothesis. Today we understand this better as the one gene-one polypeptide relationship, because many functional proteins, e.g., insulin, are fragments of much larger polypeptides, e.g., pre-pro-insulin.

1. Add these labels to the following depiction of a cell:

DNA Pre-mRNA tRNA ribosome translation polypeptide transcription

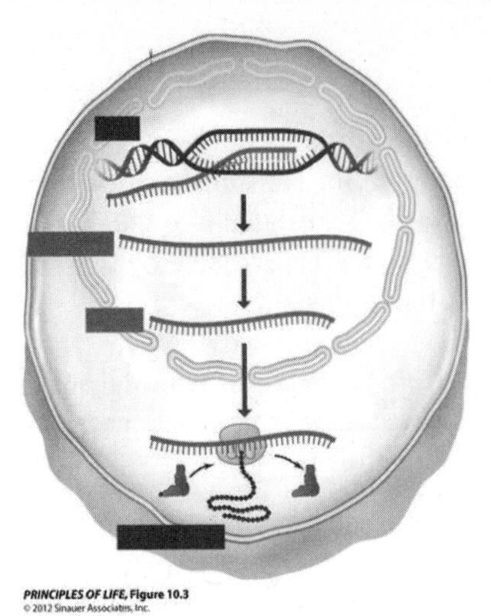

PRINCIPLES OF LIFE, Figure 10.3

2. Describe the structure and function of each of the following types of RNA: mRNA, rRNA, microRNA, and tRNA.

3. A dog was taken to a veterinary clinic because it had become increasingly inactive and lethargic. A blood test determined that the dog had very low levels of the steroid hormones produced by the adrenal glands, especially cortisol. The veterinarian told the owner that the dog has "a mutation in its cortisol gene." Explain why this characterization is inaccurate, given that steroid hormones are lipids. Speculate on a way that a genetic mutation could result in low levels of steroid hormones.

4. Describe the genetic condition of a person who has sickle-cell anemia. Use specific details about which gene is mutated, the different effects of whether the person is heterozygous or homozygous for the mutation, and include information about the polypeptide that has been altered by the mutation.

____

____

____

____

____

____

____

____

____

____

Concept 10.2 shows how DNA expression begins with its transcription to RNA**.** Transcription of DNA begins with the association of an enzyme called RNA polymerase with the DNA template. The necessary "ingredients" for transcription include free nucleoside-triphosphates (ATP, GTP, CTP and UTP) that will be incorporated into the RNA strand as it is built by the enzyme RNA polymerase moving along the DNA sequence undergoing transcription. The phases of transcription are: initiation → elongation → termination. As transcription proceeds, each base in the DNA template strand pairs with its complementary nucleoside phosphate, which is then incorporated into the RNA. Here is a short example of that complementarity:

3'-T-C-A-A-G-T-5' sequence in a DNA strand will result in
5'-A-G-U-U-C-A-3' sequence in the complementary RNA strand.

The start of transcription along a strand of DNA is determined by the presence of a promoter sequence in the DNA, which includes a transcription initiation site. The next part of the sequence specifies the "start" codon UGA in the RNA. Elongation proceeds until a "stop" codon (UAA or UAG) is specified. While still in the nucleus, pre-mRNA receives a G-cap on its 5' end, and a poly-A tail on its 3'end; these modifications enhance RNA stability and for mRNA, facilitate its exit from the nucleus and its binding to ribosomes to start translation.

5. Describe two possible consequences for a gene's expression resulting from a mutation in the promoter region of that gene.

____

____

____

____

____

5. Describe two possible consequences for a gene's expression resulting from a mutation in the stop codon of that gene.

____

____

____

____

____

6. Describe what happens to the “intron” and “exon” sequences of pre-mRNA when it undergoes processing.

7. Describe the functions of the GTP cap and the poly-A tail on mRNA.

Concept 10.3 says that the genetic code in RNA is translated into the amino acid sequences of proteins. The ribosomes “translate” the mRNA code by using it to determine which amino acid gets placed at which position in newly synthesized proteins. More specifically, the mRNA code consists of sequences of triplets of ribonucleotides (three bases in length) called codons. Each mRNA codon determines which amino acid will be placed at added to a growing polypeptide.

8. For this fragment of an mRNA molecule

5’ AUG UUU CAG CGA GGA UGA 3’

use the table to characterize the peptide product from translating this sequence.

First letter	Second letter: U	C	A	G	Third letter
U	UUU Phenyl-alanine	UCU Serine	UAU Tyrosine	UGU Cysteine	U
	UUC Phenyl-alanine	UCC Serine	UAC Tyrosine	UGC Cysteine	C
	UUA Leucine	UCA Serine	UAA Stop codon	UGA Stop codon	A
	UUG Leucine	UCG Serine	UAG Stop codon	UGG Tryptophan	G
C	CUU Leucine	CCU Proline	CAU Histidine	CGU Arginine	U
	CUC Leucine	CCC Proline	CAC Histidine	CGC Arginine	C
	CUA Leucine	CCA Proline	CAA Glutamine	CGA Arginine	A
	CUG Leucine	CCG Proline	CAG Glutamine	CGG Arginine	G
A	AUU Isoleucine	ACU Threonine	AAU Asparagine	AGU Serine	U
	AUC Isoleucine	ACC Threonine	AAC Asparagine	AGC Serine	C
	AUA Isoleucine	ACA Threonine	AAA Lysine	AGA Arginine	A
	AUG Methionine; start codon	ACG Threonine	AAG Lysine	AGG Arginine	G
G	GUU Valine	GCU Alanine	GAU Aspartic acid	GGU Glycine	U
	GUC Valine	GCC Alanine	GAC Aspartic acid	GGC Glycine	C
	GUA Valine	GCA Alanine	GAA Glutamic acid	GGA Glycine	A
	GUG Valine	GCG Alanine	GAG Glutamic acid	GGG Glycine	G

PRINCIPLES OF LIFE, Figure 10.11

9. Describe how the result would change if the above sequence were altered so that 6th base from the 5' end was switched to G.

____

____

____

10. Describe how the result would change if the above sequence were altered so that 19th base from the 5' end was switched to A.

____

____

____

11. Even though the amino acid serine is specified by UCU, UCC, UCA, UCG, AGU and AGC, is it claimed that "the code" is not ambiguous. Explain.

____

____

____

____

12. Explain the comment, "the genetic code is nearly universal," in terms of evolutionary ancestry.

____

____

____

____

13. Explain why the human gene for insulin can be inserted into an E coli bacterium and the bacterium can then produce human insulin.

____

____

____

____

14. Distinguish between the consequences of "silent" and "nonsense" mutations.

____

____

____

____

Concept 10.4 describes how translation of the genetic code is mediated by tRNA and ribosomes. The amino acids incorporated into newly synthesized proteins are carried on specialized RNA carriers known as "transfer RNA" (tRNA). Each tRNA molecule includes a triplet sequence of RNA that will be complementary to, and will bind with, a particular codon of mRNA, triggering the delivery of the tRNA's amino-acid cargo to the growing polypeptide. After that delivery is complete, the remnants of the tRNA leave the ribosome, and the mRNA code is then moved along to the next codon, thus specifying the next amino acid to get delivered and covalently peptide-bonded into the new and growing polypeptide.

15. Explain the relationship between codons and anticodons.

____

____

____

____

____

16. Describe the activity of a single ribosome, and then describe how more than one ribosome can be active in that same process.

____

____

____

____

____

____

17. For the ribosome shown to the right, state the anticodon sequence for the next tRNA to bind (at the bottom of the figure) and then use the table from question 8 to find what will be the next amino acid incorporated into the peptide.

____

____

____

____

____

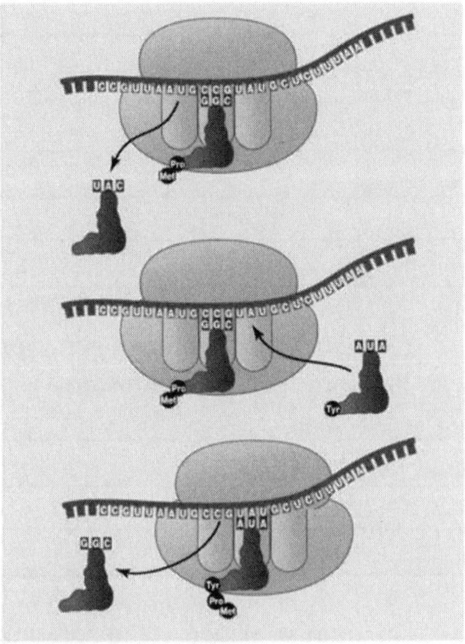

PRINCIPLES OF LIFE, Figure 10.16 (Part 2)
© 2012 Sinauer Associates, Inc.

18. Compare the actions of RNA polymerase and ribosomes.

____

____

____

____

____

Concept 10.5 looks at how proteins are modified after translation. As soon as the mRNA has been translated all the way to its "stop" codon (UAA, UAG or UGA), the polypeptide will be released from the ribosome and is either used directly, as is, or is modified by other parts of the cell. Three major categories of post-translational modification include the following. First, cutting a long polypeptide into smaller proteins is called proteolysis, and is accomplished by proteolytic enzymes. Second, carbohydrates might be added onto the polypeptide, especially those proteins that play a role in cellular identity. Third, many proteins are covalently modified by phosphorylation, thus altering the shape and the function of the protein.

19. Secreted proteins, including those that act as hormones, typically interact with two membrane-bound organelles prior to secretion. Describe the activities of the two organelles in terms of preparing the polypeptide for secretion out of the cell.

____

____

____

____

____

20. Human insulin has 52 amino acids and yet it is a product of the INS gene, which has nearly 40,000 base pairs. Explain how this is possible.

____

____

____

____

____

21. Add these labels to the diagram below:

RER   phospholipid bilayer   mRNA   polypeptide   ribosome.

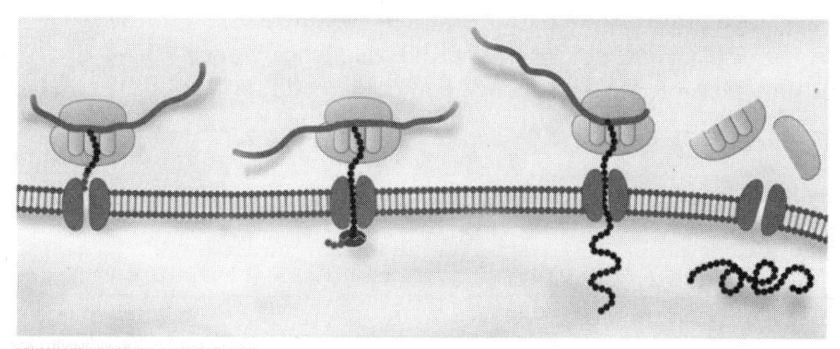

PRINCIPLES OF LIFE, Figure 10.19 (Part 2)

Science Practices & Inquiry

In the AP Biology Curriculum Framework, there is a set of 7 Science Practices. n this chapter, we will focus on two of the Science Practices:
Science Practice 1: You can use representations and models to communicate scientific phenomena and solve scientific problems. More specifically, use Practice 1.2: to describe representations and models of natural or man-made phenomena and systems in the domain.
Science Practice 6: You can work with scientific explanations and theories. More specifically, use Practice 6.4: to make claims and predictions about natural phenomena based on scientific theories and models.

Question 22 asks you to describe representations and models illustrating how genetic information is translated into polypeptides (LO 3.4) while #23 asks you to predict how a change in a specific DNA or RNA sequence can result in changes in gene expression (LO 3.6).

22. Researchers have determined that a short chain polypeptide signal (nuclear localization signal, NLS) comprised of eight amino acids must be attached to a protein if it is to enter the nucleus. When the NLS-protein complex docks with a pore in the nuclear membrane the signal causes the pore in the nuclear membrane to open.

The sequence for this NLS peptide is: -Pro-Pro-Lys-Lys-Lys-Arg-Lys-Val-

A. Write out a mRNA strand that could produce this sequence.
B. What is the minimum number of DNA nucleotides needed to produce this sequence?
C. Explain why methionine is not part of the above sequence.
D. Write out the DNA sequence that this mRNA sequence comes from.
E. Is this the only possible DNA sequence that could code for this NLS? Explain your answer.

23. Below are the results of an experiment used to determine the function of the NLS:

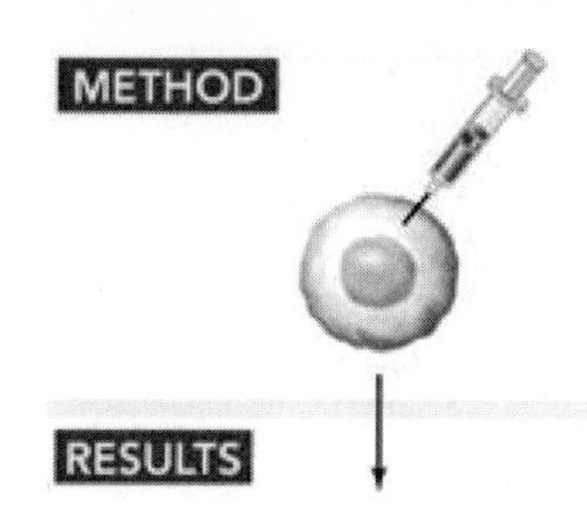

In scenario Y to the left, an NLS-protein-red-dye complex that was injected into a cell is later found in the nucleus. In scenario Z, the protein-red-dye complex, lacking the NLS peptide, is injected into the cell, and it is later found only in the cytoplasm.

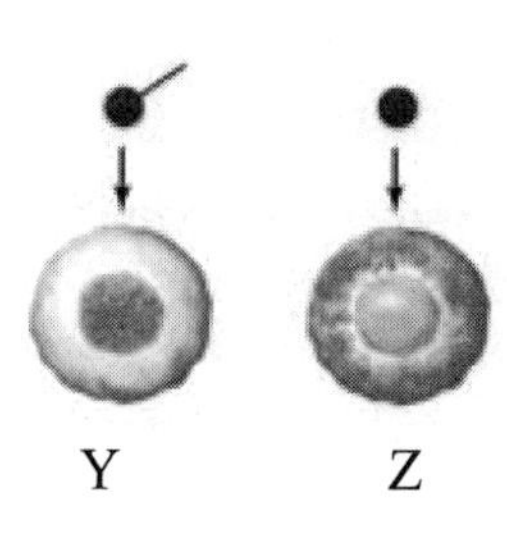

For each of the three scenarios below, A, B and C, predict where the protein-NLS complex will be found when injected into a cell, and explain why.

(A) A cytosolic protein, normally found only in the cytoplasm, is bound to the NLS peptide.

____

____

____

____

____

(B) A nuclear protein is attached to a mutated form of the NLS missing the final valine.

____

____

____

____

____

(C) A nuclear protein is attached to a NLS that was produced from the DNA sequence:
TAC-GGG-GGT-TTT-TTC-TTC-GCT-TAC-CAC-stop

____

____

____

____

____

Chapter 11: Regulation of Gene Expression

Chapter Outline

11.1 Several Strategies Are Used to Regulate Gene Expression
11.2 Many Prokaryotic Genes Are Regulated in Operons
11.3 Eukaryotic Genes Are Regulated by Transcription Factors and DNA Changes
11.4 Eukaryotic Gene Expression Can Be Regulated after Transcription

The DNA code is the transcription template for making RNA, which is then processed to make mRNA. The mRNA code is translated at the ribosomes to guide the synthesis of a polypeptide or a protein. Most of the cells of an organism carry the full set of that organism's DNA code.

The expression of different genes in the DNA in different cells is what leads to the differences between liver cells, skin cells, bone cells or any of the thousands of different kinds of cells in a human. In addition to encoding cell types, DNA encodes the messages necessary to repair cells, maintain homeostasis, regulate cell death, and address a host of other activities. Inducible genes, operons, transcription factors, post transcriptional factors and the cell's environment all work in concert in a dynamic fashion to maintain homeostasis and serve the cell's functions.

Chapter 11's coverage of the AP Biology Curriculum Framework folds together three of the Big Ideas.

Big Idea 2 states that the utilization of free energy and use of molecular building blocks are characteristic of processes fundamental to life. Specifically, Chapter 11 looks at regulation of DNA and gene expression:

2.c.1: Organisms use negative feedback mechanisms to maintain their internal environments and respond to external environmental changes.

Big Idea 3 states that living systems store, retrieve and transmit information essential to life processes. Specifically, Chapter 11 continues with DNA, but now with the regulation of DNA:

3.b.1: Gene regulation results in differential gene expression, leading to cell specialization.

Big Idea 4 states that biological systems interact in complex ways. Continuing with the theme of regulation, Chapter 11 looks at the effects of the environment as well:

4.c.2: Environmental factors influence the expression of the genotype in an organism.

Chapter Review

Concept 11.1 examines how gene regulation can involve positive and/or negative regulation. Transcription factors are regulatory proteins that bind to the DNA to prevent transcription or to activate it. Viruses, in particular bacteriophages, provide a convenient model to study this as they infect bacteria and turn them into virus factories.

1. It has long been known that there is probably a genetic link for alcoholism. Researchers studying rats have begun to elucidate this link. Briefly describe the genetic mechanism found between alcoholism and rats.

____

____

____

____

____

2. In the diagram below, label and identify the 5 potential points (arrows) for the regulation of gene expression.

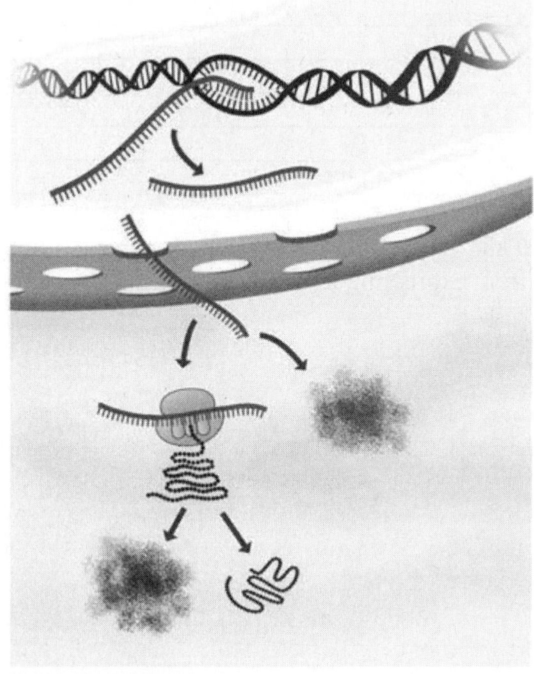

PRINCIPLES OF LIFE, Figure 11.1

3. Explain the primary difference between constitutive and inducible genes, and provide an example of each.

____
____
____
____
____
____

4. Explain why viruses are not considered to be cellular organisms.

____
____
____

5. Describe and explain the claim that the 4 types of viruses are distinguished by difference in their genetic material.

____
____
____
____

6. Describe the lytic and lysogenic phases of viral reproduction.

____
____
____
____
____
____

7. Explain how bacteriophages are an example of positive regulation and of postranscriptional regulatory mechanisms.

8. Draw and label a diagram that shows how an infected cell attempts to utilize transcription terminator proteins to prevent a retrovirus from expressing itself. Include in your diagram: the nucleus, nuclear membrane, and provirus.

9. Explain how HIV counteracts the process you described in your answer to the previous question on retrovirus protection.

Concept 11.2 shows how operons are the units of transcriptional regulation in prokaryotes. They allow prokaryotes to conserve energy and resources by only turning genes on when necessary. Operons are a primary means of regulation in prokaryotes.

10. Describe the three primary parts of the *lac* operon.

11. Genetic mutations are useful in analyzing the control of gene expression. In the *lac* operon of *E. coli*, gene *i* codes for the repressor protein, *Plac* is the promoter, *o* is the operator, and *z* is the first structural gene. (+) means wild (or normal) type; (–) means mutant. Fill in the table below by writing "YES" or "NO" in box of the following table, describing the level of transcription in different genetic and environmental conditions.

Z TRANSCRIPTION LEVEL		
GENOTYPE	LACTOSE PRESENT	LACTOSE ABSENT
i– Plac+ o+ z+		
i+ Plac+ o+ z–		
i+ Plac– o+ z+		
i+ Plac+ o– z+		

11. The Trp operon is a repressible operon. Explain what is meant when the Trp operon is described as a repressible operon, and how this regulatory function is important to a bacterial cell.

____

____

____

____

12. Describe two features of bacteria that make them especially useful for studying the mechanisms of gene regulation.

____

____

____

____

____

____

Concept 11.3 shows how eukaryotic gene regulation is often much more complex than operons. While operons are sometimes found (in eukaryotes), genes are also regulated at multiple other points as a gene is transcribed to RNA. In addition to cellular factors regulating a gene, epigenetics or environmental factors also regulate genes. These environmental factors can be passed down through multiple generations.

13. Explain why transcription factors are found more commonly in eukaryotes than in prokaryotes.

____

____

____

____

____

____

14. Describe all the necessary components that must be present before RNA polymerase II can transcribe a segment of DNA.

____

____

____

____

____

____

15. In the diagram below, label: DNA, TATA box, transcription initiation site, promoter, transcription factors, and RNA polymerase II.

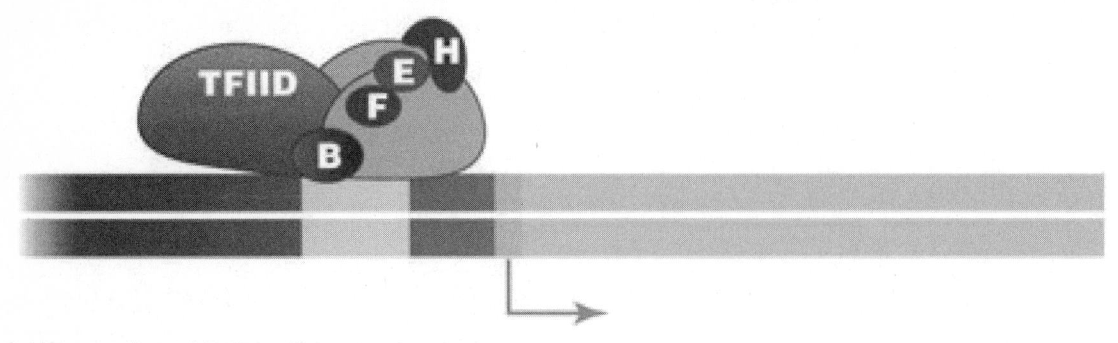

PRINCIPLES OF LIFE, Figure 11.10 (Part 2)

16. Describe the chemical matching between a transcription factor and the DNA sequence to which it can bind.

____

____

____

____

____

17. When a transcription factor binds to DNA, there is an induced fit. What is an induced fit and how does it occur?

____

____

____

____

____

18. Discuss the functional importance of coordinated gene expression.

____

____

____

____

____

19. Epigenetic changes to DNA alter gene expression without changing DNA sequences, and these changes are passed to offspring. Describe two different ways that epigenetic changes to DNA can occur, and describe how these changes can be inherited.

____

____

____

____

____

____

____

20. Early in life, identical twins are often difficult to tell apart and behave very similarly. As they age, subtle differences begin to appear and as they reach middle age, they often have distinct differences. Explain how this can occur at a molecular level with their DNA.

____

____

____

____

____

____

Concept 11.4 looks at what happens after transcription, when gene regulation continues with alternative splicing and translational controls. After a protein is produced, its longevity in the cell is also regulated as proteasomes degrade proteins after they have reached the end of their usefulness to a cell.

21. In *Drosophila*, sex is determined by a gene that has four exons, which we will designate 1, 2, 3, and 4. In the female embryo, splicing generates two active forms of the protein, containing exons 1 and 2, and 1, 2, and 4. However, in the male embryo, the protein contains all four exons (1, 2, 3, and 4) and is inactive. Draw a diagram similar to Figure 11.16 that represents this process.

22. Explain how humans have approximately 24,000 genes, yet have at least 85,000 mRNA's.

____

____

____

____

____

____

____

____

____

23. In the figure below, label: targeted protein, ubiquitin, and proteasome.

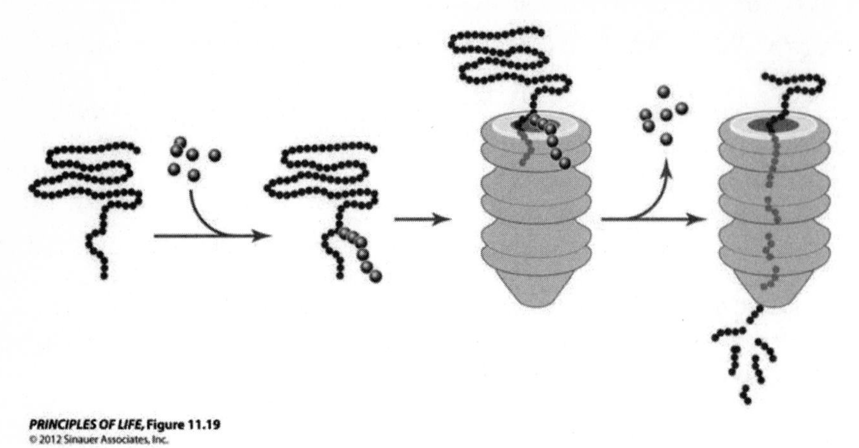

PRINCIPLES OF LIFE, Figure 11.19

24. Explain the role of ubiquitin in the figure in the preceding question.

____
____
____
____
____

25. Explain why miRNA's are considered to be a gene silencing mechanism.

____
____
____
____
____

26. Complete the table below for the different types of gene regulation:

	Location in cell	Molecule(s) acted on	Example
operons			
transcription factors			
translational			
epigenetics			
chromatin remodeling			
miRNA			
alternative splicing			
translational repressors			
proteasome			

Science Practices & Inquiry

In the AP Biology Curriculum Framework, there is a set of 7 Science Practices. In this chapter, we will focus on:
Science Practice 7: The student is able to connect and relate knowledge across various scales, concepts and representations in and across domains. More specifically, you will focus on Practice 7.1: The student can connect phenomena and models across spatial and temporal scales.

Question #27 asks you to describe the connection between the regulation of gene expression and observed differences between individuals in a population. (LO 3.19)

27. In an experiment, fruit flies showed unusual outgrowths on their eyes. This trait lasted for 11 - 13 generations of offspring until breeding of fruit flies resulted in normal fruit flies.

   a. Could this change have occurred due to a mutation or change in the DNA sequence? Why or why not?
   b. Explain how this change can occur and then result in normal fruit flies.

____
____
____
____
____
____
____
____
____
____
____
____
____
____
____

Chapter 12: Genomes

Chapter Outline

Having sampled some of the diverse mechanisms by which gene expression is regulated at the level of transcription, translation, and protein modification, Chapter 12 draws your attention back to the organism's full set of genes, its genome. To refer to an organism's genome is to refer to the full DNA sequence that constitutes its complete set of genes. Evolution's effects are especially apparent at the level of the phenotypes, which is based on variation in genomes.

Prokaryotes, eukaryotes and viruses are three major life forms for which there is detailed knowledge at the genomic level. Prokaryotic cells are produced via binary fission, and each resulting cell typically has the full set of genes. Cell replication is called mitosis in eukaryotes, and it includes copying the genome, so that all cells have the full genome. Viruses have the capacity to take advantage of, and even insert DNA into, the genomes of both kinds of cells. Mistakes in the copying of DNA, called mutation, provide additional possibilities for evolution in the face of environmental uncertainty. Genomic analysis allows detailed studies of gene/protein structure/function and of evolutionary pathways.

Chapter 12 includes material characterized in **Big Idea 3.** The type of heritable information in cells that effects change for them is their set of genes, i.e., their DNA or genome. The genome typically includes much more information than a single cell might use, but different information (different genes) in the genome can be activated in response to different signals found in the environment. These responses usually promote homeostasis in the organism. Also addressed:

3.A.1: DNA, and in some cases RNA, is the primary source of heritable information.
3.A.3: The chromosomal basis of inheritance provides an understanding of the pattern of passage (transmission) of genes from parent to offspring.
3.C.3: Viral replication results in genetic variation, and viral infection can introduce genetic variation into the hosts.

Chapter Review

Concept 12.1 discusses the methods for sequencing genomes and analyzing gene products. The "instructions for life" are found in every cell that has the full complement of DNA. Sequencing the genome started as an expensive and laborious process, but it has been greatly accelerated by automated techniques. Even so, the DNA molecule is typically an incredibly long molecule: human chromosome one alone has 246,000,000 base pairs in it.

Genomic analysis provides the basis for the analysis of protein structure and function, as proteins are among the direct products of gene expression. More than one protein can be the product of a gene being expressed; one big protein can be cut up into several smaller proteins. For the approximately 23,000 genes in the human genome, there are about 65,000 proteins known. Because many proteins serve as enzymes in the synthesis of lipids and other biomolecules, the analysis of the full set of active biomolecules has its own label: metabolomics.

Genomic analysis spans many fields of interest: evolutionary biology, molecular details of genetic function in health and disease, RNA functions, open reading frames, intron analysis, protein synthesis, chromosome stability, etc. Comparing many DNA sequences of many related organisms provides specific insight on the molecular basis for the evolutionary changes that took them in different directions from their last common ancestor. Recall that certain parts of the DNA sequence code specifically for the start and stop of transcription, and you will appreciate how far genomic analysis has advanced in such a short amount of time.

1. Add these labels to the diagram below:

terminator of transcription	promoter of transcription
centromere	telomere
RNA polymerase	mRNA

add a labeled bracket for the "open reading frame"

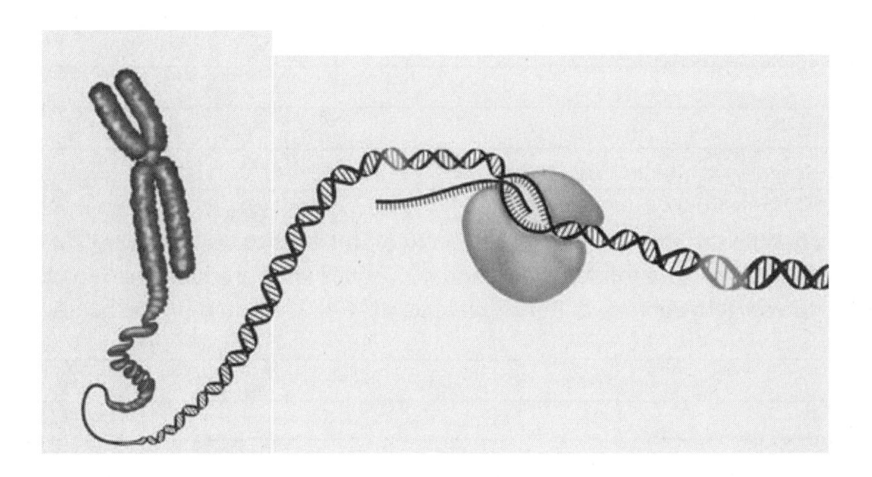

2. Add these labels to the diagram to the right:

mRNA	phenotype
genome	metabolome
proteome	genes

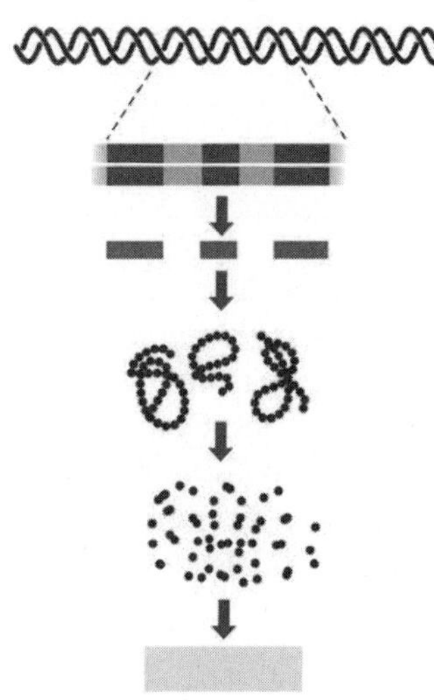

PRINCIPLES OF LIFE, Figure 12.5

Concept 12.2 dictates that prokaryotic genomes are relatively small and compact. Organisms are broadly described as either eukaryotes or prokaryotes, the latter including bacteria, such as *E. coli*, and all of the other microbes that lack membrane-bound organelles. Genomic packaging in the prokaryotes is typically minimal and compact, taking place on a single chromosome. In fact, the first genome to be fully sequenced was that of a prokaryote and the first to be assembled by "synthetic" biologists was also prokaryotic in most of its characteristics.

Very few of the genes in prokaryotes have introns, suggesting efficiency in expression. Genomic analysis of different prokaryotes has revealed the evolutionary expansion of key gene functions such as molecular uptake from the environment and metabolism of fuel molecules. It is also become apparent that some "sets" of base sequences move around in the DNA and can be loaded onto plasmids, greatly enhancing biotechnology approaches to understand life. Most prokaryotes have single copies of genes, i.e., are haploid in all stages.

3. Suppose that you are studying two strains of related bacteria. One is non-pathogenic but has antibiotic resistance, and the other is pathogenic but lacks antibiotic resistance. Speculate on the possible consequences in these two strains if you were to rely only on antibiotic soaps to "sterilize" your hands and equipment as you go back and forth between bacterial strains.

____

____

____

____

____

____

____

4. From a genomic viewpoint, discuss the differences between these two prokaryotes: *Escherichia coli* has 4,288 protein-coding genes, 243 energy-metabolism genes, and 427 genes with products that uptake molecules from the environment, whereas *Mycoplasma genitalium* has only 482, 31, and 34, respectively.

____

____

____

____

____

____

____

____

Concept 12.3 looks at eukaryotic genomes, which are large and complex. The packaging of the genome in eukaryotes is necessarily complex, owing to its large size compared to that of prokaryotes, and eukaryotes generally have many more genes. The DNA of eukaryotes is packaged in structures clearly recognized as chromosomes. Compared to prokaryotes, much more of the DNA in eukaryotes is not transcribed and/or translated and thus serves purposes other than coding for proteins, e.g., telomeres. Eukaryotes are diploid in a significant part of their life cycle, with two copies of each gene present in the diploid cells. Among animals, developmental genes in the genome are broadly shared, reflecting evolutionary history.

5. Develop an argument explaining that the claim that "DNA codes for proteins" is too limited a statement to provide a good description of "life's instruction book."

____

____

____

____

____

____

____

____

6. For each, briefly compare the genomic characteristics of prokaryotes and eukaryotes.

	prokaryotes	eukaryotes
complexity		
DNA location		
chromosomes		
RNA splicing		
size (# of base pairs)		
# of gene copies		
Proportion of DNA		
that is translated		
# of genes		

7. Discuss the claim that viruses directly increase variation in the genomes of prokaryotes and eukaryotes. Include a discussion of transposons in your answer.

____

____

____

____

____

____

8. In analyzing the genomes of diverse eukaryotes, many stretches of "gene duplication" have been found. Assume an original DNA sequence is associated with an essential protein, and that the duplicate sequence is no longer 100% identical to the original sequence. Discuss gene duplication as an evolutionary characteristic.

____

____

____

____

____

____

Concept 12.4 describes the many applications of the human genome sequence. Human DNA includes some 6 billion base pairs, packaged in 23 pairs of chromosomes. Genomic diversity is small; humans are much more alike than they are different, with fully 97% of DNA sequences remaining identical from one person to the next.

9. A newspaper story reported that the genomes of persons with African-American ancestry have fewer genetic components for resistance to malarial infection, compared to the genomes of persons with African-only ancestry. Describe methodology that could have been used to generate these results.

10. Following up on genome findings in the previous question, discuss the suggestion that the above result implies that in the past 300 years, African-Americans have evolved along a pathway different from that of Africans.

11. Assume you are able to measure the height of each of the individual members of a population of 100 people whose individual genomes have been fully sequenced. How would you determine how height is passed on from one individual to another?

12. Now assume you are able to measure intelligence in that same population. Discuss the additional challenges in reaching satisfactory conclusions. What tests might you utilize to determine intelligence? Are there different forms of intelligences?

Science Practices & Inquiry

In the AP Biology Curriculum Framework, there is a set of 7 Science Practices. In this chapter, we will focus on **Science Practice 6:** The student can work with scientific explanations and theories. More specifically, Practice 6.4: The student can make claims and predictions about natural phenomena based on scientific theories and models.

Question 13 provides you **with** enough information to support the claim that humans have the necessary technology needed to manipulate heritable information (LO 3.5).

13. Discuss two technologies that researchers can use to manipulate heritable information. Identify the technology, explain how it is used, and then give an example of its use.

Chapter 13: Biotechnology

Chapter Outline

In this chapter we examine how humans genetically modify organisms. From domesticating dogs and cats to selecting those crops that produce the best yield, humans have been manipulating genomes for thousands of years. More recently, humans have been able to insert the DNA of one organism into another, thus creating recombinant DNA. The use of restriction enzymes, ligase, and gel electrophoresis are all techniques associated with recombinant DNA that you may have practiced in your laboratories. When recombinant DNA is inserted into living organisms, the result is transformed or transgenic organisms. The majority of soy products in modern agriculture are derived from transgenic soy bean plants that have genes for protection from herbicides.

Biotechnology also allows the manipulation of an organism's own genes to answer questions in biology. How does a set of 23 pairs of chromosomes code for a human being? Which genes are turned on and turned off in any given cell? How can we block the expression of a gene, such as the uncontrolled growth of cancer cells? From replacement organs to controlling cancer, producing new high yielding crops and breeding organisms that can clean up toxins from the environment: these are all projects within the scope of biotechnology.

Within the AP Biology Curriculum Framework, Chapter 13 focuses primarily on Big Idea 3. The Big Ideas are a means for organizing the vast amount of information in biology. It is important that while this chapter is only featured in one of the Big Ideas, you still make connections across the other Big Ideas. For instance, how is the expression of DNA regulated and how can humans control this expression?

Big Idea 3 states that living systems store, retrieve, and transmit information essential to life processes. Specifically, Chapter 13 discusses biotechnology and how we can manipulate genomes, through:

3.a.1: DNA, and in some cases RNA, is the primary source of heritable information.

Chapter Review

Section 13.1 examines how restriction enzymes isolated from bacteria are used to cut DNA fragments, producing fragments with sticky ends. These fragments can be separated by size using gel electrophoresis. After cutting open a plasmid with the same restriction enzyme, a DNA fragment can be inserted into the plasmid and sealed into place by using the enzyme ligase.

1. Explain why the brewing of beer is considered to be biotechnology.

____
____
____

2. Restriction enzymes are sometimes called the immune system of bacteria. Explain how restriction enzymes protect bacteria from viruses.

____
____
____

3. How does a bacterium protect its own DNA from being cleaved by restriction enzymes?

____

____

____

4. Explain how "sticky ends" are be used to join DNA from two different organisms.

____

____

____

5. Describe the type of atomic interaction that hold together two fragments of DNA after they have been joined with sticky ends. What characteristics can strengthen this interaction?

____

____

____

6. Describe how DNA is rearranged by meiosis. Compare that description to a description of how recombinant DNA is produced.

____

____

____

____

____

____

7. Plasmids are small circular pieces of DNA. A particular plasmid (pUC 19) is cut with the AvaII restriction endonuclease (enzyme). The pUC 19 plasmid has a size of 26,867 base pairs. The Ava II cuts the plasmid at two locations: 1837 bp and 2059 bp.

a. The circle below represents a plasmid. Draw a map of this plasmid showing the location of these cuts and the size of the fragments.

b. On the gel below, sketch the results that you would expect to see if the fragments were run using gel electrophoresis. Label which end is positive, and which is negative.

8. A DNA molecule of 12,000 bp (12 kilobase, or kb) is cut by restriction enzymes as follows:

CONDITION	SIZES OF FRAGMENTS (KB)
Enzyme A	2, 10
Enzyme B	2, 10
Enzymes A+B	2, 8

a. Sketch the results of the three cuts on the electrophoresis gel below.

b. Indicate on a linear map where each enzyme cuts the DNA.

Section 13.2 shows how recombinant DNA is inserted into an organism to express the DNA. There are many techniques for doing this, including the use of plasmids as vectors and transformation of bacteria. Resistance to antibiotic genes and green fluorescent genes are often used as markers to screen for organisms that have been transformed.

9. What is the advantage of using yeast over bacteria as a model organism for the introduction of recombinant DNA?

10. Explain the benefits of using a viral vector in place of a plasmid vector.

11. What is a reporter gene? Give an example.

____
____
____
____
____
____
____
____

12. Outline the steps that a researcher would need to follow if he or she wished to insert a foreign piece of DNA into a plasmid.

____
____
____
____
____
____
____
____
____

13. A researcher has a small gene that she wants to insert into *E. coli* for expression. This gene has been cut on both ends with the EcoRI restriction enzyme. The researcher obtained a plasmid vector that includes two functional genes:

- the β-galactosidase gene (*lacZ*) that codes for an enzyme that can convert the colorless substrate X-gal into a bright blue product, and
- the gene for resistance to the antibiotic ampicillin (*amp*).

The EcoRI restriction enzyme site on the lacZ gene was cut to insert the small gene of interest. The researcher grew *E. coli* bacteria with the following treatments:

- with no plasmid added
- with plasmid alone
- with recombinant plasmid (gene added inside the lacZ gene site).

a. Draw a diagram of the uncut and the recombinant plasmid below showing how the different genes will look.

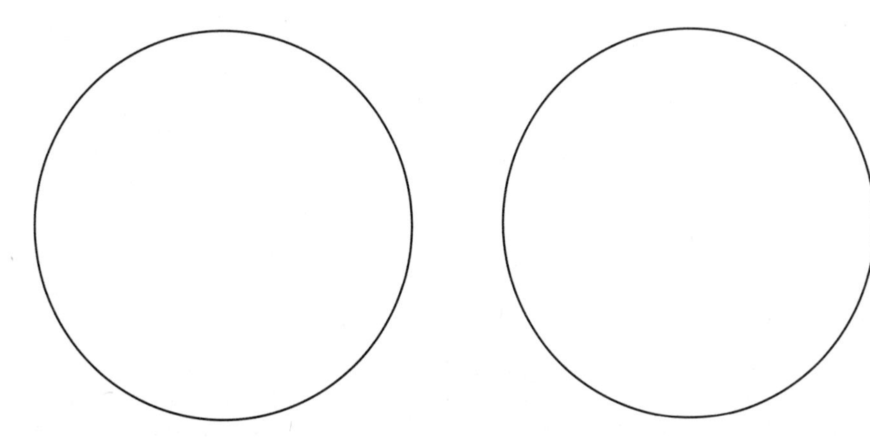

b. Complete the table below with the results the researcher will see after growing the *E coli* bacteria on media containing X-gal and the antibiotic ampicillin.

Treatment	Selection by ampicillin (colonies present or absent)	Color of colonies if present (blue or white)
Grown with no plasmid		
Transformed with plasmid		
Transformed with recombinant plasmid		

Section 13.3 reviews some of the molecular methods for manipulating gene expression including the selection or creation of DNAs for amplification, the detection of expressed genes, and the artificial regulation of gene expression.

14. Use the figure at the right to explain how knock-out genes are produced. Label FIVE parts of the figure (A, B, C, D and E) as you refer to them in your discussion.

15. How does antisense RNA regulate the expression of DNA?

Section 13.4 looks at applications of biotechnology and is the final focus of this chapter. Examples include the production of human insulin and human growth hormone by bacteria, the improvement of agricultural crops, and bioremediation techniques to clean up environmental disasters.

16. Identify three reasons why biotechnology is advantageous over more traditional plant breeding techniques.

____
____
____
____
____
____

17. Many people oppose the insertion genes from one organism onto another to produce genetically modified organisms (GMOs). Describe THREE concerns about GMOs.

____
____
____
____
____
____
____
____
____
____
____
____

Science Practices & Inquiry

In the AP Biology Curriculum Framework, there is a set of 7 Science Practices. In this chapter, we will focus on **Science Practice 6: The student can work with scientific explanations and theories**. More specifically, Practice 6.4: The student can *make claims and predictions about natural phenomena* based on scientific theories and models.

Question #18 asks you to draw on your knowledge of proteins. First you need to recall how proteins are assembled by a cell, including how their shape is formed during their production. It may be helpful to review chaperonins before answering the first part of this question. Additionally, recall the theme of structure and function as you work on this question. Part b asks you to apply your knowledge of biotechnology techniques. There are many different tools you could use to answer this question; it is important to explain what property of the molecule is being tested or analyzed by the technique. (LO 3.5 & 3.6)

18. A fellow student tells you that organic chemists frequently will try to substitute one functional group for another on a large molecule to try and make minor changes to the molecule. Cocaine, for example, has four functional groups that can be substituted for or deleted to make different pharmaceutical drugs of different potency. So if chemists can make new drugs this way, why can't biologists simply change one amino acid in a protein to try and make new proteins?

a. Explain why changing one amino acid in a protein is not as easy as changing a functional group on a large molecule.

____
____
____
____
____
____
____
____

b. If it were possible to change an amino acid in a protein, identify two biotechnology techniques that could be used to determine if a change had occurred or to determine the properties of the new molecule.

____
____
____
____
____
____
____
____

Chapter 14: Genes, Development, and Evolution

Chapter Outline

14.1 - Development Involves Distinct but Overlapping Processes
14.2 - Changes in Gene Expression Underlie Cell Differentiation in Development
14.3 - Spatial Differences in Gene Expression Lead to Morphogenesis
14.4 - Gene Expression Pathways Underlie the Evolution of Development
14.5 - Developmental Genes Contribute to Species Evolution but also Pose Constraints

Most organisms start life as a single cell. The interplay between the genes and the materials inside the cells sets in motion a precisely regulated pattern of development from which a mature organism takes shape. The materials found inside the cell depends on the location and time of that cell's appearance in the developing organism. In this way, spatial and temporal differences in gene regulation result in spatial and temporal patterns of organ and limb development. These patterns have the result that everything develops at the right time and in the right place. Spatial and temporal differences in development both influence, and are influenced by, the evolution of species differences.

Developmental biology is strong support for Big Idea Two, noting that molecular building blocks allow organisms to grow, to reproduce and to maintain dynamic homeostasis. Specifically,

2.e.1: Timing and coordination of specific events are necessary for the normal development of an organism, and these events are regulated by a variety of mechanisms.

Understanding the networks and cascades of signaling that direct developmental biology, strongly supports Big Idea Three, which is based on information flow in living systems. Here,

3.b.2: A variety of intercellular and intracellular signal transmissions mediate gene expression.

3.d.2: Cells communicate with each other through direct contact with other cells or from a distance via chemical signaling.

Finally, Big Idea Four, biological systems interact with complexity, is a fundamental underpinning in developmental biology. A bigger picture emerges, in that

4.a.3: Interactions between external stimuli and regulated gene expression result in specialization of cells, tissues and organs.

Chapter Review

Concept 14.1 introduces how development involves distinct but overlapping processes. The zygote is the single cell formed by the union of the gametes in sexually reproducing species, and mitosis is the process that gives rise to ever-more cells. Determination of any individual cell's "fate" happens early in development, and groups of cells can differentiate to become specific structures with specific functions. Morphogenesis proceeds, resulting in the formation of body parts and organs. Growth typically occurs over a much longer part of the organism's lifespan.

Cells that are labeled "totipotent" can differentiate to become any part of an organism, whereas cells that are "multipotent" have a more limited array of developmental possibilities.

1. The schematic drawing below shows possible manipulations to non-human embryos. In scenario A, shown to provide you the background to understand scenarios B and C, the oval-shaped embryo shown at the top normally gives rise to the embryo form shown below it. Briefly explain what is shown in scenarios B and C, and compare these two scenarios regarding their impact on our knowledge of "fate determination."

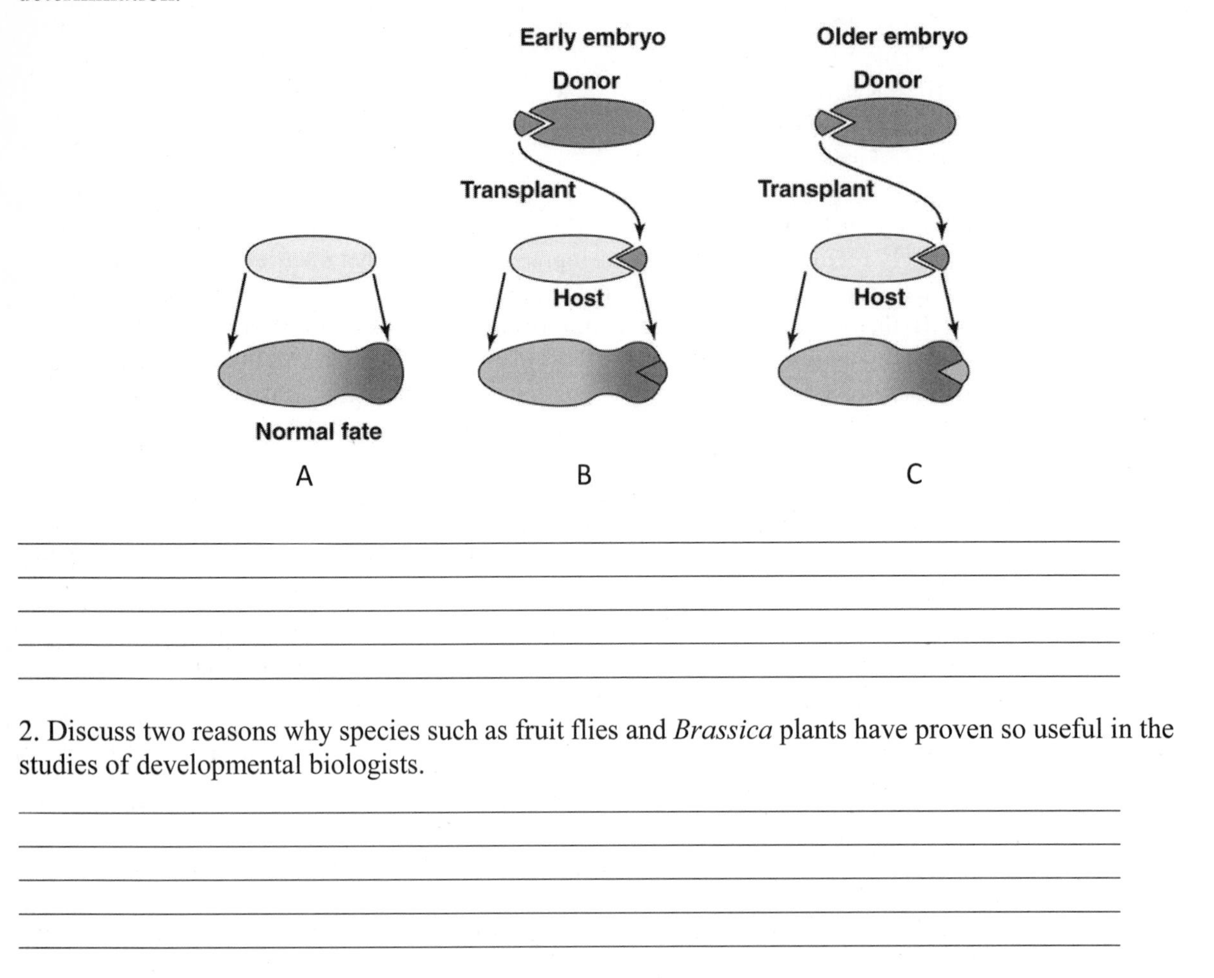

2. Discuss two reasons why species such as fruit flies and *Brassica* plants have proven so useful in the studies of developmental biologists.

3. The team that cloned Dolly the sheep used a nucleus from a mammary epithelium (ME) cell. They also tried cloning by transplanting nuclei from fetal fibroblasts (FB) and embryos (EC), with these results:

	Number of attempts that progressed to each stage		
Stage	ME	FB	EC
Egg fusions	277	172	385
Embryos transferred to recipients	29	34	72
Pregnancies	1	4	14
Live lambs	1	2	4

A. Calculate the percentage survival of eggs from fusion to birth. What can you conclude about the efficiency of cloning?

B. Compare the efficiencies of cloning using different nuclear donors. What can you conclude about the ability of nuclei at different stages to be totipotent?

Concept 14.2 discusses how changes in gene expression underlie cell differentiation in development. What guides different cells to their specific patterns of development? All cells contain the same set of instructions (DNA), but it is clear that different cells, for example a neuron in the brain and a cell in the liver, have many different proteins. This observation should suggest to you that a history of differential gene expression is what results in the different developmental fates of these cells. The explanation for how this occurs is that each cell has its own specific history of which transcription factors, gene inducers, and "morphogens" were present during its existence. Therefore, different genes can be expressed or expressed at different times, depending on the cell's local environment. The developing cells are contained in a networking cascade of interacting cells, giving rise to the genetic determination of structural development.

4. The French flag model of early development (see figure) holds that chemical gradients of morphogens across developing tissues influence regional gene expression leading to specific patterns in the emerging organism. A morphogen called Sonic hedgehog plays an important role in the formation of the vertebrate nervous system. Briefly describe some testable ideas about what the French flag model is helpful for understanding developmental studies of nerves in vertebrate embryos.

PRINCIPLES OF LIFE, Figure 14.12 (Part 1)

5. Explain how the identification of genetic "switches" that turn on and turn off gene expression has proven exceptionally useful for understanding developmental biology.

Concept 14.3 describes how spatial differences in gene expression lead to morphogenesis. Detailed observations show that different gene-expressing inducers are present in different parts of the developing organism. But that's not the whole explanation, as the loss of cells by programmed cell death (apoptosis) is also very important for the determination of form and function. As a general example, a hand in a human develops as a flipper-like mass, and then selective apoptosis of columns of cells leads to the emergence of the fingers. In the tiny nematode, *C. elegans*, the differentiation of the adult's 959 cells was preceded by the development of 131 cells that underwent apoptosis after serving a brief developmental role. The identities of many of the signals that initiate programmed cell death are known, and it is a common developmental strategy that many more cells are "born" than persist in the adult forms of organisms.

6. Discuss the observation that Hox genes are present in all animals and have a positional expression pattern during development that guides cell fates.

____

____

____

____

____

7. A fruit fly with the *Antennapedia* mutation grows a leg where its antenna ought to grow. Explain this "homeotic" mutation, in general terms, including changes in the expression patterns of Hox and related genes.

____

____

____

____

____

Concept 14.4 shows how gene-expression pathways underlie the evolution of development. The evolutionary perspective on developmental biology has advanced largely through the observations that diverse organisms use the same or very similar regulatory mechanisms during development. Much of the diversity we see among forms of different organisms is the result of small changes in the timing or place of the expression of regulatory gene networks. For example, the observation that chimpanzees and humans are so very closely related in genotype but not phenotype is the result of many similarities and differences in species-specific patterns of gene expression. An additional example is that a longer interval for the expression of a particular gene can result in the elongation or other change in a limb, as in the long neck of the giraffe.

8. Discuss the claim that orthologous (evolutionary related) regulatory genes show similar patterns of expression along the axis between the head and the abdomen in insects and vertebrate animals. Be sure to include a discussion of what category of function (signals, transcription factors, filaments, etc.) is present in the products of these genes.

____

____

____

____

____

9. The phylogenetic tree shown here was derived from observations of the number of legs among invertebrate animals. The five groups organized at the top of the drawing have six legs each, and the four groups below them have 8 or more legs. Explain the genetic implications of the small circle drawn on the tree at the key branch point for the evolution of changes in the number or location of legs.

PRINCIPLES OF LIFE, Figure 14.19

____

____

____

____

____

____

10. Control of eye formation during development of many animals is under the control of a genetic switch involving a transcription factor. Here are partial DNA sequences for the control gene from two organisms:

Mouse Pax6 gene: 5‘-GTATCCAACGGTTGTGTGAGTAAAATT-3‘
Fruit fly eyeless gene: 5‘-GTATCAAATGGATGTGTGAGCAAAATT-3‘

A. Calculate the percentage of identity between the two DNA sequences.

B. Use the genetic code to determine the amino acid sequences encoded by the two regions, and calculate their percentage of similarity.

C. The fruit fly and mouse evolved from a common ancestor about 500 million years ago. Comment on your answers to A and B in terms of the evolution of developmental pathways.

Concept 14.5 explains how developmental genes contribute to species evolution but also pose constraints. The "genetic toolkit" describes a limited set of developmentally-important genes upon which natural selection has acted to generate variability in form and function. Examination of the genetic toolkit reiterates the conclusion that evolution acts upon genes that are already present, rather than inventing anew.

11. Historically, it was thought that any two species with many differences in form and function must have huge differences in the genes that regulate their development patterns. Discuss whether or not this hypothesis has been widely supported or weakened by the data collected during more recent studies of gene expression during development.

12. Stickleback fish that live in the ocean develop a pelvic spine that helps protect them from predation, since the spine makes it difficult for predators to swallow them. A key gene has been identified that is required for the development of the spine Suppose that two freshwater populations of sticklebacks from different regions were found to never develop the pelvic armor. Develop two hypotheses that could be tested to determine if the loss of the armor is due to identical genetic changes in the gene or to different changes in the gene. For the latter, comment on the concept of parallel phenotypic evolution in your answer.

Science Practices and Inquiry

Understanding the concepts of developmental biology requires facility in working models (Science Practice 1) that provide explicit explanations of what is happening where during development. By necessity, the physical models give rise to theoretical explanations of what takes place during development (Science Practice 6, e.g. learning objective 4.25). Finally, keeping track of models and theories across the time scales of development links space and time in complex ways, thus supporting Science Practice 7.

PRINCIPLES OF LIFE, **Figure 14.20**

13. The drawing shows wings of pterosaurs, birds and mammals. Draw circles around these subsets of skeletal entities in the drawing:
  humerus,
  radius and ulna,
  metacarpals, and
  phalanges

Next, link your circles together with lines to show the homologous parts.

Discuss how these similarities demonstrate that the wing was not a completely novel "invention" of nature each time it developed.

Chapter 15: Mechanisms of Evolution

Chapter Outline

15.1 - Evolution Is Both Factual and the Basis of Broader Theory
15.2 - Mutation, Selection, Gene Flow, Genetic Drift, and Nonrandom Mating Result in Evolution
15.3 - Evolution Can Be Measured by Changes in Allele Frequencies
15.4 - Selection Can Be Stabilizing, Directional, or Disruptive
15.5 - Genomes Reveal Both Neutral and Selective Processes of Evolution
15.6 - Recombination, Lateral Gene Transfer, and Gene Duplication Can Result in New Features
15.7 - Evolutionary Theory Has Practical Applications

Evolution explains the interrelatedness of all of the different species of microbes, plants and animals. Some commonalities of life include the genetic code, the similarity of developmental genes, and similar biochemical processes (glycolysis) across the phyla. In this chapter we examine the mechanisms of change beginning with Charles Darwin.

Evolution's factual basis is that we know that organisms have changed and are still changing today. Evidence of evolutionary change comes from analysis of fossils, biochemistry, homologous structures, biogeography, and direct observation of change. Today, as we observe active evolution in the development of antibiotic resistant bacteria and pesticide-resistant insects, we ask the question, "How have these changes occurred over long periods of time?" This is the theoretical side of evolutionary study. It is important to remember that a theory is not just a random thought or idea. Rather, a theory is a well-developed idea, it is repeatedly tested with experiments, and it provides a cohesive framework for analysis.

The basis of evolutionary theory, explaining how populations change over time, was first proposed by Darwin and Wallace. The success of their ideas hinged upon the idea of natural selection. Darwin observed that there is a range of variation in any species and that not all members of each species survive to reproduce. Only those members of a species that are well adapted survive well enough to reproduce and pass their genes on to the next generation. Thus, only some variation is passed from generation to generation, resulting in change over time or evolution.

A primary inventive source of new variation in a population is mutation of DNA. In addition, genetic variation results from meiosis, non-random mating, gene flow, and genetic drift. These changes can be estimated by counting the frequencies of the alleles of traits in a population. This idea led two researchers, Hardy and Weinberg, to develop the Hardy-Weinberg theorem. They proposed that in order for a population to "not evolve" over time, five conditions must exist: no mutations, random mating, no gene flow, a very large (infinite) population size, and the environment must not play a role in determining survival. The Hardy-Weinberg conditions are not obtained in nature, except in contrived settings set up by humans, so evolution proceeds.

Population changes over time frequently follow distinct patterns, including stabilizing, directional, or disruptive selection. These changes are assessed by observations of phenotypes or observable behaviors. Other changes can be seen in the genes of organisms or genomes. Sequencing DNA or proteins and looking at mutation rates can be use to research the evolution of genomes. These techniques have provided many new insights into evolutionary theory and the relationship of different organisms including the splitting of prokaryotes into two groups and creating a new level of classification called the Domain.

Chapter 15 has ideas that are found primarily in Big Idea 1. The Big Ideas are a means for organizing the vast amount of information we know as biology. It is vitally important that you continually work to understand these **Big Ideas** and that you work to relate together the different **Enduring Understandings**. While chapter 15's emphasis is primarily Big Idea 1, it also includes some of Big Idea 4 as well.

Big Idea 1 recognizes that evolution ties together all parts of biology. In Chapter 15 we look at the mechanisms of evolutionary theory including:

1.a.1: Natural selection is a major mechanism of evolution.
1.a.2: Natural selection acts on phenotypic variations in populations.
1.a.3: Evolutionary change is also driven by random processes.
1.a.4: Biological evolution is supported by scientific evidence from many disciplines, including mathematics.
1.c.3: Populations of organisms continue to evolve.

Big Idea 4 examines the idea that biological systems interact in complex ways. Included in Chapter 15 is how the variance in a population affects reproductive fitness:

4.c.3: The level of variation in a population affects population dynamics.

Chapter Review

Concept 15.1 introduces evolutionary theory and the ideas of Charles Darwin.

1. Explain in evolutionary terms why a new and different flu vaccine is developed each year.

2. Explain the meaning of "theory" in the context of atomic theory and evolutionary theory.

3. Charles Darwin noted that the species of the temperate regions in South America were more similar to the species of the tropical regions of South America than they were to the species of the temperate regions in Europe. Explain how this observation guided his evolutionary thinking.

4. In addition to observing biological specimens, Darwin also read a book about geology by Charles Lyell. How did Lyell's concepts of geological time and space influence Darwin's biological considerations of life on earth?

5. Describe TWO ways that the discovery of genes and chromosomes affected evolutionary theory.

____
____
____
____

Concept 15.2 examines the primary mechanisms of evolution including mutation, gene flow, genetic drift, and nonrandom mating.

6. Explain the phrase "individuals do not evolve, populations do."

____
____
____
____

7. Describe the origin of new genetic variation in the genetic code.

____
____

8. Use the three terms: allele, gene frequency, and gene pool to describe the fact that 10% of the Japanese population has blood type AB.

____
____
____
____
____
____
____
____
____

9. Describe the ONE difference and ONE similarity between artificial selection and natural selection, and explain how each influenced Darwin's writings on evolution.

____
____
____
____
____
____

10. Darwin made two key observations (below). Explain how each observation supports his concept of natural selection.

1. Populations have a large amount of variation within them.
2. Most individuals that are born do not survive to reproduce.

____
____
____
____
____
____

11. In northern Canada, there are two large herds of caribou that seldom meet. Predict the amount of gene flow between these caribou populations.

12. Explain how the bottleneck principle relates to conservation problems for endangered species.

13. While exploring outdoors, a fellow student asks "Why does that male cardinal have such a brilliant red color?" Use your knowledge of sexual selection to explain why male cardinals are willing to risk being more visible to predators.

Concept 15.3 describes how we can measure evolutionary change with the Hardy Weinberg equilibrium principle.

14. Identify the five principles of the Hardy Weinberg theorem that must be true if a population is in genetic equilibrium.

15. Assume that a local population of birds experiences immigration of additional birds of the same species from another country each winter, and that there is a limited amount of breeding during the winter. For each of the five principles of the Hardy Weinberg theorem, explain why it cannot be true for the local population is not evolving.

16. Evolutionary theory has the challenge of explaining why many populations of organisms appear to be stable and unchanging from year to year and many of their genotypes are not significantly changing from Hardy Weinberg expectations. Explain how this can occur.

____
____
____
____
____

Concept 15.4 considers the interaction of genes with each other and with the environment to create shifts in a population based on selection.

17. A biologist finds a population of small arthropods on a Pacific island with white sand beaches in between black lava flows. Most of the arthropods are either dark gray or very light gray, but less than 10% of the population is an intermediate gray color. What type of selection is this? Explain why this would happen.

____
____
____

18. In lizards, it has been shown that there is an optimum size egg for survival: eggs that are too big or too small are not adaptive. Explain why this could happen, using the correct term to describe this pattern of selection.

____
____
____

Concept 15.5 looks at the specific processes that operate at the level of genes and genomes. One of the most important examples you should be familiar with from this section is the heterozygote advantage.

19. Discuss the claim that a heterozygote advantage helps maintain genetic diversity within a population.

____
____
____
____
____

20. Explain how a mutation can be considered to be a neutral mutation that does not directly affect the phenotype of an individual.

____
____
____
____

21. Much of the DNA for a eukaryote is noncoding DNA. Several years ago this DNA was called "junk" DNA. Recently this noncoding DNA has been shown to have several important functions. Identify two possible functions of noncoding DNA.

____
____
____
____
____

Concept 15.6 discusses how new genes with novel functions can form in a population.

22. Sexual reproduction enhances natural selection because it generates diverse combinations of genes within a population. Not all species reproduce sexually, and these reveal some of the drawbacks of sexual reproduction. Discuss THREE drawbacks of sexual reproduction.

____
____
____
____
____
____

23. Two "endosymbiotic" organelles in eukaryotes resulted from lateral gene transfer. What new benefits did these organelles confer on the recipients?

____
____
____

24. Give an example of gene duplication in an organism and discuss how this has benefitted this organism.

____
____
____
____
____

Concept 15.7 examines how knowledge of evolutionary theory is used to combat disease, benefit agriculture and study protein function.

25. Explain why effective farmers use a different herbicides each different year on their fields when spraying for weeds.

____
____
____

26. Discuss how genomic databases allow for better detection and treatment of diseases.

____
____
____

Science Practices & Inquiry

In the AP Biology Curriculum Framework, there is a set of 7 Science Practices. In this chapter, we will focus on **Science Practice 2: The student can use mathematics appropriately.** More specifically, practice 2.2: The student can *apply mathematical routines* to quantities that describe natural phenomena.

Questions 27 asks you to apply mathematical methods to data and predict what will happen to the population in the future (LO 1.3). Most of the science practice questions up to this point have been in the style of free response questions. The redesigned AP Biology exam will have a multiple choice portion and a set of grid in items that require your use of mathematics towards biological concepts. An equation sheet will be provided that will have the commonly used equations like the Hardy-Weinberg principle used in the question below.

27. Cystic fibrosis is an autosomal recessive genetic disorder. This disease occurs in 0.4 out of 1,000 children born in the United Kingdom. Calculate the percent of carriers in the UK.

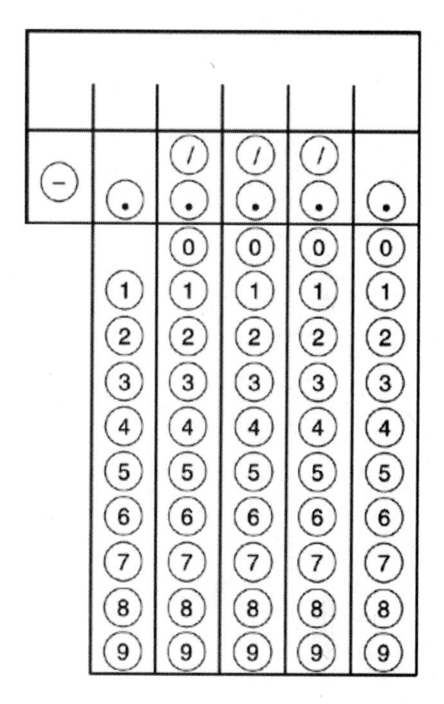

Chapter 16: Reconstructing and Using Phylogenies

Chapter Outline

16.1 All of Life is Connected through Its Evolutionary History
16.2 Phylogeny Can Be Reconstructed from Traits of Organisms
16.3 Phylogeny Makes Biology Comparative and Predictive
16.4 Phylogeny Is the Basis of Biological Classification

It is challenging to tell the full story of the amazing diversity of life, because so many attributes of living organisms are so different that the task seems hopelessly complex. Complications in the story of life include an incredibly long passage of time, far removed from our daily experience, and an incomplete record of all changes. Nonetheless, there are many, many similarities between organisms, both living and dead, and these similarities provide clues used by biologists to trace out the many branches in the vast tree of life.

Much as a family tree traces out a family's generations over time, phylogenetic trees trace out evolutionary history. A family tree includes the names of family members, whereas a phylogenetic tree is built using the units of taxonomic categories, such as orders and species. The analysis of phylogenetic trees trace out evolutionary history, and they permit us to make predictions about what might be possible as life responds to the never ending changes in environmental and other selection pressures.

Chapter 16's emphasis is evolutionary biology, i.e., Big Idea 1: The process of evolution drives the diversity and unity of life. Elaborating on this theme, you should try to prepare yourself to examine the concepts and facts showing that:

1.a.4: Biological evolution is supported by scientific evidence from many disciplines, including mathematics.
1.b.1: Organisms share many conserved core processes and features that evolved and are widely distributed among organisms today.
1.b.2: Phylogenetic trees and cladograms are graphical representations (models) of evolutionary history that can be tested.

Chapter Review

Concept 16.1 introduces how all of life is connected through its evolutionary history. Try to imagine the long personal history that led to your presence working through this book. If you were to write out your family tree, the logic and simplicity of your family pedigree would become clear. So it goes with the history of relationships that preceded today's diverse plants and animals. The evidence is particularly compelling when it comes to examining genomes, which are a kind of "molecular registry" of evolution.

Relationships within a particular level of taxonomic organization are depicted using simple phylogenetic trees. Typically, these "trees" are presented sideways, with most branching on the right side of the tree, and time (usually in unspecified intervals) along the horizontal axis of the tree. The branch points in phylogenetic trees indicate occasions where one group developed a substantial difference from its ancestral group, justifying its placement outside of a straight linear sequence of genetic inheritance.

1. Use the phylogenetic tree shown at the right to complete the following.

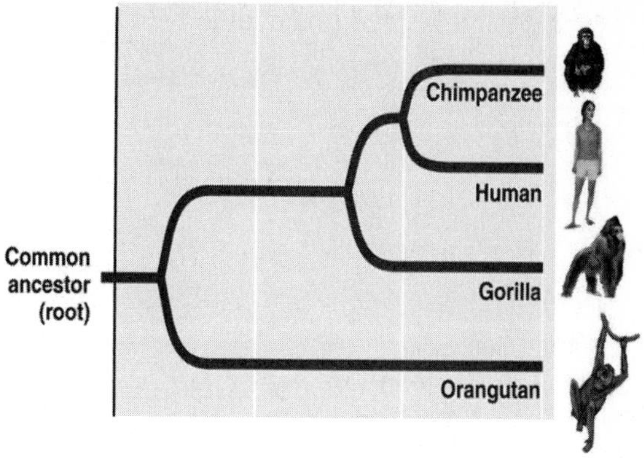

a. Explain how many clades are indicated:

____
____
____

b. Explain which branch point occurred most recently:

____
____
____

Common ancestor (root)
Chimpanzee
Human
Gorilla
Orangutan

PRINCIPLES OF LIFE, In-Text Art, Ch. 16, p. 316 (4)

c. Discuss the reasons why one of the organisms on the tree is interested in its evolutionary relationships to the other three organisms.

____
____
____
____

2. Homologous structures are of great use in determining evolutionary relationships. Explain the homology that is present in each pair of items below.

a. The spinal cord of a shark and that of a chimpanzee.

____
____
____
____

b. The wing of a bird and that of a bat.

____
____
____
____

c. A vision-related gene that is identical in fruit flies and mice.

____
____
____
____

Concept 16.2 examines how phylogeny can be reconstructed from traits of organisms. The physical and other traits of organisms help us organize groups of organisms with regard to their evolutionary relationships. Such analyses are built upon assumptions that there hasn't been any convergent evolution (that is, a "new" trait arose only once in the set of organisms selected for analysis), and that traits of interest were not lost over the evolutionary interval under consideration.

Information that is used to construct phylogenetic trees and cladograms includes: morphology (physical features of traits, e.g., flowers), patterns seen during development, e.g., *hox* genes in vertebrate and invertebrate development (Chapter 14), behaviors such as similar mating calls in varied frog populations, and molecular sequences in genes of interest, especially mitochondrial and chloroplast genes.

3. Use the data presented in the table below

TABLE 16.1 Eight Vertebrates and the Presence or Absence of Some Shared Derived Traits

TAXON	DERIVED TRAIT							
	JAWS	LUNGS	CLAWS OR NAILS	GIZZARD	FEATHERS	FUR	MAMMARY GLANDS	KERATINOUS SCALES
Lamprey (outgroup)	–	–	–	–	–	–	–	–
Perch	+	–	–	–	–	–	–	–
Salamander	+	+	–	–	–	–	–	–
Lizard	+	+	+	–	–	–	–	+
Crocodile	+	+	+	+	–	–	–	+
Pigeon	+	+	+	+	+	–	–	+
Mouse	+	+	+	–	–	+	+	–
Chimpanzee	+	+	+	–	–	+	+	–

PRINCIPLES OF LIFE, Table 16.1

to label the small circles with the derived trait shown in the diagram below.

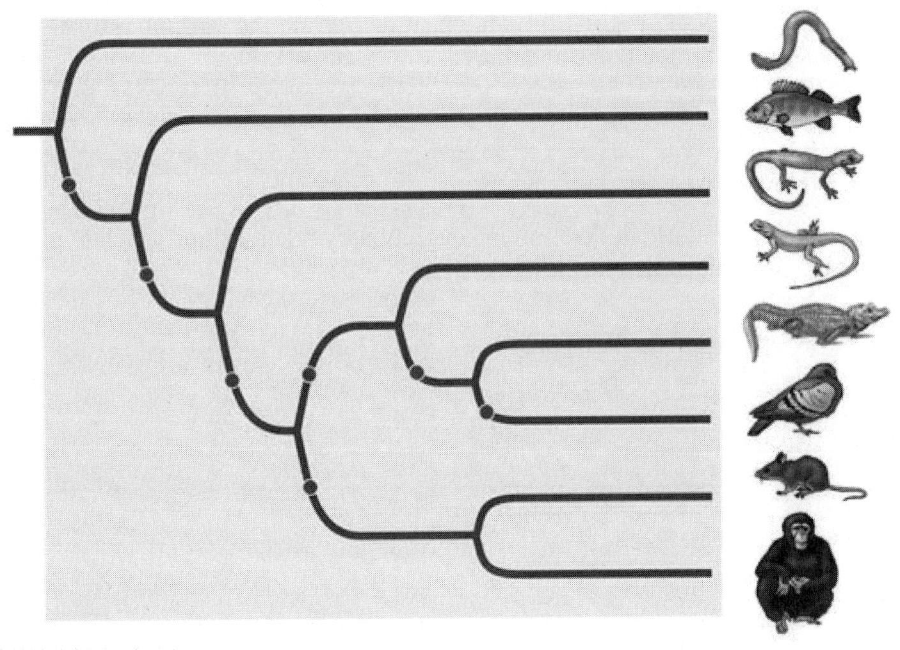

PRINCIPLES OF LIFE, Figure 16.3

4. "Parsimony" is a term used to describe how the analysis completed in the above question would resolve the complications of adding more traits and more organisms to phylogenetic analysis. Describe what "parsimony" means in this context.

____

____

____

____

5. Though their body shapes are similar, snakes and worms have substantially different evolutionary histories. Discuss how this convergence might have occurred and suggest additional data that could be sought in determining the differences in evolutionary heritage.

____

____

____

____

6. Explain why phylogenetic trees based on gene sequences are more accurate in showing evolutionary relationships than are phylogenetic trees based on morphology.

____
____
____
____

Concept 16.3 explores how phylogeny makes biology both comparative and predictive. Phylogenetic analysis of organisms that appear to be similar has often shown that they arrived at such similarity via convergent evolution rather than simply being closely related to one another. Knowing phylogenetic pathways helps us see the broad scope of evolutionary change, but it also reveals some of its limitations. Such knowledge is used to choosing genes from wild plants to engineer into domestic plants and in analyzing patterns of pathogen diversification. Therefore, phylogeny is of considerable value in solving the problems faced by our species.

7. Explain how protein analyses can help to meet the challenge of determining the approximate time that an evolutionary change took place; i.e., describe and discuss the "molecular clock" of protein change.

____
____
____
____

8. Female swordtail fish (*Xiphophorus*) show a mating preference for males with long swordtails, suggesting that sexual selection for longer tails over evolutionary time has occurred. Even though the males of a related species of platyfish do not have long tails, females of that species demonstrate a preference to mate with male platyfish that have been fitted with artificial long tails. Discuss what the females' mating preference suggests about the phylogenetic background of both groups of fish.

____
____
____
____

9. Discuss how studies of the "molecular clock" allowed an estimate of the specific year of origin for the human immunodeficiency virus.

____
____
____
____

Concept 16.4 shows us how phylogeny has become the basis of biological classification. Making sense of the immense biological diversity requires us to examine the relationships of all organisms. Phylogenetic analysis is really evolutionary history, and with it, we know more about the possible directions evolution may take in the future, with or without our help. The development of taxonomic rules will result in the most parsimonious and accurate description of evolutionary consequences: speciation.

10. Describe the Linnean binomial nomenclature of your own species.

____
____
____
____
____

11. Place these taxonomic terms in correct sequence, from most inclusive to most specific.

order, genus, family, species, classes, phyla, kingdoms

____

____

____

12. Explain why it is essential that each group of organisms included in gene bank databases has an accurate and specific taxonomic identity.

____

____

____

13. In a recent newspaper, an article about humans described humans as the genus *sapiens*. Explain why this is inaccurate and why *sapiens* cannot be used by itself.

____

____

____

Science Practices & Inquiry

In the AP Biology Curriculum Framework, there is a set of 7 Science Practices. In this chapter, we will focus on **Science Practice 5: The student can use mathematics appropriately.** More specifically, practice 5.3: The student can evaluate the evidence provided by data sets in relation to a particular scientific question.

Question 14 provides an occasion to work on evaluating evidence and finding patterns and relationships. We will examine a number of characteristics of animals that spend most of their time in water, and attempt to determine their evolutionary relationships (LO 1.9).

14. Living in water would seem to suggest shared evolutionary heritage. We will survey traits of four animals with aquatic lifestyles: sponges, fish, newts and whales. Sponges do not have a circulatory system, fish have a two-chambered heart, aquatic newts have a three-chambered heart, and whales have a four-chambered heart. Sponges, fish and newts are poikilothermic (cold-blooded) and whales are homeothermic (warm-blooded). Sponges reproduce both asexually and sexually; when reproducing sexually, sponges shed gametes externally. The fish and newt shed their gametes externally; internal fertilization, internal development (gestation, or pregnancy), and nursing of offspring are the reproductive pattern of whales. Sponges do not have true tissues, while fish, newts, and whales all have true tissues. Produce a table showing the presence of these traits in a table and then draw a phylogenetic tree.

Chapter 17: Speciation

Chapter Outline

17.1 - Species Are Reproductively Isolated Lineages on the Tree of Life
17.2 - Speciation Is a Natural Consequence of Population Subdivision
17.3 - Speciation May Occur through Geographic Isolation or in Sympatry
17.4 - Reproductive Isolation Is Reinforced When Diverging Species Come into Contact

Chapter 17 begins with an examination of species and what defines a species. Using the biological species concept, we can define a species as a group of organisms that are actually or potentially interbreeding naturally and produce fertile offspring. This concept does have limitations and is not particularly useful for examining fossils or bacteria and asexual organisms. There are many other definitions of species that can be used for these situations. Species can change slowly over time as they respond to their environment and its changes. However, the formation of a new species typically requires some form of an isolating event that prevents the flow of genes from one group to another. When this happens, new species can be the result.

Isolating events can be as simple as a population migrating on either side of a mountain range or a tectonic plate separating two groups of a species. Allopatric speciation is the result of a physical separation between groups that cuts off gene exchange between the groups. This is the most common mode of speciation in animals. Sympatric speciation appears to be less common, but it occurs when a new species arises without physical separation between the groups that diverge to form distinct species. There is lots of evidence for sympatric speciation among plants.

After a barrier to gene flow is established, differences in gene frequencies and in types of mutations can begin to build up in the separated populations. On a molecular level, genetic incompatibility can create two groups that cannot produce offspring with each other even if they try and mate. Recall that a zygote is the first new cell of an sexually-produced organism. Prezygotic- and postzygotic- isolating mechanisms serve to further reinforce the genetic isolation between species. Prezygotic mechanisms serve to prevent the union of sperm and egg to form a zygote. But there are times when a viable zygote does form between two different species. Postzygotic mechanisms result when the resulting hybrid is weak, malformed, or sterile.

Chapter 17 is one of the few chapters in this book that pulls solely from one Big Idea, evolution, in the AP Biology Curriculum Framework. Chapter 17 examines the development of new species, which is part of Big Idea #1.

Big Idea 1 states that: The process of evolution drives the diversity and unity of life. In Chapter 17 we look at speciation including

1.c.1: Speciation and extinction have occurred throughout the Earth's history.
1.c.2: Speciation may occur when two populations become reproductively isolated from each other.
1.c.3: Populations of organisms continue to evolve.

Chapter Review

Concept 17.1 discusses the definition of species. There are many different definitions of a species, but all have in common the effort to describe biological diversity and to define the most descriptive level of classification, the species.

1. Below are pictures of four small fishes found in the eastern United States.

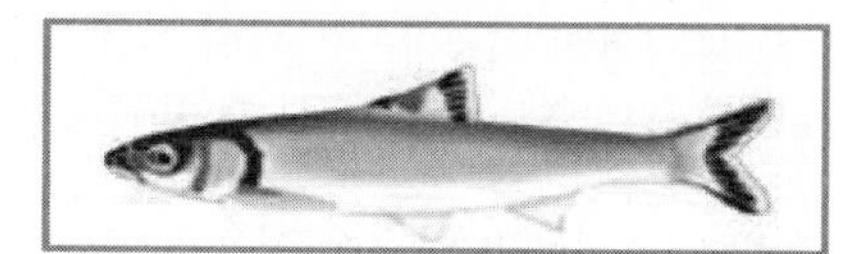

For each of the three species concepts below, describe what types of evidence would be used to distinguish which of the above fish are separate species.

a. Morphological species concept:

____

____

____

b. Biological species concept:

____

____

____

c. Lineage species concept:

____

____

____

2. Which of the three species concepts in question 1 is most useful for describing asexually reproducing organisms? Explain why you chose that concept.

____

____

____

3. Tigers and lions can be mated in zoos to produce offspring. Why are these two organisms considered to be different species?

____

____

____

Concept 17.2 examines the idea of how the accumulation of mutations among several genes in the genetic make-up of a population can result in reproductive incompatibility with a similar population. The end products of genes must be able to work with each other. If multiple gene changes create incompatibility, then reduced fitness or even lethal changes may result.

4. After two groups of a species become isolated, different mutations can quickly accumulate in each of the separated groups. Discuss how the resulting combinations of genes may have deleterious effects when a zygote is formed from by mixing gametes of the separate groups together.

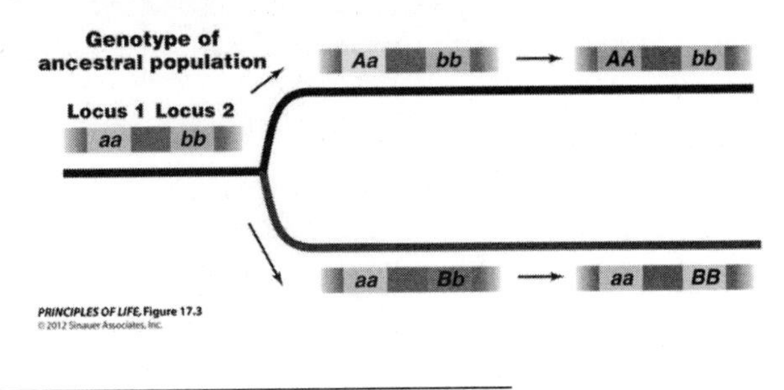

____

____

____

____

____

____

____

5. The diagram below shows modifications of the chromosomes of an original species (A) after two sub-groups (B & C) were separated.

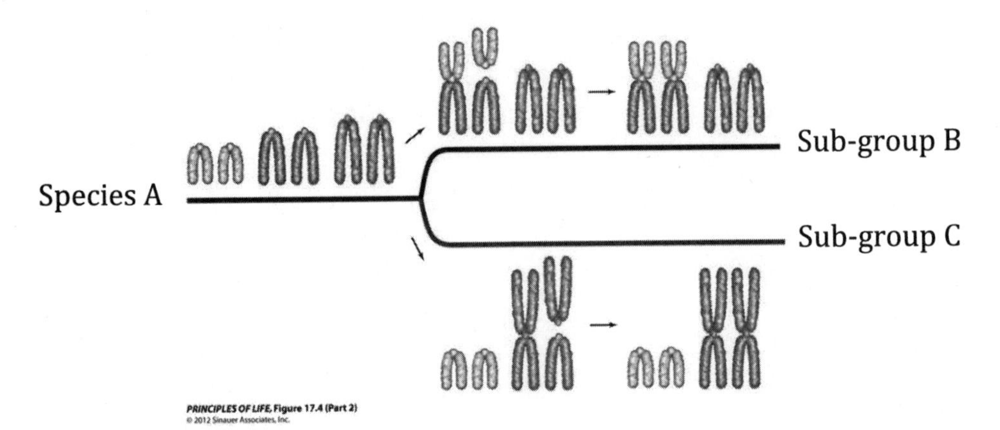

Assume that all of these groups are sexually reproducing species that forms gametes via meiosis. Explain what would happen when the gametes of sub-group B are united with those of sub-group C. Drawing the result might help you explain it.

____

____

____

____

____

____

Concept 17.3 looks at how groups of a population can split and become divided over time either through geographic isolation (allopatric speciation) or without physical isolation (sympatric speciation).

6. The opening of this chapter described the cichlids of Lake Malawi.

a. For the cichlids that populate the sandy areas of the lake versus those that populate the rocky areas of the lake, does this represent allopatric or sympatric speciation? Explain why.

b. Some of the cichlids in the rocky area evolved into plankton eating fish while others in the rocky area became algae eating fish. Does this suggest allopatric or sympatric speciation? Explain why.

7. Distinguish between autopolyploidy and allopolyploidy, and explain how sympatric speciation can be the result of each condition.

8. Describe how road construction could lead to allopatric speciation. Would this new road be predicted to have the same effect for all animal species in a given area?

Concept 17.4 shows how prezygotic and postzygotic processes reinforce the isolation of separate species.

9. Define "zygote," and discuss the processes that lead to its formation.

10. Your textbook gives several examples of pre- and post- zygotic isolating mechanisms. For each of the mechanisms given below, give another example not found in chapter 17 and explain how the isolating mechanism keeps the two species from hybridizing.

a. Mechanical isolation (orchids and wasps):

b. Temporal isolation (leopard frogs):

____

____

____

c. Behavioral isolation (frog calls, coloration of cichlids in Lake Malawi, columbine flowers and pollinators, *Phlox*):

____

____

____

d. Habitat isolation (cichlids of Lake Malawi, *Rhagoletis* flies):

____

____

____

e. Gametic isolation (sea urchins):

____

____

____

f. Low viability of hybrid adults (*Bombina* toads):

____

____

____

g. Hybrid infertility (mule and horse):

____

____

____

11. For many years, two populations of orioles in the United States were considered to be separate species. Now, most bird books list them as one species. What evidence would researchers need to have collected to make this change?

____

____

____

Science Practices & Inquiry

In the AP Biology Curriculum Framework, there is a set of 7 Science Practices. In this chapter, we will focus on these:

Science Practice 4: The student can plan and implement data collection strategies appropriate to a particular scientific question.

Practice 4.1: The student can *justify the selection of the kind of data* needed to answer a particular scientific question and

Practice 4.2: The student can *design a plan* for collecting data to answer a particular scientific question.

Science Practice 6: The student can work with scientific explanations and theories.

Practice 6.4: The student can *make claims and predictions about natural phenomena* based on scientific theories and models.

In parts a and b of question #12, you will use data from two real populations, and based on models of types of selection, explain why these groups are separate species (LO 1.22).

Part c of question #12 asks you to design a plan for collecting data to investigate a scientific claim that speciation and extinction have occurred throughout the Earth's history (LO 1.21) and to justify the selection of data that address questions related to reproductive isolation and speciation (LO 1.23).

12. Domestic horses (*Equus caballus*; 2n=64) and donkeys (*Equus asinus*; 2n=62) when inter-bred produce either the mule or hinny, both of which are infertile hybrids (2n=63).

Two groups of horses were found to have different chromosomal numbers. The Przewalski horse (*Equus przewalskii*, 2n = 66) is an ancient horse formerly found only in Mongolia; this horse is now thought to be extinct in the wild but survives in many zoos. Crosses between the domestic horse (*E. caballus*, 2n = 64) and the Przewalski horse create normal fertile offspring.

a. Explain why the offspring of horses and donkeys are infertile. Be sure to include in your answer a discussion of meiosis and gametes.
b. Even though domestic and Przewalski horses create fertile offspring, they are considered different species. Explain why.
c. Imagine a time in the future when the Przewalski horse has been introduced to Mongolia and is thriving in large numbers. Populations of domestic horses live in areas bordering the Przewalski horse range. Design an experiment to determine if these two groups of horses are one or two different species. What type of data would need to be collected to answer this question?

Chapter 18: The History of Life on Earth

Chapter Outline

18.1 - Events in Earth's History Can Be Dated
18.2 - Changes in the Earth's Physical Environment Have Affected the Evolution of Life
18.3 - Major Events in the Evolution of Life Can Be Read in the Fossil Record

It is time to think about time. Understanding the impressive evolutionary changes of plants and animals requires you to think far beyond the lifespan of a human being, as this story started very long ago, was punctuated by five major extinctions, and it will continue on for a very long time!

To organize our thinking about evolutionary time, the earth's geological history has been mapped out into four major intervals called eras, starting 4,500,000,000 years ago (4.5 bya). Each era is subdivided into periods. While it is true that intense cataclysms shaped life on earth, e.g., volcanoes, meteors, continental crashes, etc., it is also true that life shaped the earth, especially the earth's atmosphere, yielding more evolutionary opportunities. Puzzling through the evidence, the fossils, the geological changes, the calculations, allows us to see some of life's experimentation, including the dead ends, and, perhaps, the future opportunities.

The Big Idea of Chapter 18 is, of course, number one: evolution. This chapter shows that:

1.a.4: Biological evolution is supported by scientific evidence from many disciplines, including mathematics.
1.c.1: Speciation and extinction have occurred throughout the Earth's history.

Chapter Review

Concept 18.1 explains how events in earth's history can be dated. Imagine having an immense trash can into which you place all of your disposable items throughout your lifetime. Someone could, carefully, sort through the layers, or strata, in the trash, and find out some of what you did last week, some of what you did two years ago, and some of what you did when you were 5 years old. In so doing, they could put together a picture of how you changed over the years. So it is with fossils in ocean sediments—the oldest forms of life are in the bottom layer, the newer forms are closer to the top, and analyzing the layers accumulated over the years tells us much about the history of life.

Temperature of a corpse is an important clue to an investigator studying a murder scene, as the cooling rate of a recently deceased individual has been thoroughly studied in forensic science. Just as heat energy dissipates when it is no longer being produced, elemental isotopes decay in a measureable way over time, albeit a much longer interval of time, and provide important clues that "date" rocks and other materials of interest to those focused on analyzing the past.

1. Radioisotopic elements in igneous rock decay at predictable rates, as shown in the table below.

Radioisotope	Half-life (years)	Decay product	Useful dating range (years)
Carbon-14 (^{14}C)	5,700	Nitrogen-14 (^{14}N)	100 – 60,000
Potassium-40 (^{40}K)	1.3 billion	Argon-40 (^{40}Ar)	10 million – 4.5 billion
Uranium-238 (^{238}U)	4.5 billion	Lead-206 (^{206}Pb)	10 million – 4.5 billion

PRINCIPLES OF LIFE, Figure 18.1 (Part 2)

Use the table to the right to suggest which radioisotopic elements should be measured to determine the age of samples is believed to be taken from each of the following periods. Explain your answers.

a) Permian:

____
____
____
____
____
____
____

b) Quaternary:

____
____
____
____
____
____

TABLE 18.1 Earth's Geological History

RELATIVE TIME SPAN	ERA	PERIOD	ONSET
	Cenozoic	Quaternary	2.6 mya
		Tertiary	65 mya
	Mesozoic	Cretaceous	145 mya
		Jurassic	200 mya
		Triassic	251 mya
Precambrian	Paleozoic	Permian	297 mya
		Carboniferous	359 mya
		Devonian	416 mya
		Silurian	444 mya
		Ordovician	488 mya
		Cambrian	542 mya
	Precambrian		900 mya
			1.5 bya
			3.8 bya
			4.5 bya

Note: mya, million years ago; bya, billion years ago.

2. Suppose that the sample you dated in question 1, part a), was found in igneous rock 2 meters above a dinosaur fossil. Explain how this additional information would affect your assessment of the geological age of the "Permian" sample.

____
____
____
____

3. Suppose that the sample you dated in question 1, part b), was found in rock 2 meters above a dinosaur fossil. Explain how this additional information would affect your assessment of the geological age of the "Quaternary" sample.

____
____
____
____

Concept 18.2 shows how changes in the earth's physical environment have affected the evolution of life**.** The movement of the continental landmasses has been an important influence on the distribution of plants and animals. Just as sheets of ice can move around on the ocean, plates of the surface or crust of the earth move around on a bed of very hot molten rock. You probably know these "continental drift" movements can cause earthquakes, but they are also instrumental in determining where volcanoes occur, and what direction ocean currents flow, thus influencing temperature across the earth. In addition to its own restlessness, the earth has been struck by extraterrestrial objects of varying sizes, including a meteorite big enough to cause mass extinctions. Furthermore, the organisms on the planet, especially plants, have had major effects on the amount of oxygen in the atmosphere.

4. Discuss the climate changes associated with large reductions in sea level that coincide with large extinctions of marine life.

____
____
____
____

5. Describe the hypothesis of current global climate change and discuss how an unusually cold winter fails to refute the hypothesis.

____
____
____
____

6. Iridium is fairly uncommon in the clay soils of earth but is relatively more abundant in a thin layer of Cretaceous-Tertiary boundary sample (65 million years ago) across large areas of earth, a time when a mass extinction took place. Discuss this finding.

____
____
____
____

7. Describe how life on earth has impacted the concentration of oxygen in the atmosphere at these milestones over the past two billion years.

2 billion years ago:____
____
____

1 billion years ago:____
____
____

500 million years ago:____
____
____

250 million years ago:____
____
____

Concept 18.3 shows how major events in the evolution of life can be read in the fossil record. Fascination with dinosaurs is wide spread, likely because the artistic models of these animals, re-constructed based on fossil evidence, reveals animals that are so large and so strange that it is difficult to imagine them. But fossils are evidence for many other kinds of plants and animals that can factor into solving the puzzles of evolutionary history. Unfortunately, the fossil record is only a tiny representation of the past flora and fauna of the planet, since very few dying organisms end up in conditions appropriate for fossilization processes to occur.

8. Around 500 million years ago, the Cambrian "explosion" in the number plants and animals on earth was in bloom. However, 150 million years later, at the end of the Devonian period, a massive extinction occurred. This was followed by the "invasion of the land," as amphibians began to move on to land and into drier habitats. Discuss the changes that produced the opportunity for the land invasion by animals.

____
____
____
____
____
____

9. In addition to your answer to #8, describe the types of fossil evidence of animal morphology that would support the hypothesis that the invasion occurred at that time.

____
____
____
____
____
____

10. Describe the primary geologic factors at work for each of these milestones of evolution.

250 million years ago:____
____
____

225 million years ago:____
____
____

175 million years ago:____
____
____

100 million years ago:____
____
____

Science Practices & Inquiry

In the AP Biology Curriculum Framework, there is a set of 7 Science Practices. In this chapter, we will focus on Science Practice 5: The student can perform data analysis and evaluation of evidence. More specifically, practice 5.1: The student can analyze data to identify patterns or relationships.

Question number 11 asks you to analyze data related to questions of speciation and extinction throughout the Earth's history (LO 1.20).

11. The graph shows the number of families of organisms over time.

a) Determine which times appear to be major extinctions, and label them with arrows.

b) Explain how a major extinction will impact the surviving species, that is, those that do not go extinct.

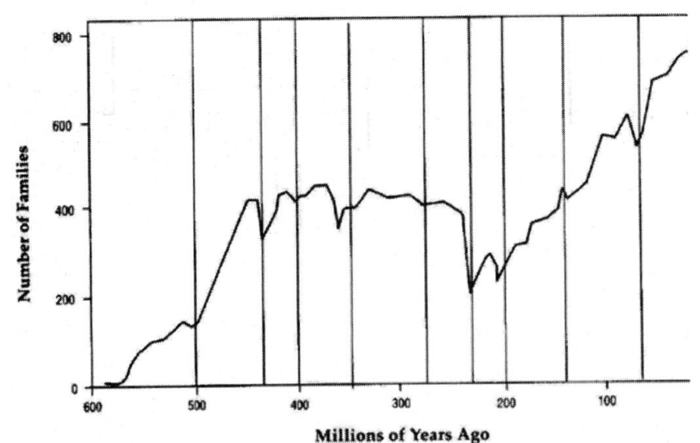

Chapter 19: Bacteria, Archaea, and Viruses

Chapter Outline

19.1 - Life Consists of Three Domains That Share a Common Ancestor
19.2 - Prokaryote Diversity Reflects the Ancient Origins of Life
19.3 - Ecological Communities Depend on Prokaryotes
19.4 - Viruses Have Evolved Many Times

In this chapter we begin our study of the diversity of life. The chapter opens with an introduction to the three domains of life. There are obvious distinctions between the eukaryotes and prokaryotes. Although you cannot see Bacteria and Archaea individuals without a microscope, the differences between the Bacteria and the Archaea are equally dramatic. The major differences are biochemical in nature with some differences in the size and structure of ribosomes and the structure of their membrane lipids.

As the AP Biology® exam is stressing taxonomy and classification less and less, the important lessons of this chapter include the evolutionary lineages of organisms from a prokaryotic predecessor and the idea that phylogenetic trees show evolutionary history. You will not be tested on whether a particular bacteria stains Gram positive or negative, but you should understand the major biochemical processes that are conserved through the phyla such as glycolysis. Prokaryotes ran the gamut of diversity when it comes to metabolic processes, including chemoheterotrophy (like humans) to photoheterotrophy. Pay close attention to how different organisms produce ATP and obtain carbon atoms for anabolic reactions.

Additionally, the prokaryotes play many vital roles in ecosystems. Early prokaryotes liberated enough molecular oxygen gas into the atmosphere to allow the evolution of aerobic respiration. Without nitrogen fixing bacteria, there would be insufficient amounts of nitrates available for plants to thrive. The cycling of nutrients by bacteria as decomposers is critical to many ecosystems. As symbionts and pathogens, bacteria are also important to larger organisms: bacteria help cattle digest grasses and also produce vitamins B_{12} and K in our intestine.

Chapter 19 concludes with a look at viruses. Viruses are considered to be non-living by many, while others classify them as acellular organisms. Regardless, viruses do replicate very efficiently, mutate frequently, and are an important group of pathogenic "organisms" to study. Viruses are classified by their type of genetic material and whether their genetic material is single or double stranded. Given its impact on human populations, the HIV virus is an important example of a retrovirus to understand.

Chapter 19 spans Big Ideas 1 and 2 of the AP Biology Curriculum Framework. The Big Ideas are a means for organizing the vast amount of information we know as biology. It is vitally important that you continually work to understand these **Big Ideas** and across different **Enduring Understandings**.

Big Idea 1 recognizes that evolution ties together all parts of biology. Specifically in Chapter 19, we see an introduction to classification and the evolutionary relationships of bacteria to eukaryotes.

1.b.1: Organisms share many conserved core processes and features that evolved and are widely distributed among organisms today.
1.b.2: Phylogenetic trees and cladograms are graphical representations (models) of evolutionary history that can be tested.

Big Idea 2 states that the utilization of free energy and use of molecular building blocks are characteristic fundamental of life processes. Specifically, Chapter 19 includes:

2.e.3: Timing and coordination of behavior are regulated by various mechanisms and are important in natural selection.

Chapter Review

Section 19.1 examines the commonalities and differences across the three domains of life. Evolutionarily, the last common ancestor of all three domains probably lived on Earth approximately 3 billion years ago. Over the last 3 billion years, organisms have evolved a diversity of metabolic pathways and differentiated into three major groups.

1. Describe and discuss four things that are common to all living organisms.

a)____

b)____

c)____

d)____

2. Explain why all living things share the features you identified in question #1.

3. Describe how binary fission in prokaryotes is similar to AND different from mitosis in eukaryotes.

4. Eukaryotes have membrane-bound organelles where specific biochemical reactions take place. How do prokaryotes accomplish these tasks without highly specialized organelles, specifically, lysosomes, mitochondria, or endoplasmic reticula?

5. Describe and discuss TWO types of evidence supporting the idea that all three domains of life have a common ancestor.

6. Explain the similarity of genes found in eukaryotes to those of the Archaea and the Bacteria.

7. There are a multitude of different prokaryotes on the Earth today. State your agreement or disagreement with this statement, "Bacteria are primitive life forms that are more than 3 billion years old." Explain your answer.

8. Below are two possible cladograms showing the relationships of 4 species. What would be needed to determine the relationship between B, C, & D?

A

B

C

D

A

B

C

D

Section 19.2 looks at the taxonomy and classification of prokaryotes. Much of this will not be on the AP Biology exam. Understanding the extreme habitats different bacteria flourish in can give us insights into shared biochemical pathways and some useful biotechnology tools.

9. The keyboard you are typing on, or the pen you are holding, may or may not have any Archaea and/or Bacteria on it. Explain why or why not.

10. Describe the importance of spores to bacteria that can form them.

11. Most animal classification involves physical features or behavioral characteristics. Describe the key feature that was used originally to separate Archaea and Bacteria into separate domains.

12. Frequently one encounters unfamiliar words in science textbooks. For instance p. 377 in this sentence in Principles of Life states: "These ether linkages are a synapomorphy of archaea." Very few introductory biology students know this word and it is not defined for you well in context.

a. Define the word "synapomorphy" as used here.

b. Synapomorphy occurs when a trait is shared by two or more ancestors and their common ancestor, but not the ancestor of the common ancestor. Complete the phylogenetic tree below showing how this might look with two species sharing a common trait. Use shaded circles to show a shared trait and unshaded circles to show the lack of this trait.

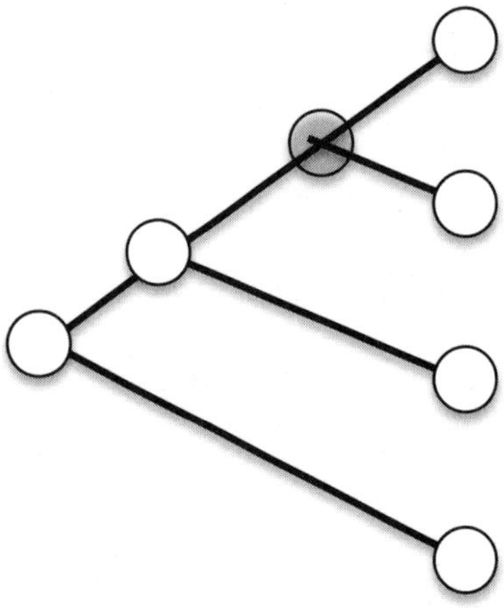

13. Describe how the membranes of archaea are unique among prokaryotes.

Section 19.3 shows how prokaryotes flourish almost everywhere and are important members of ecosystems.

14. Discuss and compare the metabolism of obligate and facultative anaerobes, and then classify humans using this terminology.

15. Complete the table below:

Category	Energy source	Carbon source	Example
Photoautotrophs			
Photoheterotrophs			
Chemolithotrophs			
Chemoheterotrophs			

16. Use the categories in the table above to classify the following:

____________________ An organism that lives in complete darkness and cannot utilize CO_2 but is able to produce its own sugars.

____________________ Humans

____________________ A fungus decomposing a rotting tree.

____________________ A plant absorbing CO_2 and producing sugars

17. Compare the reactions of nitrogen-fixing bacteria with those of nitrifying bacteria.

18. In an aquarium, fish excrete ammonia as a waste product. Describe the types of bacteria that convert the ammonia to nitrates.

19. A fellow student complains that there are too many bacteria in the world and we should spray all surfaces in the classroom with antibiotic spray.

a. How would you explain to the student why very few of the bacteria would be able to hurt him/her?

b. If all surfaces in the room were sprayed with the same antibiotic spray every day, what effect would there be on the bacterial populations?

Section 19.4 focuses on viruses and their importance to other organisms. Evolutionarily viruses probably evolved multiple times as either escaped genetic elements or as highly reduced parasites.

20. Viruses are generally classified into 4 large groups, DNA or RNA, and single- stranded or double-stranded genetic material. Explain why most believe that these groups are not monophyletic.

____
____
____
____

21. Explain why some biologists view viruses as "highly reduced parasitic organisms."

____
____
____
____

22. Explain quorum sensing and explain why it is important to bacteria from the aspect of natural selection. (Hint, you may wish to reread the opening page of the chapter).

____
____
____
____

Science Practices & Inquiry

In the AP Biology Curriculum Framework, there is a set of 7 Science Practices. In this chapter, we will focus on **Science Practice 6:** The student can use representations and models to communicate scientific phenomena and solve scientific problems. More specifically, practice 6.1: The student can *justify claims with evidence.*

Questions 22 & 23 ask you to justify the scientific claim that organisms share many conserved core processes and features that evolved and are widely distributed among organisms today (LO 1.16).

22. Draw a diagram showing how a retrovirus infects a host cell. Label all the different forms of genetic material that are utilized.

23. Most biologists now believe that viruses evolved from other living organisms recently and are not an ancient form of life. Describe evidence that viruses have evolved from other living organisms.

____
____
____
____
____
____
____
____
____
____
____
____
____
____
____

Chapter 20: The Origin and Diversification of Eukaryotes

Chapter Outline

20.1 - Eukaryotes Acquired Features from Both Archaea and Bacteria
20.2 - Major Lineages of Eukaryotes Diversified in the Precambrian
20.3 - Protists Reproduce Sexually and Asexually
20.4 - Protists Are Critical Components of Many Ecosystems

Chapter 20 explores an early question challenging evolutionary biologists: How did living organisms gain such complexity? As you encounter diverse organisms, here are some questions to think about. What does it look like? How does its metabolism work? What is its mode of reproduction? These broad functional questions will help you identify the biological characteristics of most use.

In Chapter 20, the primary theme is Big Idea number one, evolution, with further emphasis that:

1.b.2: Phylogenetic trees and cladograms are graphical representations (models) of evolutionary history that can be tested.

Chapter Review

Concept 20.1 shows how eukaryotes acquired features from both Archaea and Bacteria. Having explored the single-celled Archaea and Bacteria in Chapter 19, this section reviews the cellular transitions hypothesized to have occurred leading from simple cells of Archaea and Bacteria to the more complex cells of the Eukarya. First, it appears that size limits on diffusion and metabolism were barriers to single cells with smooth surfaces, and the in-folding of the cell's surface resulted in an increased surface area for exchange between the cell and its environment. Second, the DNA of prokaryotes, bound to a small piece of membrane, became encircled by that membrane, leading to the eukaryotic nucleus. Third, ever-larger cells engulfed other, smaller cells, and their life cycles became united as they prospered together – the endosymbiotic theory that explains eukaryotic complexity.

1. Describe the hypothesized (non-endosymbiotic) relationship between the diffusion requirements of cells and the development of Golgi apparatus, the smooth endoplasmic reticulum, and the rough endoplasmic reticulum.

____
____
____

2. Describe the hypothesized endosymbiotic relationship between a bacterium and the eukaryotic mitochondrion. Include in your discussion the role of the bacteria in detoxification metabolism.

____
____
____
____

3. Describe the hypothesized endosymbiotic relationship between the metabolism of a cyanobacterium and the first photosynthetic eukaryotes.

____
____
____
____
____

4. Describe the structural characteristics suggesting a formerly independent lifestyle for:

a) Mitochondria____

b) Chloroplasts____

5. Compare and discuss the DNA of Bacteria and Archaea with the DNA of eukaryotes.

6. Compare and discuss primary and secondary endosymbiosis.

7. Discuss this proposal: "The evolutionary development of photosynthetic eukaryotes was likely preceded by the development of mitochondria." Explain the reasons for your argument.

Concept 20.2 explains how major lineages of eukaryotes diversified in the Precambrian. Protists are the eukaryotes of interest in this chapter. "Protista" is not a formal taxon, but the term is used to suggest that these organisms are not the traditional plants, animals and mushrooms we see in everyday life. Even though memorizing taxonomy is not required by the AP-Biology Curriculum Framework, the chapter includes some phylogenetic trees that you should be able to read and understand (without memorizing).

8. Freshwater *Paramecium* would explode if you deactivated a specific organelle. Name the structure and describe how it works.

9. In spite of their amazingly diverse forms, diatoms follow one of two simple "body plans." Describe these plans and discuss whether humans follow either plan.

10. A fellow student, using a high-powered microscope, discovers a unicellular organism in a water sample taken from the creek behind your school. Suggest some visual landmarks she should look for in determining whether or not this is a eukaryote.

11. Explain how this phylogenetic tree shows that the term "protists" does not refer to a formal taxonomic group.

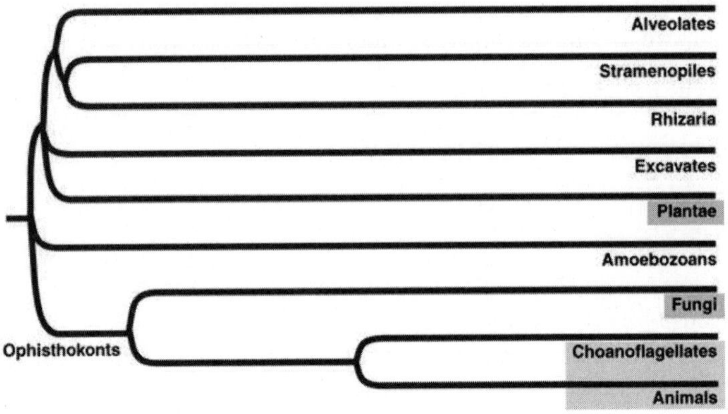

PRINCIPLES OF LIFE, Figure 20.3
© 2012 Sinauer Associates, Inc.

Concept 20.3 explores variations among protists in the capacity to reproduce sexually and asexually. Some protists reproduce asexually and other are sexual in their reproductive pattern. Recall that asexual reproduction is similar to a clone factory, wherein progeny are nearly fully identical to the parent. Some protists exchange DNA without producing offspring. Others engage in sexual reproduction, and you will see that these can include prominent diploid and haploid stages of development, yielding insight into the "the alternation of generations" characteristic of these and many other forms of more complex eukaryotes.

12. Explain what the statement means that some protists have sex without reproduction.

13. Explain what the statement means that some protists have reproduction without sex.

14. Explain alternation of generations. Is this part of your life pattern?

Concept 20.4 explains how protists are critical components of many ecosystems. Protists are incredibly bountiful on earth, but their small size easily escapes our attention. However, their byproducts can be abundantly obvious, whether at the beach, or deep in a petroleum well.

15. "Tidal blooms" occur in the ocean when certain plankton species become especially abundant. Explain how toxins produced under "bloom" conditions can make it into the human food chain, even though we don't intentionally eat protists.

16. Describe the relationship between protists and petroleum.

Science Practices & Inquiry

In the AP Biology Curriculum Framework, there is a set of 7 Science Practices. In this chapter, we will focus on **Science Practice 3**: The student can engage in scientific questioning to extend thinking or to guide investigations within the context of the AP course. More specifically, practice 3.1: The student can pose scientific questions.

Question 17 asks you to pose scientific questions about a group of organisms whose relatedness is described by a phylogenetic tree or cladogram in order to (1) identify shared characteristics, (2) make inferences about the evolutionary history of the group, and (3) identify character data that could extend or improve the phylogenetic tree. (LO 1.17)

17. The phylogenetic tree (gene tree) to the right shows the evolutionary relationships of rRNA gene sequences isolated from the nuclear genomes of humans, yeast, and corn; from an archaeon (*Halobacterium*), a proteobacterium (*E. coli*), and a cyanobacterium (*Chlorobium*); and from the mitochondrial and chloroplast genomes of corn. Use the phylogenetic tree to answer the following questions.

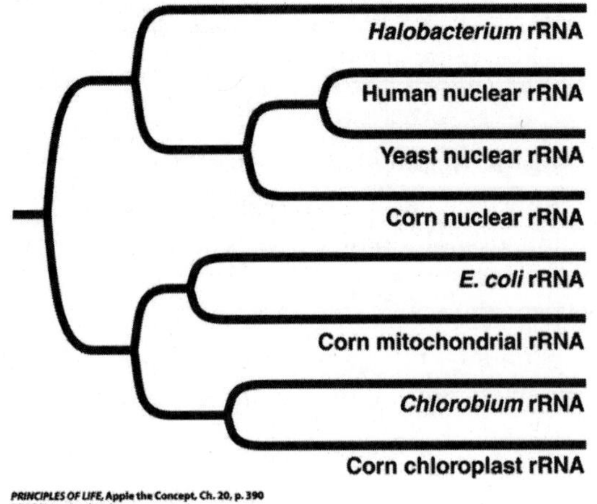

PRINCIPLES OF LIFE, Apple the Concept, Ch. 20, p. 390

a. Why aren't the three rRNA genes of corn one another's closest relatives?

____

____

____

____

____

____

b. Explain where the rRNA genes from human and yeast mitochondrial genomes would be placed on the gene tree.

____

____

____

____

____

____

c. What other characteristics might you analyze that could improve the phylogenetic tree?

____

____

____

____

____

____

Chapter 21: The Evolution of Plants

Chapter Outline

21.1 - Primary Endosymbiosis Produced the First Photosynthetic Eukaryotes
21.2 - Key Adaptations Permitted Plants to Colonize Land
21.3 - Vascular Tissues Led to Rapid Diversification of Land Plants
21.4 - Seeds Protect Plant Embryos
21.5 - Flowers and Fruits Increase the Reproductive Success of Angiosperms

In this chapter we examine the evolution of land plants. The opening page of the chapter provides an excellent overview of the relationships among the different groups of plants, presenting the major groups in a phylogenetic tree. The evolutionary relationships among plants and the survival of plants on land are important biological concepts, even though the taxonomy and the diverse reproductive patterns in plants in Chapter 21 are unlikely subjects for the AP Biology exam. The evolution of plants, however, including the features that allowed them to colonize the land, is worthy of your attention.

The evolution of a cuticle, stomata, gametangia, a protected embryo, protective pigments, and thick spore walls allowed the colonization of land. Later features of vascular plants include the xylem and phloem and tracheids. The evolution of the seed brought the gymnosperms and related plants to dominance. With the evolution of the flower and fruits, the angiosperms had a major advantage in reproduction over other plants. A key feature of angiosperm reproduction is double fertilization.

The most important part of this chapter is that of evolutionary relationships found in **Big Idea 1** which recognizes that evolution ties together all parts of biology. In Chapter 21 we look at the evolutionary history of land plants:

1.b.2: Phylogenetic trees and cladograms are graphical representations (models) of evolutionary history that can be tested.

Chapter Review

Section 21.1 describes the evolution of land plants from an early green algae.

1. When observing an orchid with a very long flower, Charles Darwin predicted the discovery of a moth with a very long proboscis. Several years later, a sphinx moth that pollinates the flower was discovered. Explain the logic that Darwin used to make his prediction.

____
____
____
____
____
____
____
____
____

2. Examine the diagram below. The black dots represent new evolutionary features that gave the plants a reproductive advantage. Label each black dot with its feature.

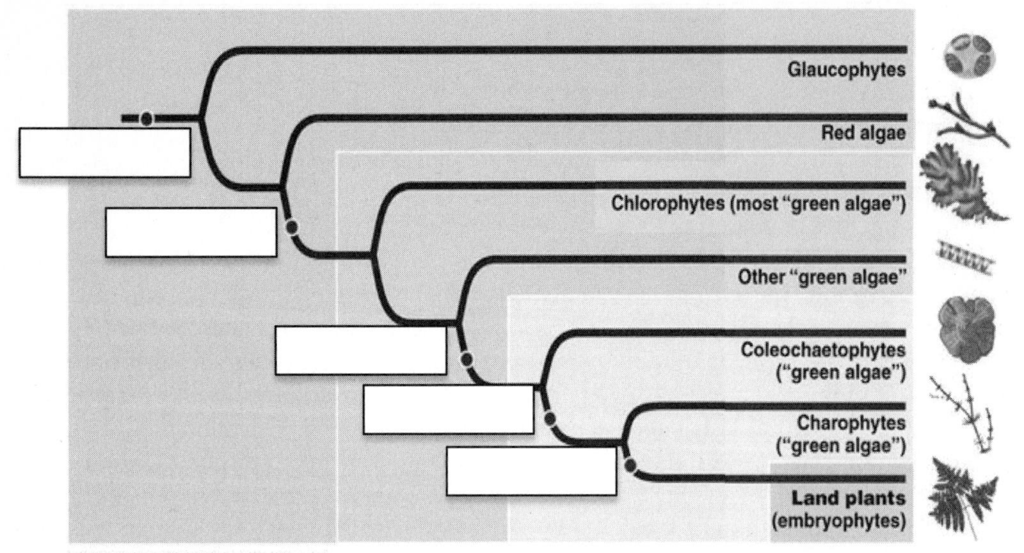

PRINCIPLES OF LIFE, Figure 21.1 (Part 1)

3. Recent genome analyses have revealed remarkable similarities between the human and green algal proteins involved in cilium assembly and function. The discovery that these proteins are strikingly similar even after hundreds of millions of years of evolutionary divergence suggested the use of green algae as experimental organisms for research on diseases that affect human cilia. Propose an idea why these proteins have changed so little over hundreds of millions of years of evolution.

____

____

____

____

____

____

Section 21.2 examines the evolution of land plants from algae living in marginal habitats along shorelines.

4. A plant's life on land is much more difficult than life in the ocean or a pond. Describe four aspects of life on land that make terrestrial environments difficult places for plants to grow.

a) ____

____

b) ____

____

c) ____

____

d) ____

____

5. Describe the two key processes needed for the alternation of generations.

a) ____

____

____

b) ____

____

____

6. For plants whose gametes are produced by mitosis, explain the role of meiosis in the alternation of generations.

____

____

____

____

____

____

7. A fellow student tells you that he/she has discovered a new species of moss that grows over six feet tall. Explain why this moss is probably classified within a different group of plants.

____

____

____

____

____

____

Section 21.3 describes the evolution of vascular tissue in land plants.

8. Explain the function of the two vascular tissues in plants.

a) ____

____

____

b) ____

____

____

9. The evolution of a cellulose-digesting enzyme in fungi likely occurred after the evolution of land plants. Explain how coal and oil form and why they are considered nonrenewable resources.

____

____

____

____

____

____

Section 21.4 studies the evolution of the seed and its importance to the gymnosperms and angiosperms.

10. Explain the role of the sporophyte and the gametophyte in the reproduction of an angiosperm, and describe where you would find each in a blooming lily plant.

11. Pollen grains are integral components in angiosperm reproduction. Describe pollen grains and discuss how pollen can survive for many years.

Section 21.5 looks at the reproductive advantage of the fruit and the flower to the dominant plant form today, the angiosperms.

12. Is pollen unique to angiosperms or gymnosperms?

13. Draw a cladogram (phylogenetic tree) showing the evolution of pollen grains with these two groups of plants and their ancestors. Clearly show in your cladogram when pollen evolved.

14. Draw a cell that undergoes mitosis three times without cytokinesis. Show each mitotic division (you do not need to draw each phase of mitosis.) Show the number of nuclei present in the final cell.

15. Explain the difference between pollination and fertilization in plants. Draw a cross section of a flower and show where each occurs.

Science Practices & Inquiry

In the AP Biology Curriculum Framework, there is a set of 7 Science Practices. In this chapter, we will focus on Science Practice 1: The student can use representations and models to communicate scientific phenomena and solve scientific problems. More specifically, practice 1.1: The student can *create representations and models* of natural or man-made phenomena and systems in the domain.

Question 16 asks you to create a phylogenetic tree or simple cladogram that correctly represents evolutionary history and speciation from a provided data set. Below is a set of data about a group of plants that you will use to create a cladogram (LO 1.19).

16. Below are some characteristics found among four groups of plants.

	Movement to land	Presence of chlorophyll a	Vascular system	Inversion in chloroplast DNA	Seeds	Flowers
A	+	+	+	+	+	+
B	+	+	+	+	+	–
C	+	+	+	+	–	–
D	+	+	–	–	–	–

a. Using the data, draw a phylogenetic tree for these organisms.
b. Compare your phylogenetic tree to those in the chapter to see if you can determine the identities of A, B, C, and D.

Chapter 22: The Evolution and Diversity of Fungi

Chapter Outline

22.1 - Fungi Live by Absorptive Heterotrophy
22.2 - Fungi Can Be Saprobic, Parasitic, Predatory, or Mutualistic
22.3 - Major Groups of Fungi Differ in Their Life Cycles
22.4 - Fungi Can Be Sensitive Indicators of Environmental Change

The opening of this chapter demonstrates the relevance of understanding natural selection as we use more and more antibiotics in daily life. When Alexander Fleming discovered that mold growing as contamination on his bacterial growth plates inhibited the growth of the bacteria, he launched the discovery of the first antibiotic drug, penicillin. It took almost 20 years for chemists to isolate the new antibiotic and create a stable version that could be used in modern medicine, but once they had, penicillin ushered in a new age of modern medicine, greatly reducing the number of complications and deaths from infections. Subsequently, the attraction to blindly killing as many bacteria as possible led us to our current state of antibiotic overuse. As a result, many new forms of antibiotic-resistant bacteria have become abundant through natural selection.

The evolutionary relationships among different organisms and fungi are important biological concepts, even though the taxonomy and the diverse reproductive patterns of fungi in Chapter 22 are unlikely subjects for the AP Biology exam. The evolution of fungi as absorptive heterotrophs, and the symbiotic relationship of fungi with other organisms, however, are worthy of your attention. The vocabulary used to describe fungi might be new to you, but this is a chance to apply what you've already learned about other living organisms: form arises during development, and serves to meet the metabolic (vegetative) and reproductive functions of the organism.

The scope of Chapter 22 includes Big Ideas 1 and 2 in the AP Biology Framework. The Big Ideas are a means for organizing the vast amount of information we know as biology. It is vitally important that you continually work to understand these **Big Ideas** and across different **Enduring Understandings**.

Big Idea 1 recognizes that evolution ties together all parts of biology. In Chapter 22, we note that:

1.a.2: Natural selection acts on phenotypic variations in populations
1.b.2: Phylogenetic trees and cladograms are graphical representations (models) of evolutionary history that can be tested.

Big Idea 2 states that the utilization of free energy and use of molecular building blocks are characteristic fundamental of life processes. Specifically, Chapter 22 includes:

2.e.2: Timing and coordination of physiological events are regulated by multiple mechanisms.

Chapter Review

Concept 22.1 introduces the fungal lifestyle. Most fungi are absorptive heterotrophs meaning they secrete digestive enzymes onto their intended food, releasing nutrients that get absorbed for use. The primary "body" of a fungus is called the mycelium that is made up of individual filaments called hyphae.

1. Explain the following terms of fungal structure: hyphae, mycelium, septa.

____

____

____

____

2. Explain how heterotrophy by humans is similar to that by fungi.

3. Explain how heterotrophy by humans is different from that by fungi.

4. The hyphae of many fungi are multinucleate. Explain how they solve the problem of being multicellular, with comparison to multicellular plants and animals.

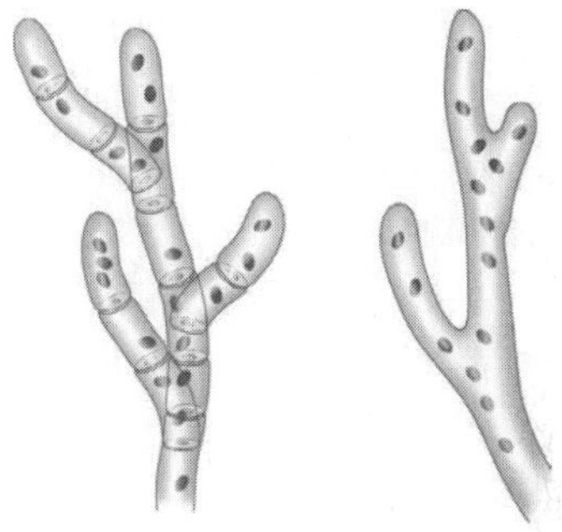

Concept 22.2 compares feeding patterns among different fungi. Many fungi (saprobes) feed on organic material in dead animals and plants, releasing carbon to supply the carbon cycle. Other fungi are parasitic. Several fungi have mutualistic relationships with other organisms, such as lichens and mycorrhizae. Lichens are a symbiotic relationship between an algae and a fungi, and mychorrhizae are an association between fungi and vascular plants.

5. Lichens are mutualistic relationships between fungi and algae. Explain the contributions of each partner in this relationship and how this allows lichens to colonize and thrive on bare rock.

6. Not all fungi are as pleasant or benign as lichens or mycorrhizae. The fungus shown to the right is a predatory fungi (*Arthrobotrys dactyloides*) that traps small nematode worms in soil. Explain how this fungus is still an absorptive heterotroph.

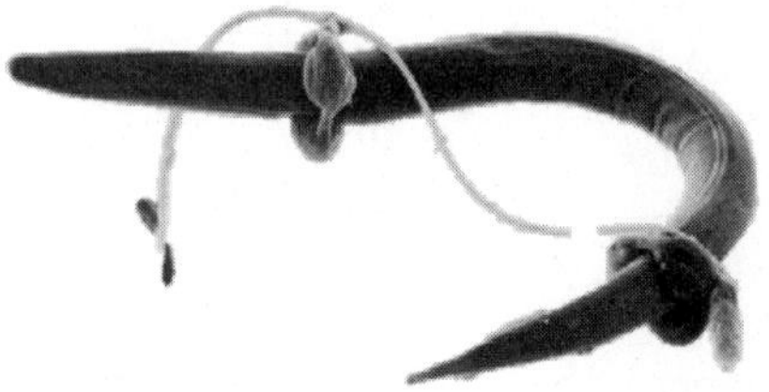

7. Many plants have a mutualistic relationship with mycorrhizae. A gardener starts some seeds in a pot with new sterile soil that has all the needed nutrients to grow plants, but finds that the plants are small and stunted compared to the seeds that were sown in a nearby garden. How would you explain this phenomenon to the gardener?

Concept 22.3 compares the many and varied life cycles of different fungi. It is not important to memorize the life cycles of the various groups for the AP exam, but do realize that fungi respond to external signals that regulate their physiological responses.

8. Explain when and where a mycologist should go looking for the dikaryotic mycelium of a club fungi if the sought species is responsive to and dependent on environmental moisture for its well-being.

Use the lifecycle diagram of a fungus below to answer questions 9 and 10.

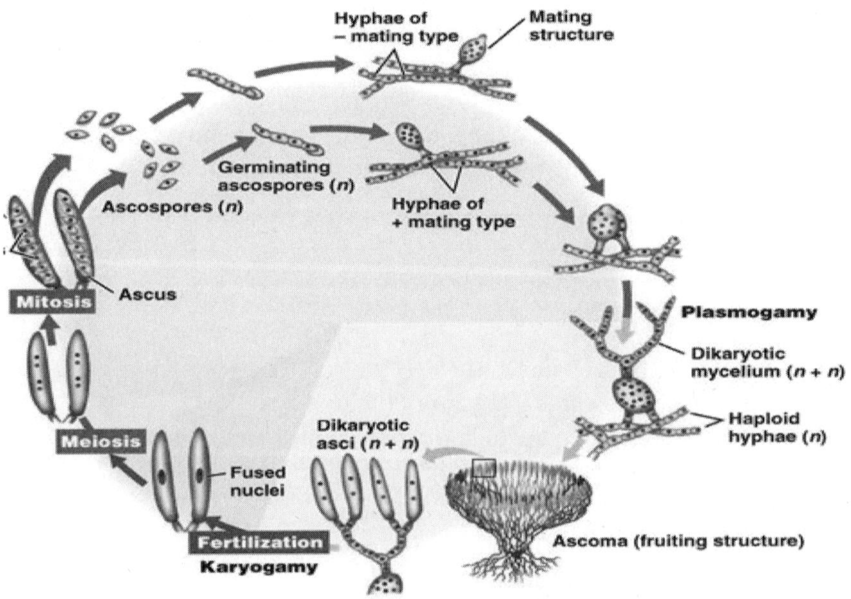

9. Describe the life cycle in terms of how much time the fungus is in haploid or diploid form.

10. Describe the function of the ascospores.

____

____

____

Concept 22.4 describes how fungi can be used to study environmental problems and how to solve them.

11. An environmental scientist has decided to utilize lichens to monitor air quality in a city. Propose a method for doing this, and explain your logic.

____

____

____

____

____

____

12. A forester proposed that the difficulty and expense of reforestation projects after tree harvest might be reduced by retaining viable communities of mycorrhizal fungi. Explain how this could be done.

____

____

____

____

____

____

____

____

____

Science Practices & Inquiry

In the AP Biology Curriculum Framework, there is a set of 7 Science Practices. In this chapter, we will focus on **Science Practice 1:** The student can use representations and models to communicate scientific phenomena and solve scientific problems. More specifically, practice 1.1: The student can create representations and models of natural or man-made phenomena and systems in the domain.

Questions 13 asks you to create a phylogenetic tree or simple cladogram that correctly represents evolutionary history and speciation from a provided cladogram and your prior knowledge. (LO 1.19).

13. Below is a phylogenetic tree showing the relationships between choanoflagellates (protists), animals, and fungi. Draw a new phylogenetic tree showing the relationships between plants, humans, bacteria, and fungi. Label the shared derived characteristics for each group.

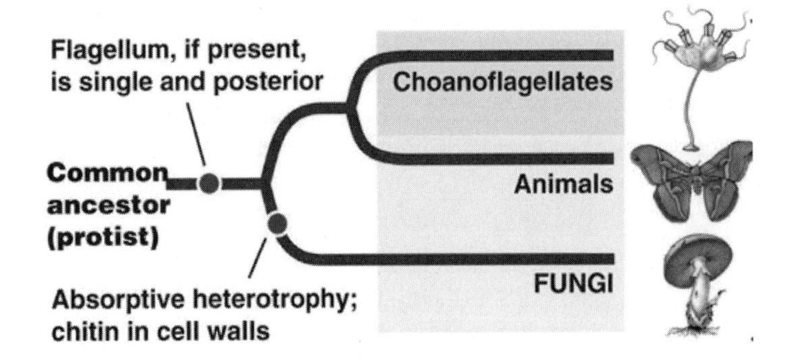

Chapter 23: Animal Origins and Diversity

Chapter Outline

23.1 - Distinct Body Plans Evolved among the Animals
23.2 - Some Animal Groups Fall Outside the Bilataria
23.3 - There are Two Major Groups of Protostomes
23.4 - Arthropods Are Diverse and Abundant Animals
23.5 - Deuterostomes Include Echinoderms, Hemichordates, and Chordates
23.6 - Life on Land Contributed to Vertebrate Diversification
23.7 - Humans Evolved among the Primates

Chapter 23 looks at the evolutionary relationships among the animal groups. The chapter begins with an overview of animals, their phylogeny, and their basic developmental patterns and body plans. Beginning with the sponges, the chapter reviews each of the major phyla, their characteristics and evolutionary developments. This is a lengthy chapter with many names and vocabulary terms.

Most of this chapter is outside of the scope of the AP Biology Curriculum Framework. However, while illustrative examples and excluded content will not be assessed on the AP Biology Exam, they may be provided in the body of exam questions as background information for a concept or science practice being assessed. Do not spend your time memorizing the many phyla and characteristics found within this chapter. Rather, familiarize yourself with the terms and ideas so when you see them in the context of a paragraph or diagram you are asked to analyze, you will recognize the terms and ideas and be able to provide a reasonable answer. Some examples of these terms and ideas can be found in questions 8, 9, and 11 below.

The most important part of this chapter is that of evolutionary relationships found in **Big Idea 1,** which recognizes that evolution ties together all parts of biology. Chapter 21 summarizes the major branches of the evolutionary history of animals, using a familiar tool:

1.b.2: Phylogenetic trees and cladograms are graphical representations (models) of evolutionary history that can be tested.

Chapter Review

Section 23.1 begins with an overview of the animal phyla and their characteristics.

1. Discuss the difficulties that scientists face when trying to classify the multitude of different insect taxa.

____
____
____
____
____

2. Tree canopies in tropical forests are rich locations for collecting previously unclassified species of insects. Explain why there is a high diversity of insects in the tropics, basing your logic on natural selection as the mechanism for evolutionary diversification.

____
____
____
____
____

3. Describe and discuss two reasons that the animal phyla are considered to be monophyletic.

Use the phylogenetic tree below to answer questions 4 through 6.

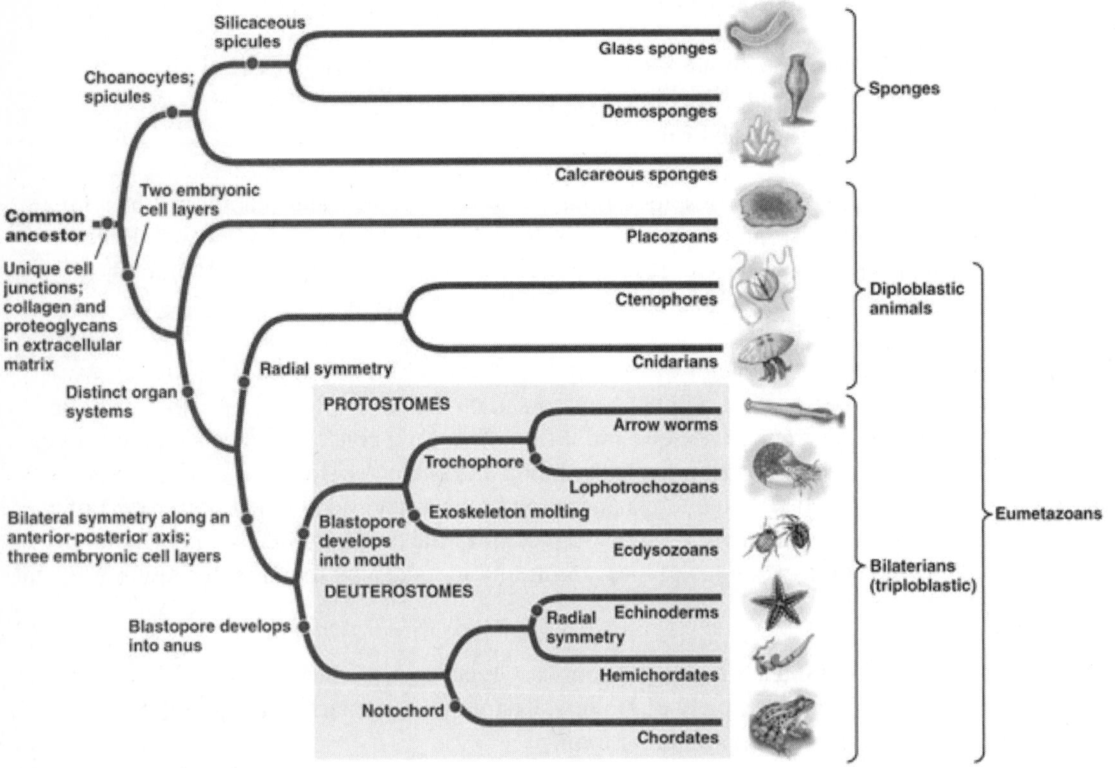

PRINCIPLES OF LIFE, **Figure 23.1**
© 2012 Sinauer Associates, Inc.

4. Based on the phylogenetic tree shown above, identify THREE characteristics found in both echinoderms and ctenophores.

5. Based on the phylogenetic tree shown above, identify FOUR characteristics found in both chordates and arrow worms.

6. Discuss the differences and similarities in your answers to questions 4 and 5.

7. Describe the primary difference between radial and bilateral symmetry.

8. Describe the primary difference between protostomes and deuterostomes, and explain which group includes humans.

9. Describe the primary difference between pseudocoelomates and acoelomates.

10. Discuss the principle evolutionary advantage provided by the development of segmentation in the body plan of animals.

Section 23.2 reviews the major animal groups outside of the Bilataria.

11. The group Bilataria (or bilaterians) composes a very large group of animals. Discuss the characteristics of the Bilataria and explain why sponges are not included in the Bilataria.

12. Many sponges reproduce asexually and sexually. Describe how a sponge reproduces asexually.

Section 23.3 studies the two major groups of protostomes, the lophotrochozoans and the ecdysozoans.

13. The onychophorans are often considered to be a transitional animal group, leading to the evolution of the arthropods. To the right is a drawing of an onychophoran. Describe TWO evolutionary adaptations you would expect to see in the drawing if this organism really is an arthropod.

Livingstone © BIODIDAC

Image Source: BIODIDAC

____
____
____
____
____
____

14. New anatomical features evolved in rotifers that assist food gathering and digestion. Identify these new features and discuss how they increased digestive capacity and reproductive success.

____
____
____
____
____
____

15. In a handful of rich soil, you could find several hundred or even thousands of nematodes. Identify and briefly describe three benefits of nematodes to humans and/or the environment.

____
____
____
____
____
____
____

Section 23.4 discusses the arthropods, a group that might include a billion billion (10^{18}) living individuals on the planet at this moment. There are nearly two million species of insects among the Arthropoda, and the consequences of evolutionary change are impressive.

16. Describe how jointed appendages have contribute to the evolutionary success of arthropods in colonizing terrestrial habitats.

____
____
____
____
____
____

17. Distinguish between complete and incomplete metamorphosis. From an evolutionary perspective, explain which one likely evolved first.

Section 23.5 considers the diversity of the three major clades of the deuterostomes.

18. Protostomes and deuterostomes differ in a key characteristic, and this difference forms the basis for the names "protostome" and "deuterostome." Explain.

19. Chordates have three derived features that are apparent at some point in development. However, many chordates, including sea stars and sea squirts, do not show these three features during all stages of life. Describe the three features and describe appearance in the lives of sea stars and sea squirts.

20. Echinoderms have several distinctive features, including an internal skeleton and a water vascular system. Describe each and comment on its contribution to the success of these organisms.

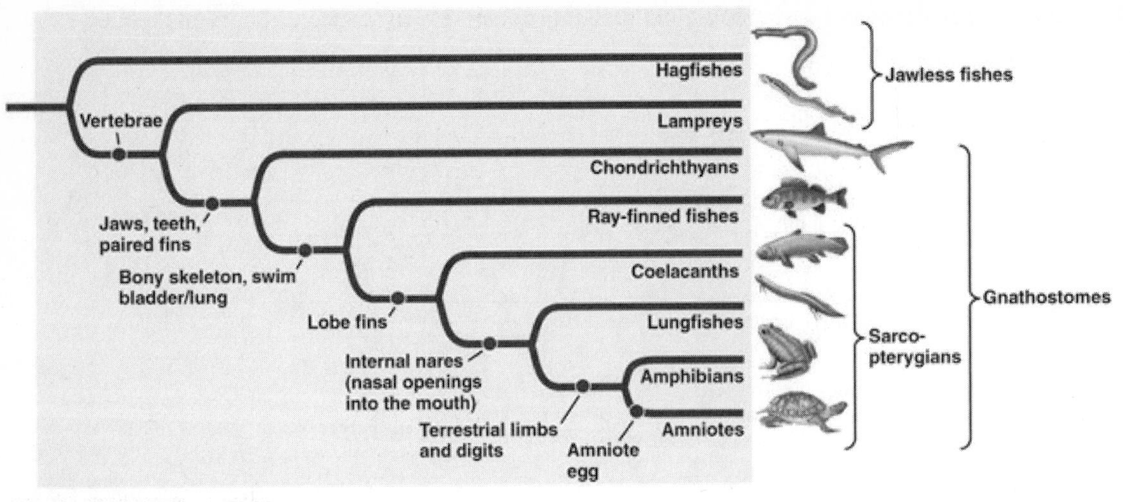

PRINCIPLES OF LIFE, **Figure 23.36**

21. The hagfish and lampreys were lumped together as "jawless fishes" in early classification efforts. Based on the diagram above, explain why they are no longer classified together.

22. Discuss the claim that the cartilaginous fish (chondrichthyans) differ from the ray-finned fish by three features of substantial evolutionary benefit.

Section 23.6 describes the vertebrates and their evolutionary adaptations that allowed them to colonize land.

23. Discuss TWO of the reproductive adaptations that allowed animals to successfully move out of the water and on to land.

24. Discuss the TWO key features that distinguish mammals from other vertebrate phyla.

25. Describe the three patterns of reproduction in the three classes of mammals.

26. Birds were recently reclassified as a taxonomic group more closely related to reptiles and dinosaurs than mammals. The placement of turtles is still fairly uncertain. Explain why some classifications are not always perfect and are subject to change.

Section 23.7 provides an overview of primate evolution.

27. Discuss the adaptive importance of each of the following hominid characteristics:

Brain size

Bipedal motion

Grasping fingers

Tool use

28. One of the biggest misconceptions of evolution is that humans evolved from apes. Explain the evolution of these two groups and why they are so similar.

29. Complete the chart below. Your teacher may ask you to expand this chart to include other phyla or features such as body cavities, proto- or deuterostomes, symmetry, sensory organs, skeletal or support systems, or movement.

Phylum	Example	Symmetry	Digestion	Excretion	Respiration	Reproduction	Nervous	
Porifera								
Platyhelminthe								
Nematoda								
Mollusca								
Annelida								
Arthropoda								
Echinodermata								

30. Using the chart on the previous page, discuss the evolution of the nervous system from the sponges to primates.

Science Practices & Inquiry

In the AP Biology Curriculum Framework, there is a set of 7 Science Practices. In this chapter, we will focus on Science Practice 1: The student can use representations and models to communicate scientific phenomena and solve scientific problems. More specifically, practice 1.1 The student can *create representations and models* of natural or man-made phenomena and systems in the domain.

Question 31 asks you to create a phylogenetic tree or simple cladogram that correctly represents evolutionary history and speciation from a provided data set. (LO 1.19) Below is a set of data about a group of animals that you will use to create a cladogram.

31. Below are some characteristics found among four groups of animals.

	Spiral Cleavage	Chitonous Outer Skin	Specialized Segments	Bilateral Symmetry
Mollusks	+	–	–	+
Annelids	+	–	–	+
Onychophorans	–	+	–	+
Arthropods	–	+	+	+

a. Using the data, draw a phylogenetic tree for these organisms.
b. Figures 23.1 and 23.36 shown earlier clearly label the common features for each group. Label the four features of this diagram showing who shares them.
c. Mollusks and annelids are clearly different phyla even though the chart above does not differentiate between them. Identify two features that separate them.

Chapter 24: The Plant Body

Chapter Outline

24.1 - The Plant Body Is Organized and Constructed in a Distinctive Way
24.2 - Meristems Build Roots, Stems, and Leaves
24.3 - Domestication Has Altered Plant Form

There is no part of human existence that is not touched by plants, and people have studied plants extensively over the millennia. Chapter 24 describes the development and the anatomy of plants, introducing many vocabulary terms. Apply what you've already learned about other living organisms: form arises during development, and serves to meet the metabolic (vegetative) and reproductive functions of the organism. Many plants are unique from animals in having parts called root and apical meristems that remain embryonic and can grow throughout life.

Chapter 24 includes content that is relevant for consideration of Big Idea 1, evolution, especially in its consideration of the ways that humans have artificially selected plants over time to increase agricultural yield and increase food quality.

1.a.2: Natural selection acts on phenotypic variations in populations.

Chapter 24 also includes material that fits in Big Idea 4, emphasizing the interactions with biological organisms, especially:

4.a.4: Organisms exhibit complex properties due to interactions between their constituent parts.

Chapter Review

Concept 24.1 reviews the fundamental parts of plants, with a focus on the development and anatomy of flowering plants. The cells of plants are surrounded by cell walls and the cells are usually totipotent in terms of development potential. The root system, which develops from the root apical meristem, is largely underground, and it supplies the plant with water and soil-borne nutrients needed for growth and maintenance. The shoot system develops into the stems, leaves, and flowers. Plants have three tissues systems: i) the dermal system is the outer covering of the plant, and it is in contact with the outside world, ii) the ground tissue system, where photosynthesis takes place and where the supporting mechanism for the plant is found, and iii) the vascular system, which includes the xylem distribution network for mineral and water delivery from the roots, and the phloem network, which distributes carbohydrates, the products of photosynthesis, throughout the plant for use as fuel molecules or for storage in starch.

1. Below are some of the anatomical parts that a water molecule might pass along, in its delivery from the soil to a location where it becomes involved in the reactions of photosynthesis in a shoot. Put the parts in the correct root-to-shoot sequence.

leaf vascular tissue, root dermal tissue, leaf ground tissue, root ground tissue

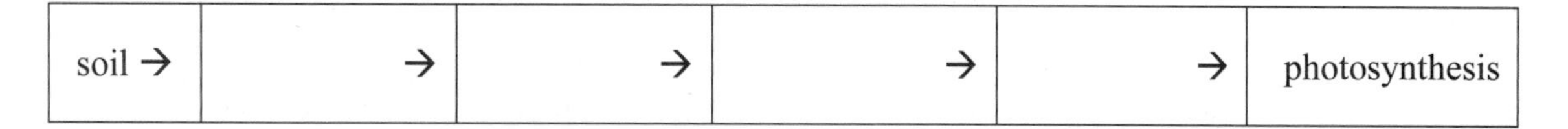

soil →	→	→	→	→	photosynthesis

2. Briefly explain, in general terms, how of the protein components in root cells come to be different from the proteins found in leaf cells.

3. Discuss the phenomenon that cells taken from the growing root tip on a carrot plant can be grown in culture and then coaxed, with growth factors and nutrients in a supportive medium, to make a full carrot plant, including roots, stems, leaves and flowers.

4. Discuss the mechanism by which an individual species of plant can grow to very different shapes when grown in different environmental conditions.

5. The roots, stems and leaves of a plant are considered to be organs. Explain with a specific example how their coordinated interactions provide essential biological activities.

Concept 24.2 reviews the mechanisms of development and differentiation used by plants. An elongating young plant undergoes primary growth via activity in its apical meristems, while a thickening older plant, e.g., a tall tree, undergoes secondary growth via its lateral meristems.

6. Compare the long-term growth potential in an apical meristem with the long-term growth potential of a newly emerged leaf.

7. Discuss the observation that desert-dwelling plant species growing in arid conditions produce a thicker extra-cellular matrix (cuticle) than do plant species growing in a very moist environment.

8. Most species of deciduous trees growing in areas with harsh winters drop their leaves in the fall, and grow new leaves in springtime. Discuss what this implies for the location of apical meristem in such species.

9. Describe TWO of the selective advantages of indeterminate growth, above and below ground, for a plant such as the poison ivy vine.

10. For all multicellular organisms, coordination between systems provides essential biological activities. Explain how the vascular tissue of a plant and the mesodermal tissue of a leaf are similar to the respiratory and circulatory systems of an animal.

Concept 24.3 only scratches the surface of the ways that humans have selectively altered plants over the millennia. The predecessor of modern corn plants is pictured as a relatively simple grass, one that makes a considerably smaller energetic investment in seed production compared to a modern day ear of corn pictured to the right.

11. The yield of corn from domesticated corn plants has increased tremendously over the past 100 years. In the earlier years, and continuing today, disease susceptibility of domesticated plants has been seen to increase. Discuss this "cost" of artificial selection.

Photo Credit: Gewin V (2003) Genetically Modified Corn— Environmental Benefits and Risks. PLoS Biol 1(1): e8.

Science Practices & Inquiry

In the AP Biology Curriculum Framework, there is a set of 7 Science Practices. In this chapter, we will focus on **Science Practice 3:** The student can engage in scientific questioning to extend thinking or to guide investigations. More specifically, practice 3.3: The student can evaluate scientific questions.

Question 12 asks you to evaluate scientific questions concerning organisms that exhibit complex properties due to the interaction of their constituent parts (LO 4.8).

12. In an experiment that lasted 60 days, seeds were sprouted and grown in wet or arid conditions, and then above-ground (shoots) and below-ground (roots) mass was measured every 10 days.

Table 1: Shoot and Root Mass (grams) of Three Species of Trees over 60 days

	10 days	20 days	30 days	40 days	50 days	60 days
wet -						
shoots	3.2	3.3	3.8	4.4	6.2	8.1
roots	2.4	2.5	2.9	3.3	4.5	6.0
arid –						
shoots	3.0	3.3	3.5	3.8	4.1	4.4
roots	2.4	3.0	4.2	6.1	6.6	7.5

Graph the data, using four curves.

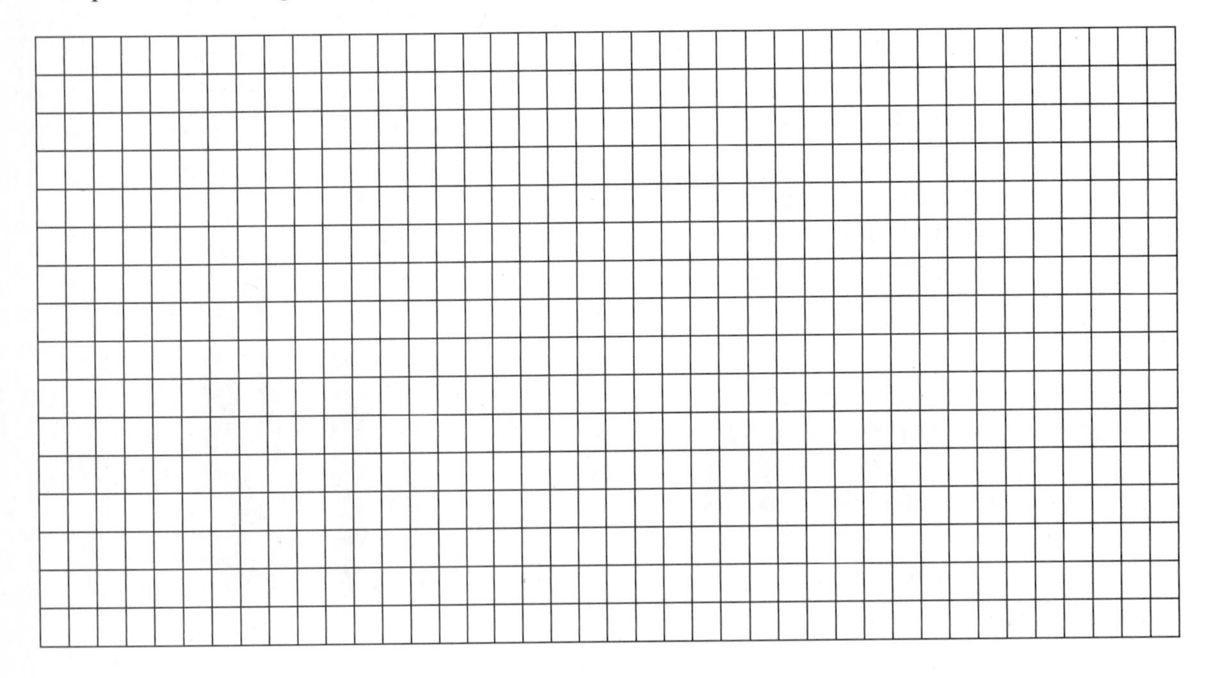

For each condition, and at each time point, calculate the ratio of root to shoot growth, which is equal to root mass divided by shoot mass.

Table 1: Calculated root to shoot ratio

	10 days	20 days	30 days	40 days	50 days	60 days
wet						
arid						

Graph these ratios on the second graph.

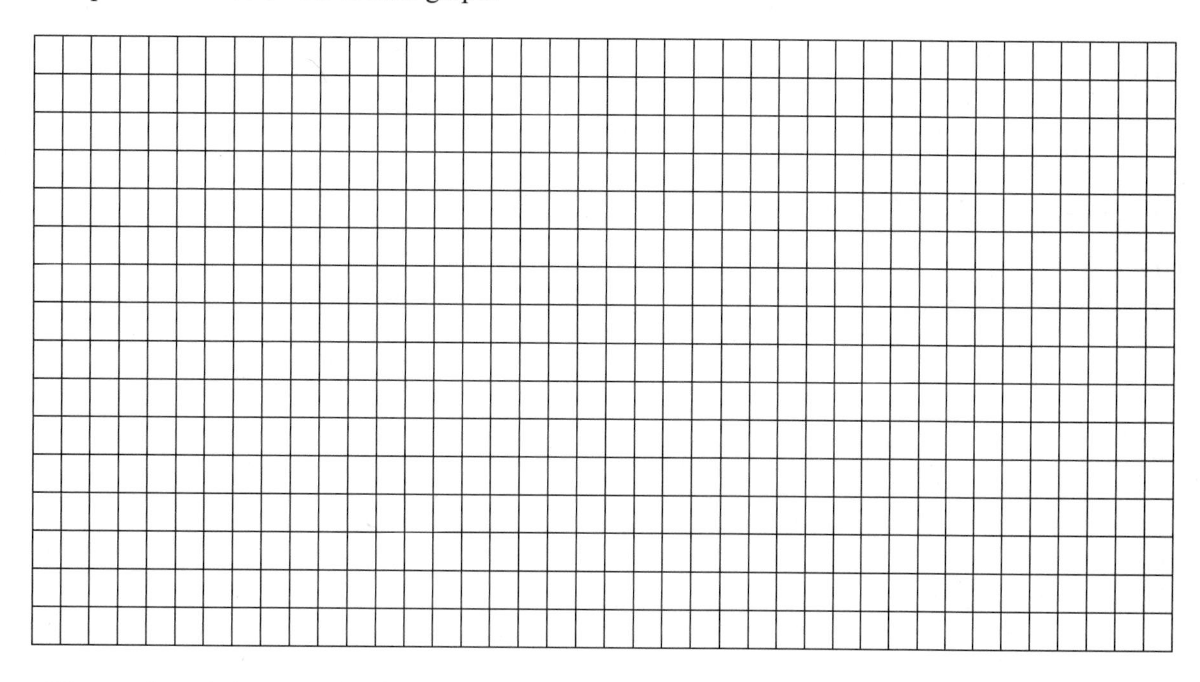

Discuss the relative growth of roots and shoots in the two experimental conditions used here, and include relative growth of root meristem and apical meristem in your answers.

____

____

____

____

____

____

____

____

Speculate, without mentioning specific chemicals, what caused the plant to grow in the different patterns you observed in your graphs.

____

____

____

____

____

____

____

____

Discuss one evolutionary advantage to the plants growing in the different patterns you observed in your graphs.

____
____
____
____
____
____
____
____

Describe which part of the plant would have the highest concentration of sugars when subjected to a chemical assay to measure carbohydrates.

____
____
____
____
____
____
____
____

Chapter 25: Plant Nutrition and Transport

Chapter Outline

25.1 - Plants Acquire Mineral Nutrients from the Soil
25.2 - Soil Organisms Contribute to Plant Nutrition
25.3 - Water and Solutes Are Transported in the Xylem by Transpiration–Cohesion–Tension
25.4 - Solutes Are Transported in the Phloem by Pressure Flow

Plants are autotrophic organisms that acquire carbon from atmospheric carbon dioxide, using energy from the sun. Oxygen is released by plants as a byproduct of photosynthesis, but oxygen is also absorbed from the atmosphere by plants and used in respiration, especially during the evening hours or during periods of darkness. Plants also acquire nutrients from soil, including nitrogen from nitrates in ground water, via the roots. Learning the details of mineral cycles is not required for success on the AP-Biology exam; however, it is important realize that human manipulation of mineral cycle via the use of agricultural fertilizers can easily disrupt such cycles, leading to undesirable environmental changes.

The majority of terrestrial plants have an interaction with fungi called mycorrhizae that promotes the absorption of nutrients and water by the plants' roots. Some plants form relationships with nitrogen-fixing bacteria that convert atmospheric nitrogen (N_2) to a useable form of nitrogen (NO_3^-) that plants can absorb. These relationships between plants and organisms are a result of signaling compatibilities between the plants and these other organisms.

Xylem and phloem are plant tissues that transport substances throughout the plant. Water is transported inside the xylem from the roots up through the leaves. The movement of water is governed by the evaporation of water form leaves providing a pull (transpiration—cohesion—tension) to move water up from the roots (recall that water molecules form hydrogen bonds that make them "stick" together, much like a long line of cars in a train). Phloem is the tissue that moves sugars produced by photosynthesis throughout the plant, thus translocating them from an area of high pressure (source) to low pressure (sink.)

Big Idea #2 is captured in Chapter 25. The Big Ideas are a means for organizing the vast amount of information we know as biology. It is vitally important that you continually work to understand these **Big Ideas** and make connections across different **Enduring Understandings**.

Big Idea 2 states that the utilization of free energy and use of molecular building blocks are characteristic fundamental of life processes. Specifically, Chapter 25 includes:

2.a.3: Organisms must exchange matter with the environment to grow, reproduce, and maintain organization.

Chapter Review

Section 25.1 describes how plants acquire mineral nutrients from the soil. Some of the minerals move into the roots as a result of ion exchange, an ATP-dependent process that "pumps" hydrogen ions (H^+) out of the root, where the hydrogen ion "trades places" with cations that are then taken up by the roots.

1. Identify three ways that the growth of plants benefits humans.

____
____
____
____

2. Identify three ways that the growth of plants benefits other organisms.

3. Explain how plants utilize each of the following:

a. carbon dioxide (CO_2)

b. oxygen gas (O_2)

c. water (H_20)

d. nickel (Ni)

e. nitrogen gas (N_2)

f. nitrogen compounds (NO_3^-, NO_2^-)

4. Design an experiment to show the effect of carbon dioxide (CO_2) uptake on plant growth. Be sure you include in your design, a hypothesis, a control, and at least two constants.

5. Epidermal root cells utilize ion exchange to dislodge cations bound to clay particles by pumping protons outside of the cell. Does this represent active or passive transport? Explain why.

___

___

___

Section 25.2 discusses how fungi aid the uptake of water and nutrients for plants, and the importance of nitrogen-fixing bacteria.

6. Researchers grew lemon seedlings in soils that contained only phosphate fertilizer, only mycorrhizal fungi, and both the phosphate fertilizer and mycorrhizal fungi together. The table below shows the mean dry weight of the seedlings after 6 months.

Phosphate Fertilizer (g)	Mycorrhizal Added (?)	Mean dry weight of seedlings (g)
0	No	1
0	Yes	10
12	No	28
12	Yes	166
24	No	20
24	Yes	210

a. Summarize the results of this experiment.
b. Identify the independent and dependent variables in this experiment.
c. Identify three other features that the experimenters needed to hold constant in this experiment.

___

___

___

___

___

___

___

___

___

___

___

___

___

___

7. Explain how mycorrhizae expand the surface area of a plant for water absorption.

___

___

___

8. The atmosphere on earth is almost 80% nitrogen gas. Explain why molecular nitrogen cannot be used directly by plants.

___

___

___

9. Describe how plants use the nitrates they absorb at their roots.

____

____

____

Section 25.3 explains how the plant absorbs water from its environment and how xylem transports water throughout the plant.

10. Draw simple representations of two water molecules and show how they form a hydrogen bond. Be sure to show their individual polarities.

11. The variables in water potential (ψ) are shown by the formula:

$$\psi = \psi_s + \psi_p$$

Describe what the two variables (ψ_s and ψ_p) of this equation represent.

____

____

____

____

12. Water potential is described in terms of the concentrations of water molecules in compartments. Describe how adding a salt to water reduces that solution's concentration of water molecules.

____

____

____

13. Explain what prevents the diffusion of ions across the cell membrane.

____

____

____

____

____

14. Explain what is wrong with the statement: Water in the root flows faster through the apoplast pathway (via cell walls) than through the symplast (via cytoplasm and plasmodesmata); thus, the majority of the water entering the xylem flows only through the apoplastic pathway.

____

____

____

____

____

15. The transpiration-cohesion-tension mechanism explains how water streams up a tall tree to its leaves where much of it lost to the atmosphere. Explain how each component in this mechanism functions:

Transpiration: ____

____

____

Cohesion: ____

____

____

Tension: ____

____

____

16. Discuss how the structure of a guard cell supports its ability to open and close a stoma.

____

____

____

____

17. Most water potential values are negative. Describe a means for generating a positive value for the water potential.

____

____

____

18. Explain what it means to say that air has a water potential of -55 MPa and soil water has a much higher water potential of -0.3 MPa. Explain why these values are negative in your answer.

____

____

____

Section 25.4 explains how phloem transports sugars through the plant.

19. Explain each of the two terms below in the context of the movement of sugars through a plant. Give an example for each.

Source: ____

____

____

Sink: ____

____

____

20. While trimming grass around trees with a weed whacker, a person gets too close to the tree and cuts a groove in the bark around the base of the tree. The tree survives for a few months, but then dies the following spring. Explain why the tree survived for a short period of time and died the next year.

____

____

____

____

____

____

____

____

____

21. Explain how the movement of phloem sap from one part of a plant to another is considered to be an example of both active and passive transport.

22. Explain how plant growth of plants depends on other organisms.

23. Describe crop rotation and explain how this practice helps farmers to produce food in a more sustainable manner.

Science Practices & Inquiry

In the AP Biology Curriculum Framework, there is a set of 7 Science Practices. In this chapter, we will focus on **Science Practice 2: The student can use mathematics appropriately.** More specifically, practice 2.2: The student can *apply mathematical routines* to quantities that describe natural phenomena.

Question #24 asks you to model quantitatively the exchange of molecules between an organism and its environment, and the subsequent use of these molecules to build new molecules (LO 2.19.) An equation sheet will be provided that will have the commonly used equations and conversions that you will need to solve problems like the one below.

TIP: When you are writing a free response question with multiple parts in it, write each part separately and clearly label each part. Part "a" of the question is labeled for you.

24. While you are comparison shopping at a food store, you find two tomatoes, one is marked organic and priced at $2.37 per pound while the non-organic tomatoes cost $1.75 per pound. (One pound = 0.45 Kg)

a. Explain any differences between the actual make up of the two tomatoes.
b. Calculate the cost for 2 kilograms of each.
c. Approximately 13 gallons of water are used to produce each tomato. If there are 4.3 tomatoes per pound, calculate how much water is used to produce two kilograms of tomatoes.
d. Explain any differences for the impact on the environment caused by the growth of these two tomatoes.
e. Which tomato should you buy? Why?

Chapter 26: Plant Growth and Development

Chapter Outline

The germination of a dormant seed followed by its growth into a leafy green plant with flowers is a beautiful thing to watch. In this area of biology, cause-and-effect thinking and careful experimentation have made great progress in understanding how the expression of the information coded in the plant genome is precisely orchestrated by environmental and hormonal signals.

For the AP Biology exam, you will not be required to know the names and mechanisms of the many plant hormones that have been identified, but some basic ideas about signaling are reinforced in this chapter. For example, hormones are chemical signals that are released by cells and then received by receptor proteins on other cells, also known as target cells. As a result of the hormone binding to the receptor protein, the biochemical reactions in the target cell are altered, typically in a way that is of benefit to the organism. Furthermore, the activation of light-sensitive chemicals in plants influences their hormonal and genetic responses, coordinating metabolism, growth and reproduction.

Chapter 26 includes topics in Big Ideas Two and Three. The Big Ideas are the fundamental concepts on which the curriculum is built. Continue to focus on the Big Ideas and expand your support for them by reviewing their many Enduring Understandings.

Big Idea 2 emphasizes chemical building blocks and energy flow in living organisms, with emphasis in Chapter 26 on:

2.e.1: Timing and coordination of specific events are necessary for the normal development of an organism, and these events are regulated by a variety of mechanisms.

Big Idea 3 emphasizes information transfer based on chemical in living organisms, with Chapter 26 providing the occasion to reinforce your knowledge that:

3.b.2: A variety of intercellular and intracellular signal transmissions mediate gene expression.

Chapter Review

Concept 26.1 starts by describing some of the factors that control seed dormancy. Dormancy is typically the result of having a tough seed coat that must be disrupted in some way, but it can also be the result of chemical inhibitors. Upon the appropriate environmental stimulation, the seed germinates to form a sprout, which sends the growing root tip toward the soil and the growing plant stem toward the open air and sunlight. Various environmental and chemical factors determine the differentiation and development of the plant from the sprout.

1. Discuss why the surface of some garden seeds needs to be physically scratched before the seed will be ready to sprout.

____
____
____
____

2. Explain why too much scratching of the seed can greatly impair its ability to sprout. Use the terms seed coat and embryo in your answer.

____
____
____
____

3. Discuss why the scratched seeds from question 1 need to be placed in a moist environment before sprouting will take place.

____
____
____
____

4. Despite the correct scratches made on the seeds from his dogwood tree, followed by proper soaking, William is frustrated to find that his dogwood seeds will not sprout. To slow down bacterial growth on the wet seeds, he decides to put the container of remaining wet seeds in the refrigerator for safekeeping. Alas, he forgets about them until three months later, when he takes them out and keeps them wet for another week. He was surprised to find this time that all of the refrigerated seeds germinated. Discuss three physical factors that might be responsible for sprouting dogwood seeds.

a.____
____

b.____
____

c. ____
____

Concept 26.2 describes some of the actions of six kinds of growth hormones in plants. While knowing the identities of specific plant hormones is beyond the scope of the AP-Biology framework, there are still important messages to remember about signaling. First, chemical signals come from a source that synthesizes them in response to an environmental cue. The signals that are released by the source cells travel to target cells, where "target" is defined by the presence of the appropriate receptor protein needed to receive the signal. The binding of the signal and the receptor activates a response in the target cell, yielding a biochemical change in the target cells such as enzyme activation or altered gene expression.

5. Charles Darwin and his son Francis grew some grass seedlings where the light source was kept on only one side of the sprout. The sprouts grew toward the light, a response called phototropism. We know today that higher levels of the plant hormone auxin causes elongation of growing plant cells, compared to such cells exposed to lesser concentration of auxin. Speculate in detail about whether auxin is produced symmetrically **or** asymmetrically in phototropism by grass seedlings. Include receptor proteins and consider auxin's effects on cell elongation in your answer.

____
____
____
____
____
____

6. The downward growth of root tips of the sprouted grass, called gravitropism, is also mediated by auxin. Speculate in detail about whether auxin is produced symmetrically **or** asymmetrically in the root tips of grass seedlings. Include receptor proteins and consider auxin's effects on cell elongation in your answer.

____

____

____

____

7. Giberellins stimulate starch hydrolysis in plants. Describe the general function of starch in plants, and then predict the expected level of gibberellin action in a germinated seed compared with that of a plant that is undergoing a high rate of photosynthesis. Explain your predictions.

____

____

____

____

8. In an experiment completed 20 years ago, researchers reported that auxin treatment of plant cells led to increased levels of mRNA. Discuss the meaning of this observation of signal transduction in terms of the mechanism of action for auxin's effects on target cells.

____

____

____

____

9. Suppose that dwarf and excessively tall plants were studied and were both found to have mutations in the same gene, one that modifies the actions of auxin; however, the sequence of the translated mRNAs and the amino acid sequence of the translation product are different. Given that a protein can have more than one different functional domain, propose an explanation for how two mutations of the same gene can have such different effects.

____

____

____

____

Concept 26.3 expands on the survey of chemical signals used by plants to coordinate the three fundamental processes apparent in all living organisms: growth, metabolism and reproduction. For example, ethylene is a gas that promotes fruit ripening and senescence in plants, and is possibly the most widely used plant hormone in the practical and applied sense. Cytokinins (inhibited elongation) and brassinosteroids (enhanced growth) are also reviewed here, along with abscisic acid (inhibited growth).

10. Discuss the science behind the old saying, "one rotten apple spoils the whole barrel."

____

____

____

____

11. Radishes and other root crops produce an abundance of cytokinins, which are signals that inhibit the elongation of the roots. Auxins are signals that are active in the elongation of roots and stems. Discuss whether or not it is likely that both types of signal would be active in the same plant.

____

____

____

____

12. In some plants, the brassinosteroids stimulate many aspects of above-ground growth. Steroid hormone responses in animals are the classic example of how a signal can alter gene expression, which requires a receptor protein for the steroid hormone to be active in the nucleus of the target cells. However, the brassinosteroid receptors of plants are found on the cell membrane. Discuss whether or not the brassinosteroids are likely to have the same mechanism of action as the steroid hormones of animals.

13. Abscisic acid is often regarded to have the opposite action of the giberellins. Given that giberellins stimulate starch hydrolysis in plants, predict the effect of abscisic acid on seed germination. Explain your prediction.

Concept 26.4 directs our discussion of plant signals outside of the plant, to introduce the non-photosynthetic effects of light on plants. (Chapter 6 explores photosynthesis.) For example, light exposure of particular wavelengths can activate or inhibit seed germination. Furthermore, flowering responses of many plants are dependent on changes in the light cycle to which they are exposed. In both cases, light-sensitive chemicals in the plants are altered upon exposure to the appropriate light, thus signaling the cells to alter their patterns of growth and activity. Many of these types of response are categorized as photomorphogenesis, a term suggesting that light exposure changes the shape of the plant.

14. Suppose that three possible pigments (compounds F, G and H) have been identified in the grass seedlings discussed in question 5, and Darwin's ancestors want to get to the bottom of the phototropism reaction. Compound F is activated by green light (500 nm), Compound G is activated by yellow light (540 nm), and Compound H is activated by red light (650 nm).

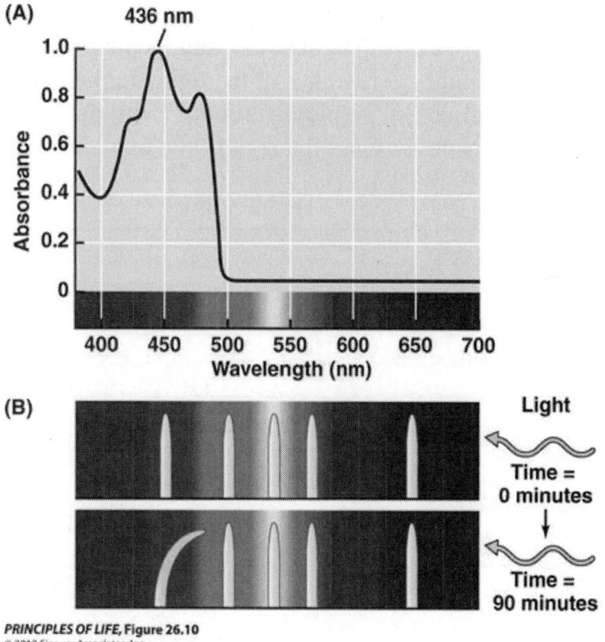

Violet = 400
Blue = 450
Green = 500
Orange = 560

PRINCIPLES OF LIFE, Figure 26.10

Given the results shown in the graph, where the slender stems in part (B) represent the phototropism or non-phototropism, select the pigment (F, G, or H) that is most justified by these data for further study as "the phototropism pigment." Explain your answer based on the results shown in the figure.

15. Phytochrome (P) is a blue-colored plant pigment that exists in alternate forms. When exposed to broad-spectrum light, it is converted to form P_{fr}. At night, or when exposed to far-red light, it is converted to P_r, where P_r is considered to be the default, inactive state of the pigment. By contrast, P_{fr} is the signal that triggers numerous light-dependent responses, e.g., seed germination, flowering and shoot development. At sunset, most of the phytochrome is in the P_{fr} form, but as the night progresses, it is slowly converted, in a time-dependent manner, to the P_r form. The P_{fr} form is an activator of transcription factors and is a transcription factor in its own capacity as well. Discuss the theory that the ratio of P_{fr}:P_r is an important signal of season in plants, and explain how this could work.

16. The flowering pattern of temperate-zone plants is often seasonal. For example, cockleburs are called "short-day-flowering plants" because their flowering stage begins when day-length shortens to less than 13 hours per day, such as might happen in the latter half of the summer. In an experiment, cockleburs were grown under a light cycle with 16 hrs of light alternating with 8 hrs of dark (16L:8D) to the stage where flowering was possible, but not yet underway. These plants were then separated into three rooms (J, K and L). Room J had its light-cycle kept at 16L:8D, and its plants did not flower. Room K had its light-cycle set to 12L:12D light cycle, and its plants flowered. Room L had its light-cycle set to 11L:6D:1L:6D light cycle (think of the lights in this third room being turned on for one hours in the middle of the "night"). The plants in room L did not flower, even though there was 12 hours of darkness every 24-hour cycle. Discuss the results and explain your discussion.

Science Practices & Inquiry

In the AP Biology Curriculum Framework, there is a set of 7 Science Practices. In this chapter, we will focus on **Science Practice 1:** The student can use representations and models to communicate scientific phenomena and solve scientific problems. More specifically, practice 1.4: The student can use representations and models to analyze situations or solve problems qualitatively and quantitatively.

Question 17 asks you to use a graph or diagram to analyze situations or solve problems (quantitatively or qualitatively) that involve timing and coordination of events necessary for normal development in an organism (LO 2.32).

17. Solve parts A and B of the problem set up in the accompanying box.

A.____________________

B.____________________

INVESTIGATION

CONCLUSION

Red light and far-red light reverse each other's effects.

ANALYZE THE DATA

Seven groups of 200 lettuce seeds each were incubated in water for 16 hours in the dark. One group was then exposed to white light for 1 min. A second group (controls) remained in the dark. Five other groups were exposed to red (R) and/or far-red (FR) light. All the seeds were then returned to darkness for 2 more days. Germination was then observed.

Condition	Seeds germinated
1. White light	199
2. Dark	17
3. R	196
4. R then FR	108
5. R then FR then R	200
6. R then FR then R then FR	86
7. R then FR then R then FR then R	198

A. Calculate the percentage of seeds that germinated in each case.

B. What can you conclude about the photoreceptors involved?

PRINCIPLES OF LIFE, Figure 26.11 (Part 2)

18. The data above demonstrate that some of the old-fashioned or heirloom varieties of certain lettuce plants only germinate when exposed to light of a certain wavelength. Put this observation in the context of evolutionary adaptation to explain why a seed's exposure to light is needed for germination to proceed (hint: think competition!).

Chapter 27: Reproduction of Flowering Plants

Chapter Outline

27.1 - Most Angiosperms Reproduce Sexually
27.2 - Hormones and Signaling Determine the Transition from the Vegetative to the Reproductive State
27.3 - Angiosperms Can Reproduce Asexually

In Chapter 27, the "alternation of generations" in angiosperm plants is the label describing their primary reproductive pattern. The "alternation" refers to a plant's progression from the diploid to the haploid form as offspring are generated. To imagine this happening in humans, sperm and/or egg cells would have to grow by mitosis into multicellular forms. This obviously does not happen with humans and most other animals, but it does in all plants. The multicellular haploid form in angiosperms is typically small, consisting of only a few cells within the flower of the sporophyte, as the multicellular diploid phase is known. Pollen is the male gametophyte and the embryo sac is the female gametophyte.

The flowering and reproduction of plants is coordinated by plant hormones. For purposes of the AP Biology exam, it is important to understand that these events are coordinated by chemical signals, but you do not need to memorize the names or specific actions of any plant hormones. In particular, the role of photoperiodism is an important concept to understand, as this is what determines the timing of seasonal flowering for many plants.

Chapter 27 notes that some plants can reproduce asexually. Asexual reproduction is efficient, especially for an organism in a predictable environment. Apomixis, the asexual production of seeds, vegetative reproduction, resulting in clones of the parent, and agricultural grafting, are three different means of asexual reproduction.

Chapter 27 has ideas that are drawn from Big Idea 2. The Big Ideas are a means for organizing the vast amount of information we know as biology. It is vitally important that you continually work to understand these **Big Ideas** and across different **Enduring Understandings**.

Big Idea 2 states that the utilization of free energy and use of molecular building blocks are characteristic fundamental of life processes. Specifically, Chapter 25 includes:

2.a.1: All living systems require constant input of free energy.
2.c.1: Organisms use negative feedback mechanisms to maintain their internal environments and respond to external environmental changes.
2.e.2: Timing and coordination of physiological events are regulated by multiple mechanisms.

Chapter Review

Section 27.1 discusses sexual reproduction and describes the alternation of generations. At the time of zygote formation (sperm and egg joining), double fertilization of nearby cells produces the nutritive endosperm that will be included in the seed.

1. Distinguish between each pair of terms below.

Perfect *vs.* Imperfect Flowers____

____

____

Complete *vs.* Incomplete Flowers____________________________________

Carpel *vs.* Stamen

Gametophyte *vs.* Sporophyte

Seed *vs.* Fruit

2. Explain why a pollen grain is considered to be an immature gametophyte.

3. Help Katie and Benson solve an argument. As they were walking, they came across a pine tree that had all of its pinecones on the upper half of the tree. Benson claimed that deer had eaten all of the lower cones. Katie claimed that the upper cones were the female cones and that they only formed on the top half of the tree as a mechanism to protect self-fertilization from pollen formed by the male cones on the bottom half of the tree. Design an experiment to determine who is correct.

4. Discuss the adaptive value of producing sweet fruits

5. Coevolution proposes that species evolve defenses even as other species evolve mechanisms to overcome defenses. The avocado fruit is buttery and sweet, yet the single seed inside is large and hard to crack open. The avocado seed presents a possible evolutionary anachronism, having evolved its size and hardness well before humans starting eating them, and there are no known *living* animals that eat it. Propose an explanation why this seed evolved its large size and became so difficult to crack.

6. Most plants do not self-pollinate, in spite of how easy this might be. Explain why.

____

____

____

____

____

7. Complete the table below:

	Humans	Lily, a perfect angiosperm flower
Site of production of male gametes		
Site of production of female gametes		
Process that produces the haploid form		
Process that produces the diploid organism		
Number of cells in the haploid form	Male _____ Female _____	Male _____ Female _____
Location of fertilization		
Number of fertilizations		

8. For each term below, identify them in the diagram to the right and briefly explain their function.

a. antipodal cells

____

____

b. polar nuclei

____

____

c. egg cell

____

____

d. synergids

____

____

e. ovary

____

____

f. ovule

____

____

PRINCIPLES OF LIFE, **Figure 27.3**

Section 27.2 describes some of the events that signal a plant to make the transition from vegetative growth to the formation of flowers. The change from producing leaves and stems with indeterminate growth to a floral meristem is caused by a cascade of gene expression, which is frequently triggered by changes in the light-dark cycle, especially in plants adapted to seasonal climates.

9. It might appear that short-day and long-night are equivalent terms to describe plants that flower in autumn. Explain why one of these terms is more accurate.

10. Many varieties of chrysanthemums are called short-day plants. A friend of yours whose house is close to a busy and well-lit street planted several short-day chrysanthemums around her front door hoping they would bloom, but they did not, even as day length shortened. Suggest a reason for the lack of flowers on her plants.

11. Briefly discuss why flowering is preceded by a "cascade" of gene expression.

12. Not all flowers utilize day or night length as the flowering cue. Identify and briefly discuss two other triggers for flowering in such plants.

13. Flowering in some varieties of the plant *Arabidopsis* is triggered by the light cycle experienced by the leaves rather than that experienced by the apical meristem. It appears that the protein called florigen, made in the leaves, mediates the flowering response. Discuss what components would be needed for such a communication system to function in this manner.

Section 27.3 explains how plants can reproduce asexually, essentially yield cloned copies of the parental stock.

14. Explain the term totipotency and its relevance to asexual reproduction.

____

____

____

15. Design an experiment to show that dandelions do not cross pollinate with other dandelions and still produce many seeds.

____

____

____

____

____

____

____

____

____

16. Complete the table below for flowering plants.

	Asexual Reproduction	Sexual Reproduction
Advantages		
Disadvantages		
Example		

17. Citrus farmers produce seedless oranges by grafting trees and by apomixis.
   a. Explain why the farmers' intervention is needed for the plants' reproduction.
   b. Discuss how you could improve upon the qualities of the seedless orange. (Hint: recall what the source of new variation is.)

____

____

____

____

____

____

____

____

____

____

____

____

Science Practices & Inquiry

In the AP Biology Curriculum Framework, there is a set of 7 Science Practices. In this chapter, we will focus on Science Practice 6: The student can work with scientific explanations and theories. More specifically, practice 6.1: The student can *justify claims with evidence.*

Questions 18 asks you to justify a scientific claim that free energy is required for living systems to maintain organization, to grow or to reproduce, but that **multiple strategies** exist in different living systems and to relate ideas from chapters 6 and 27 together as you consider the energy needs of plants (LO 2.2). You may wish to review the concept of free energy before answering the question below.

18. As plants grow they require an input of free energy.
    a. Explain where the free energy for plant growth comes from and the end result of that energy.
    b. Explain how energy transfers are necessary for growth
    c. Strawberries can reproduce asexually by runners throughout the summer, but as Fall approaches, they begin to form flowers. Explain the benefits of asexual and sexual reproduction from an energetic standpoint.

Chapter 28: Plants in the Environment

Chapter Outline

28.1 - Plants Have Constitutive and Induced Responses to Pathogens
28.2 - Plants Have Mechanical and Chemical Defenses against Herbivores
28.3 - Plants Adapt to Environmental Stresses

Plants would sometimes seem to lead an idealized and peaceful existence, transpiring water and minerals as needed, exchanging gases with the atmosphere, and generally just soaking up the sunshine to run photosynthesis. However, pathogens, predators, and environmental stress are never far away. Plants are not defenseless in the face of such attacks, and manage to maintain growth, metabolism, and reproduction, even while under attack. These defenses are described in Chapter 28.

There are three big ideas (2, 3, and 4) from the Framework that merit mention in this unit. However, the specific types of memorized information needed from this unit is fairly small, and you will not be required to know the names and mechanisms of the many plant defenses listed in this chapter. Even so, reinforce your understanding of the Big Ideas as you read the chapter and expand your review of the many Enduring Understandings.

Big Idea 2 emphasizes chemical building blocks and energy flow in living organisms, with emphasis in Chapter 28 on:

2.c.1: Organisms use negative feedback mechanisms to maintain their internal environments and respond to external environmental changes.
2.d.3: Biological systems are affected by disruptions to their dynamic homeostasis.
2.d.4: Plants and animals have a variety of chemical defenses against infections that affect dynamic homeostasis.

Big Idea 3 emphasizes information transfer based on chemical in living organisms, with Chapter 28 reinforcing that:

3.d.2: Cells communicate with each other through direct contact with other cells from a distance via chemical signaling.
3.e.1: Individuals can act on information and communicate it to others.

Big Idea 4 is that biological systems interact in complex fashion; Chapter 28 shows that:

4.c.3: The level of variation in a population affects population dynamics.

Chapter Review

Concept 28.1 begins with the general protective features on the surface of plants, including thick and waxy materials that can protect against pathogens and also have the potential to aid water conservation. These features are called "constitutive" protection, always present. Figure 28.1 directs your attention to the pathogens at the cellular level, and shows that pathogen detection can lead to changes in gene expression, including genes that aid the plant's defenses. This second category of pathogen protection is called "inducible" protection, as the pathogen's presence induced the changes.

1. Many pepper plants that can be infected by bacteria express resistance genes, but some bacteria, in turn, express genes that defeat this resistance. Discuss the "evolutionary arms race" between plants and their pathogens, including the ability of plants to recognize pathogens and the ability of pathogens to remain undetected.

2. Physical isolation at an infection site refers to plants evoking signals to kill the plant cells surrounding an area of infection. Explain how this might help the plant survive the pathogen attack

3. Some plant defenses depend on the development of systemic acquired resistance, including examples where salicylic acid plays an important role in plant defense. Assuming that changes in gene expression are needed for systemic acquired resistance, and that salicylic acid is produced locally, describe the possible chain of events leading to changes in gene expression throughout the plant.

4. Close examination of the plant leaf shown in the picture would reveal that each of the dark spots, where a fungal infection has killed the tissue, is completely surrounded by a boundary of dead tissues, called necrotic lesions. The necrotic lesions are due to the expression of plant genes for programmed cell death, also known as apoptosis. Explain why it is adaptive for the plant to create the boundary around the edge of the infected sites.

Concept 28.2 describes some of the mechanical and physical defenses found among plants. Many examples of such defenses take the form of "chemical warfare" directed against herbivorous insects. The chemical are often called secondary metabolites, and many of these are of considerable interest to humans, *e.g.*, nicotine.

5. Some plants synthesize secondary metabolites that are very similar to amino acids, but have one or more critical changes in chemical composition. Explain how these amino-acid-like compounds function in defense against herbivores by affecting protein structure in the herbivore that consumes them.

6. Find a plant that you believe has constitutive anatomical features that reduce the impact of herbivores. Describe the structures and speculate on the evolutionary origin of the feature – was it formerly a stem? A leaf?

____

____

____

____

Concept 28.3 describes the adaptations that plants use for coping with environmental stress. The discussion includes constitutive adaptations of plants as well as stress-induced changes that aid survival. The ability of plant roots to extract contaminants from soils, called phytoremediation, is being used as a new tool for cleaning up the environment.

7. In arid environments some plants persist poorly when they reach the adult form. Explain the reproductive pattern and life history of such plants, considering that it sometimes rains hard in arid environments.

____

____

____

____

8. Plants with adaptations for long-term continuous survival of the adult form in a desert or other arid condition are often called xerophytes. For each of the following items, describe one structural adaptation and discuss how this adaptation aids the survival of the xerophyte.

a. leaf size: ____

____

____

b. leaf cuticle: ____

c. water-storage systems: ____

____

____

d. roots: ____

____

____

9. Drought stress can induce the formation of abscisic acid in the roots of drought-stressed plants. Explain how this compound affects the leaves of the plant, and how this response aids in water conservation.

____

____

____

____

Science Practices & Inquiry

In the AP Biology Curriculum Framework, there is a set of 7 Science Practices. In this chapter, we will focus on **Science Practice 5:** The student can perform data analysis and evaluation of evidence. More specifically, practice 5.3: The student can evaluate the evidence provided by data sets in relation to a particular scientific question.

Question 10 asks you to evaluate data that show the effect(s) of changes in concentrations of key molecules on negative feedback mechanisms (LO 2.17).

10. The *Arabidopsis* gene *cor15a* is thought to be under the control of a promoter that responds to environmental conditions. The COR15A protein is an enzyme that appears to reduce membrane damage at cold temperatures. The levels of the enzyme coded for by this gene were measured after plants were grown at room temperature and then transferred to one of three conditions of varying environmental temperature (2°C, 19°C and 75°C); the data are shown.

	ENZYME ACTIVITY Units/g of protein		
Hours after transfer	2°C	19°C	75°C
0	0.21	0.55	0.22
12	3.21	0.35	0.14
24	5.66	0.51	0.03
36	9.65	0.45	0.06
48	10.10	0.60	0.01
72	11.44	0.49	0.00

A. Plot the data with hours after transfer on the *x* axis and enzyme activity on the *y* axis. Be sure to label the axes. Provide three labeled curves, one for each temperature.

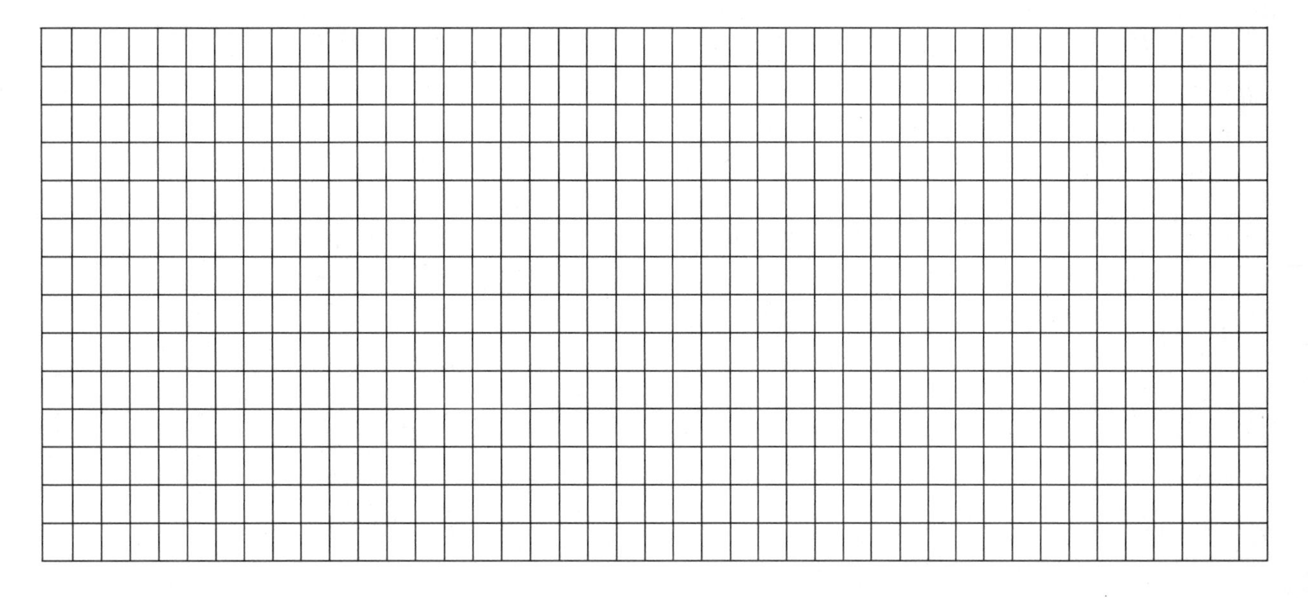

B. Write a description of the meaning of the data for this gene.

C. Explain the meaning of the data collected from plants moved to 75°C.

Chapter 29: Physiology, Homeostasis, and Temperature Regulation

Chapter Outline

29.1 - Multicellular Animals Require a Stable Internal Environment
29.2 - Physiological Regulation Achieves Homeostasis of the Internal Environment
29.3 - Living Systems Are Temperature-Sensitive
29.4 - Animals Control Body Temperature by Altering Rates of Heat Gain and Loss
29.5 - A Thermostat in the Brain Regulates Mammalian Body Temperature

Chapter 29 provides an overview of homeostasis, discussing how and why animals regulate their internal environments. Single-celled organisms and small flat or thin organisms rely on diffusion of materials across cell membranes to get nutrients in and wastes out. However, diffusion is slow over distances larger than a cell, so multicellular and complex animals rely on more elaborate means or systems to ensure homeostasis. For purposes of the AP Biology® exam, pay particularly close attention to the nervous and immune systems and their functions. Other chapters may be used as illustrative examples for different ideas or concepts as well.

Homeostasis is the dynamic maintenance of the internal environment. The key concept is that organisms have an optimal set point for physiological variables, sensors for detecting fluctuations in the variable, and effector mechanisms that help restore the variable to the optimum value. Many physiological variables are under homeostatic regulation, for example, body temperature, and the concentration of H^+ (pH), buffers, ions, gases (O_2 and CO_2), and glucose in body fluids. Communication between sensors and effectors is via nerves and hormones or other chemical signals.

The activity of each effector system is controlled—speeded up or slowed down—by signals, typically via the nervous and/or endocrine systems. Feedback from sensors guides the effectors. As the sensors monitor information, one of two types of regulation, negative or positive feedback, is activated. Negative feedback is the most common stabilizing force. Negative in this context is not a bad thing. Rather, negative feedback assures that the effectors do not overshoot the set point, much like a thermostat in a house turns off the heat after detecting that enough heat has been delivered. Positive feedback, on the other hand, amplifies a response, and is less common. Childbirth and emptying body cavities (urination) are two activities where positive feedback is needed.

Chapter 29 includes discussion about the regulation of body temperature. Temperature regulation is an illustrative idea in the new Curriculum Framework. Animals produce heat as a byproduct of metabolism, and can gain or lose heat by radiation, convection, conduction, and evaporation. Ectothermy, taking on the temperature of the environment, and endothermy, maintaining a set body temperature, can use different mechanisms to deal with heat gain and loss. As you study these mechanisms, take the "regulation viewpoint." How does the organism sense the departure from the set point, i.e., what is the physical stimulus? What is the cause-and-effect response to the stimulus? Following this scheme, you can study the regulation of most anything in the human body.

The topics in Chapter 29 span Big Ideas 2 and 4. The Big Ideas are a means for organizing biological information. It is vitally important that you understand these **Big Ideas** across their many **Enduring Understandings**. While Chapter 29's emphasis is primarily on Big Idea 2, Big Idea 4 is considered as well.

Big Idea 2 states that the utilization of free energy and use of molecular building blocks are characteristic fundamental of life processes. Specifically, Chapter 29 includes:

2.a.1: All living systems require constant input of free energy.
2.a.3: Organisms must exchange matter with the environment to grow, reproduce, and maintain organization.

2.b.2: Growth and dynamic homeostasis are maintained by the constant movement of molecules across membranes.
2.c.1: Organisms use negative feedback mechanisms to maintain their internal environments and respond to external environmental changes.
2.c.2: Organisms respond to changes in their external environments.
2.d.1: All biological systems from cells and organisms to populations, communities, and ecosystems are affected by complex biotic and abiotic interactions involving exchange of matter and free energy.
2.d.2: Homeostatic mechanisms reflect both common ancestry and divergence due to adaptation in different environments.

Big Idea 4 examines the idea that biological systems interact in complex ways. Included in Chapter 29:
4.b.2: Cooperative interactions within organisms promote efficiency in the use of energy and matter.

Chapter Review

Section 29.1 explains how multicellular organisms utilize systems of organs to maintain homeostasis. Life requires the transfer of matter and energy, and this exchange is optimized by evolutionary tuning to provide organisms with the greatest likelihood of successful reproduction. For a multicellular animal such as a human, the transfers of matter typically involve solutes moving as follows:

Environment ↔ Organism ↔ Blood ↔ Interstitial Fluid ↔ Active Cells

1. Define homeostasis and explain why it is important to maintain homeostasis.

____
____
____
____
____

2. Explain why ion concentrations and water balance in the interstitial fluid are of great homeostatic importance.

____
____
____
____
____

3. Discuss some of the benefits of having specialized tissues in multicellular organisms.

____
____
____
____
____

4. Discuss the interactions of the four separate tissue types in the stomach.

____
____
____
____
____

5. Jellyfish do not have an internal system of organs. Compare how humans and jellyfish solve the problem of getting nutrients to, and waste products away from, their internal cells.

____
____
____
____
____
____

Section 29.2 examines how sensors monitor controlled variables in the internal environment, and how this information alters the activities of relevant effectors.

6. At many of its target cells, insulin binding to its receptor proteins on the target cells activates a biochemical response that increases the uptake of glucose into the target cells. Insulin secretion from the pancreas is evoked by several factors, but especially by an increase in the local concentration of glucose. Use these terms -- set point, negative feedback, feedback information, and effector -- to explain how the lack of "insulin action" might impair health.

____
____
____
____
____
____

7. For each of the situations described below, identify the feedback as either negative or positive feedback and explain your choice.

a. Your stomach secretes an enzyme precursor called pepsinogen that is inactive. As pepsinogen is converted to the active enzyme pepsin, it triggers the conversion of other pepsinogen molecules to pepsin. A cascade effect occurs and your stomach has enough pepsin molecules to digest proteins.

____
____
____
____

b. Cells along blood vessels have sensory neurons that measure the stretch caused by blood flow against the vessel walls, thus being part of a blood-pressure monitoring system. When blood pressure decreases, thirst is stimulated to increase body hydration.

____
____
____
____

c. Upon injury to a blood vessel, the platelets change shape and clump together, releasing compounds that cause more platelets to accumulate.

____
____
____
____

8. Certain sensory cells are neurons that increase their activity when stretched. Describe a situation where information from stretch-sensitive neurons is used as negative feedback and a second example where it is used as positive feedback.

____
____
____
____
____
____

Section 29.3 reviews how animals regulate their bodies in response to temperature changes. The section also includes a discussion of Q_{10} which is a measure of the temperature sensitivity of a reaction. As a concept, Q_{10} is not an important concept for the AP exam, but it is a useful idea for comparing the sensitivity of different organisms to temperature changes.

9. Explain why physiological reactions generally slow down with cooler temperatures.

____
____
____

10. Explain why biochemical reactions speed up as temperature increases but then suddenly cease to operate at high temperatures. Be specific with your answer, simply saying enzymes denature is not enough.

____
____
____
____

11. The graph to the right shows how an endotherm (a mouse) and an ectotherm (a lizard) react to changes in environmental temperature. Summarize the information portrayed in the graph. (The ability to summarize a graph or the results of an experiment is a skill that may well be tested on the new AP Biology exam.

____
____
____
____
____
____
____
____
____

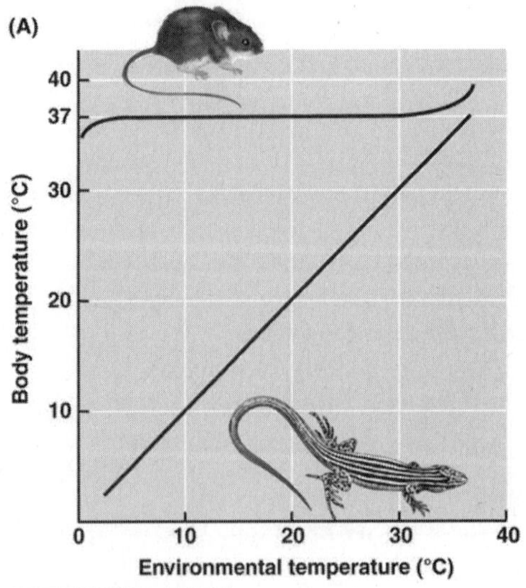

PRINCIPLES OF LIFE, Figure 29.5 (Part 1)

12. The graph to the right shows how the metabolic rates of an endotherm (a mouse) and an ectotherm (a lizard) react to changes in environmental temperature. Compare the metabolic rates of the mouse and the lizard. Why are they so different?

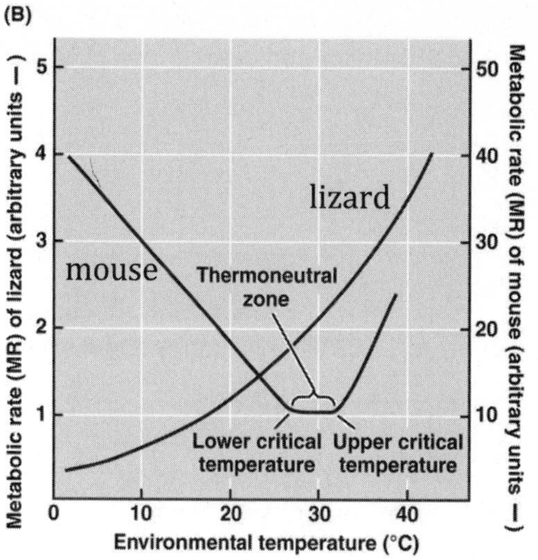

Section 29.4 explores the properties and adaptations of animals that affect body temperature. Animals exchange heat with the environment via metabolism, radiation, convection, conduction, and evaporation.

13. Explain the heat loss and heat gain as a person sits on a frozen lake, next to a hole in the ice fishing.

14. Penguins can walk miles across ice in subzero conditions. Draw a diagram of the vascular anatomy (blood vessels) in a penguin's leg detailing how heat loss from the body core is conserved, and provide a verbal description.

Draw the artery-vein relationship in a penguin's leg here:

15. Give three examples of ectotherms that demonstrate behavioral regulation of body temperature. Describe each behavior and discuss how it affects body temperature.

16. Give two examples of endotherms that demonstrate behavioral regulation of body temperature. Describe each behavior and discuss how it affects body temperature.

17. Give two examples of endotherms that demonstrate physiological (non-behavioral) regulation of body temperature. Describe each behavior and discuss how it affects body temperature.

18. Many large marine fish (tuna) conserve heat loss even as they dive to great depths in cold water. Discuss what might happen to these fish in deep water if they could not conserve heat loss as they dive.

Section 29.5 discusses how mammals regulate their body temperature.

19. Ectotherms can regulate heat loss and gain with behavioral adaptations, while endotherms can also utilize metabolic mechanisms to regulate heat loss and gain. Discuss how humans regulate body temperature, discussing one physiological mechanism and one behavioral mechanism.

20. Draw two flow charts showing the mechanisms of heat exchange in humans, one during exposure to a cold environment and one during exposure to a hot environment. Include in each diagram one behavioral response, two physiological responses, and the negative feedback systems at work, including sensors and effectors.

Science Practices & Inquiry

In the AP Biology Curriculum Framework, there is a set of 7 Science Practices. The question below combines learning objectives found within **Science Practices 5, 6 and 7.**

Science Practice 5: The student can perform data analysis and evaluation of evidence.
Science Practice 6: The student can work with scientific explanations and theories.
Science Practice 7: The student is able to connect and relate knowledge across various scales, concepts and representations in and across domains.

More specifically, practices

5.1: The student can *analyze data* to identify patterns or relationships.
6.1: The student can *justify claims with evidence.*
6.4: The student can *make claims and predictions about natural phenomena* based on scientific theories and models.
7.2: The student can *connect concepts* in and across domain(s) to generalize or extrapolate in and/or across enduring understandings and/or big ideas.

Question 21 consider the effect on a biological system at the organismal level in a scenario where the temperature is changed, and how the organisms use negative feedback for homeostasis of body temperature. You will have to make predictions about how organisms use negative feedback mechanisms to maintain the internal environment. You will need to analyze data to detect patterns and relationships between an abiotic factor and a biological system (an organism). (LO's 2.15, 2.16, 2.18, and 2.24)

21. The data below were recorded from an experiment in which the hypothalamus of a ground squirrel was cooled (**T_H**) while the animal's metabolic rate (MR) was measured. **T_H** is given in °C, and MR is given in calories per gram of body mass per minute. (From the Investigation in Figure 29.13)

Temperature of Hypothalamus (T_H)	Metabolic Rate (MR)
39.5	0.040
39.0	0.041
38.5	0.040
38.0	0.038
36.5	0.038
36.0	0.040
35.5	0.060
35.0	0.080
37.5	0.041
37.0	0.039
34.5	0.110
34.0	0.140

a. Plot the data on the graph paper supplied on the following page.
b. Summarize the trends shown in your graph.
c. Determine the threshold temperature for the metabolic-heat-production response from your graph.
d. Describe what these experimental results suggest about the function of the hypothalamus. Explain your answer.
e. Explain how the outcome of this experiment would differ if the tested animal was an ectothermic organism such as a turtle.

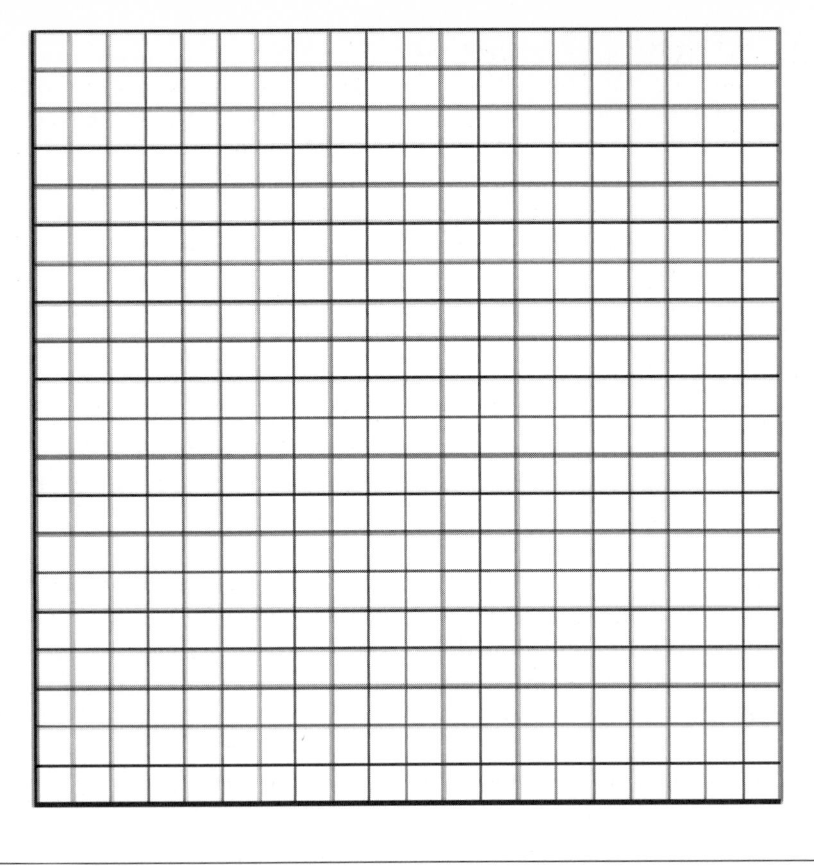

Chapter 30: Animal Hormones

Chapter Outline

30.1 - Hormones Are Chemical Messengers
30.2 - Hormones Act by Binding to Receptors
30.3 - The Pituitary Gland Links the Nervous and Endocrine Systems
30.4 - Hormones Regulate Mammalian Physiological Systems

If you have any doubts that hormones are important, you must be unconcerned about the operation of your reproductive system, you must not have any relatives or friends with *diabetes mellitus*, and you must be willing to discount the information in this chapter. In fact, hormones ARE important, and this chapter by itself includes thirteen different learning objectives in the AP-Biology Curriculum Framework.

Hormones are the main chemical messengers found in the blood. They are produced and released by endocrine glands, travel through the blood, and influence the activity of target cells; target cells are those that have the receptor proteins to bind the hormone. The binding of the hormone and its receptor produces a biochemical change in the target cell, thus accomplishing the target cell's role as an effector in a homeostatic system. We will review the nervous system as a signaling system in Chapter 34.

Big Ideas 2 and 3 from the Framework are central to understanding hormones. This material can facilitate your understanding of the Big Ideas and the many Enduring Understandings because you might already be familiar with hormones.

Big Idea 2 emphasizes chemical building blocks and energy flow in living organisms, with Chapter 30 discussing the fact that:

2.c.1: Organisms use negative feedback mechanisms to maintain their internal environments and respond to external environmental changes.

Big Idea 3 emphasizes information transfer, by expanding your knowledge that:

3.a.4: The inheritance pattern of many traits cannot be explained by simple Mendelian genetics.
3.b.2: A variety of intercellular and intracellular signal transmissions mediate gene expression.
3.d.2: Cells communicate with each other through direct contact with other cells from a distance via chemical signaling.
3.e.1: Individuals can act on information and communicate it to others.

Chapter Review

Concept 30.1 states that hormones are chemicals released by endocrine glands into the blood. Hormones serve messenger functions when the hormone reaches its receptor proteins on or in target cells. Only cells with the appropriate receptor proteins will be directly altered by the arrival of the hormone – all other (receptor-less) cells are non-target cells. Chemically, most hormones are amines, proteins or steroids. Description of two insect hormones is presented, reiterating the ancient origin of chemical signaling pathways in animals and plants.

1. Hormones can act on the cell that produces them (autocrine effect), neighboring cells (paracrine effect) and on distant cells, after passage in the blood (endocrine, or hormonal, effect). Discuss whether or not it is possible for a single hormone to work along all three effector routes.

____
____
____
____

2. Discuss the water and lipid solubility of steroid hormones and protein hormones.

____
____
____
____

3. In humans, the hormone prolactin stimulates milk synthesis in the mammary glands, but in salmon, prolactin is a signal that mediates physiological changes needed for moving from saltwater to freshwater. Discuss how it comes about that any given hormone can have very different effects in different organisms.

____
____
____
____

4. The insect pictured here is the adult, reproductive form of the genus *Rhodnius*. The juvenile form of these blood-sucking insects can live for up to a week after having been decapitated. Molting to the adult form takes place after an advanced juvenile within 7 days after it has taken a blood meal, but if the juvenile is decapitated immediately after a blood meal, the headless juvenile does not show any apparent molting changes, even a full week later, when it is still alive. Explain what this result tells us about the source of a "molting hormone" in these animals.

____
____
____
____
____
____
____

Image Source: Dr. Erwin Huebner, University of Manitoba, Winnipeg, Canada

Concept 30.2 reiterates that receptor proteins are just as important as hormones: cells that lack receptor proteins for Hormone Q cannot to show a direct response to Hormone Q. Receptor proteins are either located on the plasma membranes of target cells (especially true for the receptors of hormones that are proteins, e.g., insulin) or located in the cytosol and nucleus of the target cells (especially true for lipid-soluble hormones, e.g., steroids). In both cases, the binding of the hormone to its receptor protein is the event that alters the biochemistry of the target cell, eliciting its "effector" responses.

5. The "estrogens" are a group of steroid hormones that stimulate growth in many of their target cells. Discuss whether or not it is important to know if a "breast cancer" tumor in the mammary glands of a human expresses the gene for the estrogen receptor.

6. Epinephrine is a hormone secreted by your adrenal glands when you are profoundly startled or when you are getting a good exercise workout. This one chemical has two opposite effects in different parts of the circulation: it reduces blood flow in the gastro-intestinal tract, but it increases blood flow in your large skeletal muscles. Discuss the means by which one signal (epinephrine) can have two opposite effects (vasoconstriction versus vasodilation).

7. Discuss the "downregulation" of insulin receptors that is often noted in patients with diabetes mellitus, type II.

Concept 30.3 describes the pituitary gland. You will not need to memorize most of the hormonal details in this chapter, but work to understand conceptually how hormones and classes of hormones function. Along with the hypothalamus in the brain, the pituitary gland is a central player in orchestrating hormone secretion and allowing hormonal control mechanisms to function in homeostasis. In humans, the pituitary gland has two main parts: the anterior pituitary gland (APG), the side closest to the nose, and the posterior pituitary gland (PPG), located closer to the back of the head. Don't let their proximity to each other fool you: the APG is made up of specialized epithelial "skin" cells arising first in the roof of the embryonic mouth, whereas the PPG is a down-growth from the brain. The APG and PPG both secrete protein hormones, although those from the PPG are released by the nerve endings of neurons from the hypothalamus and those from the APG are released after the APG cells are stimulated by "releasing hormones" secreted from the hypothalamus.

8. Explain why the hormones secreted by the anterior pituitary gland are called "tropic" hormones.

9. The "stress" hormone axis is organized in the pattern shown by the diagram below. This axis includes a hypothalamic hormone (CRH) that stimulates secretion by the anterior pituitary gland (APG), an APG hormone (ACTH) that stimulates steroid synthesis (cortisol) in the adrenal glands, and adaptive responses, like glucose release from the liver and reduced immune system activity. Explain why the stress hormone axis includes a part that is in the brain (hint: think about the many forms of stress in life).

External or internal conditions
or
= Stimulation
= Inhibition
Hypothalamus
Releasing hormone
"Short loop" negative feedback
"Long loop" negative feedback
Anterior pituitary
Tropic hormone
Endocrine gland
Hormone

PRINCIPLES OF LIFE, Figure 30.9

10. Cortisol, mentioned previously in question 9, has receptors in cells of the immune system and the liver. However, cortisol receptors are also found in the hypothalamic and pituitary components of the stress hormone axis. Explain why.

Concept 30.4 provides some examples of mammalian hormone systems. To analyze a hormone system, review these four topics: the source of the hormone, the stimuli that elicit the secretion of the hormone, the cellular location of the hormone's receptor proteins, and the link between the target cell response and homeostasis. Make sure you remember that negative feedback requires receptor proteins, too.

11. Thyroid hormones boost metabolism when environmental conditions become more difficult, for example, during the cold weather of winter. Explain how boosted metabolism can aid an animal's survival as temperatures decline.

12. The condition called "goiter" is often apparent as a large swelling of the neck in the region where the thyroid gland is located. In hypothyroid goiter, the gland is greatly enlarged as it unsuccessfully attempts to make thyroid hormones. Discuss each of the following points by circling whether you think the statement is true or false and then explaining your logic.

A. Persons with hypothyroid goiter have very high levels of thyroid hormones in their blood. TRUE FALSE

B. Persons with hypothyroid goiter have very high levels of thyroid-stimulating hormone in their blood. TRUE FALSE

C. Persons with hypothyroid goiter have very high levels of thyrotropin-releasing hormone in their blood. TRUE FALSE

13. Suppose that your uncle eats ice cream, loaded with calcium, after every meal. Predict what his levels of parathormone are likely to be, relative to those of a person who eats a normal diet, and explain your prediction.

14. The adrenal glands have two sections, one more involved with acute stress, and the other more involved in prolonged stress. Explain this statement.

15. Describe the role of exposure to androgens, e.g., testosterone, in the normal development of the male and female reproductive systems.

16. Compare the normal operation of the hypothalamic-pituitary gonadal hormone axis before and after puberty in males.

Science Practices & Inquiry

In the AP Biology Curriculum Framework, there are 7 Science Practices. In the next question, you will exercise your ability to deal with Science Practice 5: The student can perform data analysis and evaluation of evidence. More specifically, the question works with practice 5.1: The student can analyze data to identify patterns or relationships.

Question 17 asks you to analyze data that indicate how organisms exchange information in response to internal changes and external cues, and which can change behavior. (LO 3.40; see "Apply the concept" on . 616 in the text).

17. The time courses of action for different hormones vary widely. Some hormones are released rapidly, establish their effects almost immediately, and then are cleared from the bloodstream within minutes. Others are released slowly and remain in the blood for many hours or even days. One way of characterizing the time course of a hormone is to measure its half-life in the blood: the length of time it takes for the blood level of a given hormone to fall to half of the baseline (maximum) following its release (or injection).

The table gives blood concentrations of thyroxine (T4) following a 600-μg injection. Plot these data on the graph space provided.

TIME (HRS)	T_4 (mg /dL)
0	7.5
6	13.7
12	12.3
24	11.1
36	10.7
48	10.3
60	9.9
72	9.5
84	9.3
96	9.1

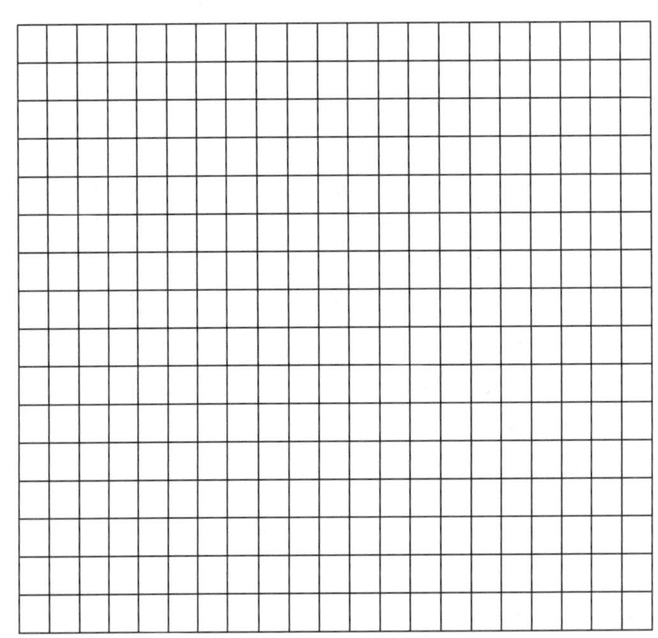

A. Before assessing your graph, consider what you know about the functioning of T4. Would you expect this hormone to have a short or a long half-life? Why?

____

____

____

____

____

____

____

____

B. Use your graph to estimate the half-life (to the nearest 0.1 hr) of T4 in the bloodstream.

Chapter 31: Immunology: Animal Defense Systems

Chapter Outline

31.1 - Animals Use Innate and Adaptive Mechanisms to Defend Themselves against Pathogens
31.2 - Innate Defenses Are Nonspecific
31.3 - The Adaptive Immune Response Is Specific
31.4 - The Adaptive Humoral Immune Response Involves Specific Antibodies
31.5 - The Adaptive Cellular Immune Response Involves T Cells and Their Receptors

To maintain health and to prevent infection, multicellular organisms can mount a coordinated defense against invasion by pathogenic organisms. Pathogenic invaders disrupt homeostasis in their hosts, sometimes even to the point of death. There are two major levels of defense: innate immunity and adaptive immunity.

Most multicellular organisms have a range of nonspecific, innate defenses against invasion. For example, salty skin with its own flora is not often a hospitable environment for additional bacterial growth. In addition, natural killer cells roam the body, attacking invaders as they are encountered and recognized chemically as invaders.

Most multicellular organisms also have an adaptive or specific response system. A major conceptual premise here is that the organism's adaptive immunity must distinguish between self and non-self, so that only the latter will be subject to attack. Activated by the chemical recognition of invaders, B and T lymphocytes are key players in adaptive immunity. Host cells coordinate defense by using chemical signals. Adaptive immunity expressly allows the organism to respond quickly and emphatically to specific pathogens that had been previously encountered and defeated, a sort of chemical memory that thwarts re-infection and forms the basis of modern vaccination practices.

Immunity is one of the three major physiological systems explicitly included on the AP Biology exam. This chapter unites many of the Big Ideas in the AP Biology curriculum including cell communication, homeostasis, and variation. There is an abundance of detail in this chapter, but memorization of many details, e.g., the structure of specific antibodies, is beyond the scope of the AP Exam**.** As you study this chapter, focus on how these ideas are united together in immunology.

Chapter 31 spans Big Ideas Two, Three, and Four. The Big Ideas are a means for organizing the vast amount of information we know as biology. It is vitally important that you continually work to understand these **Big Ideas** and across different **Enduring Understandings**.

Big Idea 2 states that the utilization of free energy and use of molecular building blocks are characteristic fundamental of life processes. Specifically, Chapter 31 shows that:

2.d.3: Biological systems are affected by disruptions to their dynamic homeostasis.
2.d.4: Plants and animals have a variety of chemical defenses against infections that affect dynamic homeostasis.

Big Idea 3 states that living systems store, retrieve and transmit information essential to life processes. Specifically, Chapter 31 shows how cell communication is utilized by cells for immunity as:

3.d.2: Cells communicate with each other through direct contact with other cells or from a distance via chemical signaling.

Big Idea 4 examines the idea that biological systems interact in complex ways. Included in Chapter 31 is the fundamental idea that:

4.c.1: Variation in molecular units provides cells with a wider range of functions.

Chapter Review

Concept 31.1 describes the primary differences between innate (nonspecific) and adaptive (specific) immunity. White blood cells, including phagocytes, leucocytes, and natural killer cells, are involved in both forms of immunity. Cytokines are chemical messengers that coordinate activities of immune-system components.

1. Identify each of these examples of immune mechanisms as innate (I) or adaptive (A):

a. ____ Stomach acid destroying a bacteria
b. ____ T-cell destroying a virus
c. ____ Skin preventing a virus from entering
d. ____ Defensins destroying invaders that have plasma membranes
e. ____ Interferons preventing viruses from spreading between neighboring cells
f. ____ Antibodies preventing reinfection by chicken pox

2. Describe the major difference between antibodies and the major histocompatibility complex (MHC) proteins.

____
____
____
____

3. White blood cells can be classified as either phagocytes or lymphocytes. Explain how these are

a. similar: ______________________________
____
____

b. different: ______________________________
____
____

Concept 31.2 focuses on the innate immune system, including the skin and mucus, lysozymes and defensins. Other innate defenses involving proteins and cellular defenses include phagocytes, natural killer cells, complement proteins, interferons, and inflammation. These examples are discussed in the context of non-specific immunity.

4. Outline the innate immune system defenses met by a pathogen as it:

a. lands on your skin.

____
____
____

b. lands in your mouth.

c. is ingested in a drink of water.

5. Describe natural killer cells and explain how they can be part of both innate and adaptive immune systems.

6. Describe how the inflammation response aids in fighting infection.

7. Many people take non-prescription drugs, e.g., antihistamines, to control allergies. Explain how these drugs can reduce an allergic response.

8. Explain how interferons are an example of cell communication. Include source and targets in your answer.

9. Synthetic cortisol-like (agonists) drugs are often injected into swollen and painful bone-joints, e.g., tennis elbow, pitcher's arm, arthritic knees. Discuss how the injected cortisol agonist affects the operations of the immune system.

10. Suppose that you were exposed to a newly synthesized "artificial" bacterium. After exposure, all signs of the bacterium from your body were gone within 24 hours. Assume further that this bacterium is novel enough that it does not share chemical identity signals with other bacteria. Decide if your immune system's victory over this bacterium was via innate or adaptive immunity, and provide explanations of some of the ways the bacterium was defeated.

Concept 31.3 introduces the adaptive immune response. The four premises for this type of response are: specificity, diversity, self-recognition, and memory. Adaptive immunity hinges on the B and T lymphocytes. In terms of specificity, activated T-cells destroy foreign invaders displaying specific antigens. In terms of diversity, B cells can synthesize specific antibodies that can recognize upwards of 10 million different antigens. This diversity is created by DNA changes and mutations after B cells are formed in the bone marrow. Self-recognition is an important premise, as the cells attacking invaders must be able to distinguish them from cells that are part of the self; the latter possess major histocompatibility complex (MHC) proteins on their surfaces as a form of chemical identification. Memory and antibodies are discussed below (Concepts 30.4 and 30.5).

11. Describe the differences between antigens and antibodies.

____
____
____
____

12. Explain how one type of bacterium can have several antigens.

____
____
____
____

13. Explain how a single molecule (polypeptide or polysaccharide) can include multiple antigen sites.

____
____
____
____

14. Severe Combined Immunodeficiency is an uncommon genetic disease in which T-lymphocyte and B-lymphocyte function is absent or greatly reduced. Describe a transplant procedure that might establish the production of lymphocytes in such a person.

____
____
____
____

15. For a person with Severe Combined Immunodeficiency, describe TWO types of disease of special concern.

____
____
____
____

16. After winter break vacation at a boarding school, many students will get sick from the flu or other viruses. Explain why older people are often more resistant to sickness than are younger people.

____
____
____
____

17. Explain how humoral immunity differs from cellular identity, using a brief example of each type

Concept 31.4 describes B cells serving as the basis of the humoral immune response. Antigen-antibody reactions are described, but for the AP-Biology exam, you are not required to memorize the detailed information presented in this section.

18. Although you have probably never been exposed to botulism or diphtheria or their toxins, you are making B cells and antibodies that can recognize these antigens, and therefore help protect you from botulism. Explain.

19. Discuss how the diversity of producing millions of different antibodies comes about, and describe how this process is similar to the idea of introns and splicing.

20. Discuss how the involvement of antibodies leads to the destruction of foreign pathogens or toxins.

Concept 31.5 describes how two types of T-cells are the basis of cellular immunity, T-helper cells and cytotoxic T cells.

21. Explain the role of T-helper cells and cytotoxic T cells, showing the key differences between their actions in the immune system.

22. Explain how a flu vaccination can induce mild flu symptoms, and yet protects against a more difficult case of the flu at a later time.

____
____
____
____
____
____

23. Some people describe allergies as an overactive immune response. Explain.

____
____
____
____
____

Science Practices & Inquiry

In the AP Biology Curriculum Framework, there is a set of 7 Science Practices. In this chapter, we will focus on **Science Practice 1:** The student can use representations and models to communicate scientific phenomena and solve scientific problems. More specifically, practice 1.1: The student can create representations and models of natural or man-made phenomena and systems in the domain.

Question 22 asks you to create representations and models to describe immune responses. (LO 2.29)

24. For each of the scenarios below, draw a diagram or flow chart showing how the immune system of a person might react to a foreign virus.

a. While walking down a crowded hallway, someone in front of you coughs without covering his/her mouth. You feel the spray of the cough on your arm but you do not get sick.
b. While swimming in the ocean Mary swallows some seawater. Later that night Mary feels ill, and runs a temperature. The next day Mary feels fine.
c. The daughter of a couple comes down with the chicken pox, but her parents do not. Both of her parents had the chicken pox as young kids. (Create two flow charts here, one for the daughter, and one for the parents.)

Chapter 32: Animal Reproduction

Chapter Outline

32.1 - Reproduction Can Be Sexual or Asexual
32.2 - Gametogenesis Produces Haploid Gametes
32.3 - Fertilization Is the Union of Sperm and Egg
32.4 - Human Reproduction is Hormonally Controlled
32.5 - Humans Use a Variety of Methods to Control Fertility

Reproduction is the "biological imperative" for any species of living organism, for without reproduction, there are no new individuals, and without new individuals, evolutionary change is no longer possible. Observe any parent's behavior with their new offspring and you will see that reproduction is also an incredible process, as new individuals come into being, fresh and full of opportunity.

Like plants (Chapter 21), animals have two possible directions in reproduction: asexually, efficiently making copies of oneself or, sexually, exchanging genetic material with a partner and producing offspring with new combinations of adaptations.

Chapter 32 includes Big Ideas Two and Three. The Big Ideas are a means for organizing biological information. Work continually to understand these **Big Ideas** across their many **Enduring Understandings**.

Big Idea 2 states that the utilization of free energy and use of molecular building blocks are characteristic fundamental of life processes. Specifically, Chapter 32 shows that:

2.d.3: Biological systems are affected by disruptions to their dynamic homeostasis.
2.d.4: Plants and animals have a variety of chemical defenses against infections that affect dynamic homeostasis.

Big Idea 3 states that living systems store, retrieve and transmit information essential to life processes. Specifically, Chapter 32 shows how cell communication is critically important in reproduction, as:

3.d.2: Cells communicate with each other through direct contact with other cells or from a distance via chemical signaling.

Chapter Review

Concept 32.1 reviews the two major reproductive options for animals: asexual and sexual reproduction. The simple but evolutionarily limiting direction is called asexual reproduction. Budding (e.g., *Hydra*), regeneration (e.g., sea stars) and parthenogenesis (e.g., whiptail lizards). Most animal reproduction occurs via sexual interactions, but a few species, e.g., aphids, can alternate between asexual and sexual reproduction.

1. Oyster fisherman used to regard sea stars as unwelcome competition for the oysters they sought to bring to market. Accordingly, any sea stars caught during oyster trawling were cut into pieces and thrown overboard to "feed the fishes." Describe how this result likely generated the opposite effect than was desired.

____
____
____
____

2. Barry and Mike agree that *Hydra* reproduce by budding, but Barry says budding requires mitosis and Mike says it occurs via meiosis. Solve their argument with your explanation.

3. Whiptail lizards are all females, but successful parthenogenesis requires that one member of a pair express the courting and mating behaviors seen in males of related but non-parthenogenic lizards. Define parthenogenesis and discuss the behavioral requirements observed in whiptails.

4. Some species of aphids alternate between asexual reproduction in springtime and summer and sexual reproduction during autumn. Discuss the benefits to aphids of this pattern, in light of the fact that the aphids thrive on fresh green vegetation and only the fertilized eggs are able to survive the cold months of winter.

Concept 32.2 describes how the meiotic production of haploid gametes by sexually reproducing animals occurs in a process called gametogenesis. Female gametes, called ova or eggs, are typically much larger than the male gametes, typically called sperm. Gametogenesis of ova is called oogenesis and of sperm is called spermatogenesis. In mammals, oogenesis is dependent on the supply of diploid primary oocytes, produced by mitosis during embryonic development, whereas diploid spermatogonia, the male equivalent of the primary oocytes, are mitotically produced throughout adult life. The meiotic progression to haploid sperm occurs within the male gonads, the testes, whereas the progression to haploid secondary oocytes occurs during each ovarian cycle (the menstrual cycle), with the final meiotic step only occurring after fertilization. The interaction of the egg with a single sperm causes the final meiotic step to progress with the resulting union of two haploid nuclei and the formation of a zygote.

5. Mitochondria appear to have once been an independent form of life that formed a symbiosis with a cell ancestral to the eukaryotic cell. Discuss your mother's and your father's likely contributions to the mitochondria in your cells and describe the events leading up to that result.

6. In humans, older age leads to diminution and then cessation of female reproductive potential, whereas older men are still capable of reproduction. Provide greater detail on the differences in reproductive potential.

7. Earthworms are "simultaneous" hermaphrodites; explain.

8. Clown fish, often found as symbionts with anemones, are "sequential" hermaphrodites; explain.

Concept 32.3 jumps to the union of sperm and egg, called fertilization. This fusion might occur externally, as in the well-studied sea urchin, or internally, as in all mammals. The molecular "recognition" of a sperm by an egg is chemically mediated for precision and accuracy. Next, the haploid nuclei are brought to close proximity, fuse to convert the former two haploid nuclei to a single diploid nucleus in what is now called a zygote, the single cell that gives rise to new life.

9. Describe THREE events that are in the mechanism leading to the "block to polyspermy" of sexually reproducing animals.

10. Discuss the primary reason that the "block to polyspermy" is necessary for the success of sexual reproduction.

11. Some animals deposit their zygotes into the external environment for development while other animals retain zygotes internally for development. Provide one example of each pattern and describe it.

Concept 32.4 moves beyond gametogenesis to consider the hormonal milieu that coordinates reproduction. What follows are many specific details on the development of human reproductive cells. You will not be responsible for memorizing each hormone its source and its target. Rather you should work to understand conceptually how the hormones are working together as a coordinated whole, through cell communication.

In males, spermatogenesis takes place in the seminiferous tubules in the testes, and begins at puberty. It requires hormonal support, in that androgens (including testosterone), and the gonadotropins (including Follicle Stimulating Hormone), are needed by the Sertoli cells in the tubules. The Sertoli cells respond to the hormones by metabolically supporting the mitotic activity of the spermatogonia and the meiotic events that follow to yield sperm. Following maturation of sperm in the epididymis of the testis, sperm are mixed with other glandular products along the reproductive tract, forming semen, which is ejaculated through the penis during orgasm.

In females, oogenesis takes place in the ovaries. Ovarian follicles surround each growing primary oocyte, supporting its growth and maturation, and then the follicle bursts at the time of ovulation, releasing the oocyte near the reproductive tract. The oocyte is swept into the oviducts, where fertilization will occur, if sperm are present. Pregnancy requires steroid-hormone support for the uterus.

12. The flow chart on the right describes the "tripartite endocrine axis" controlling male reproduction. Add the names of the hormones, targets, and results, drawing from this list:

Anterior pituitary gland (APG)
Follicle Stimulating Hormone (FSH)
Gonadotropin-Releasing Hormone (GnRH)
Hypothalamus (HT)
Inhibin (INHB)
Leydig Cells (LCs)
Luteinizing Hormone (LH)
Sertoli Cells (SCs)
Spermatogenesis
Testosterone (T)

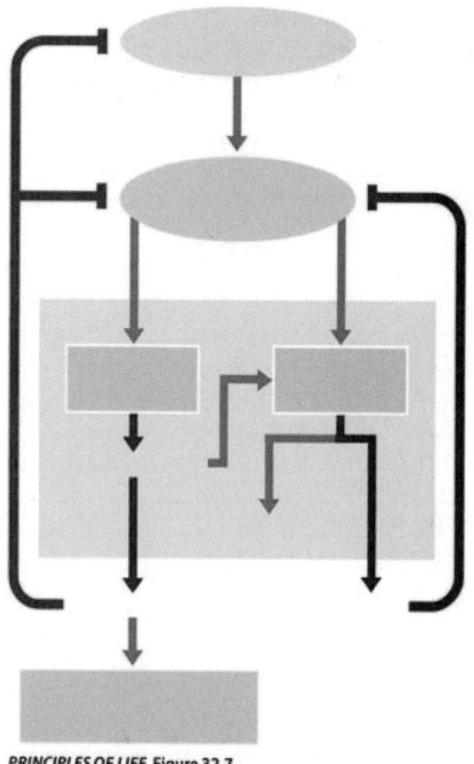

PRINCIPLES OF LIFE, Figure 32.7

13. Return to the drawing of Question 12 and add all hormone receptors using the abbreviations in this list:

Follicle Stimulating Hormone Receptor (FSH-R)
Gonadotropin-Releasing Hormone Receptor (GnRH-R)
Inhibin Receptor (INHB-R)
Luteinizing Hormone Receptor (LH-R)
Testosterone Receptor = Androgen Receptor (A-R)

14. Describe the tissue source of testosterone in males and describe how one might go about determining which cells in the body are the "target cells" for testosterone effects.

15. Negative feedback is characteristic of most endocrine pathways, including female reproductive physiology. However, there is a brief period where the feedback process is greatly changed. Describe that time and explain what happens.

16. One part of this diagram is missing, part (C), labeled "Ovarian hormones and the uterine cycle." Draw two curves in that part to show the cycles of estrogen and progesterone synthesis and release, and explain how these hormones cause the physical changes that occur in the reproductive tract.

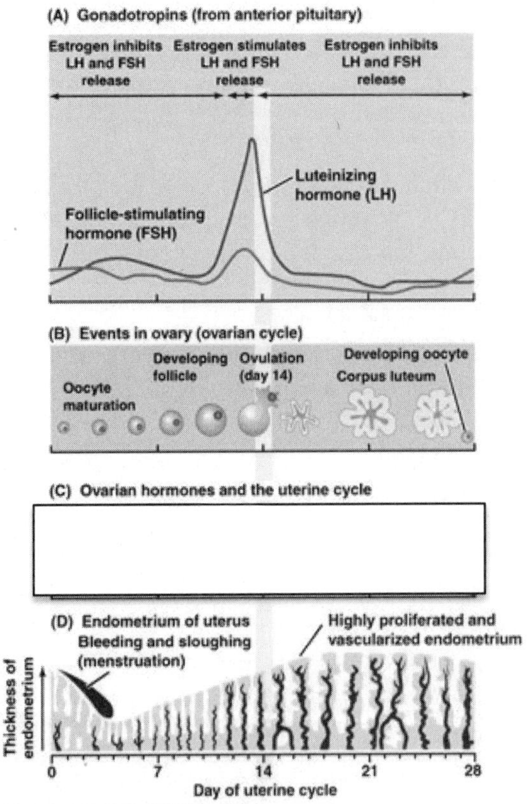

(A) Gonadotropins (from anterior pituitary)

Estrogen inhibits LH and FSH release | Estrogen stimulates LH and FSH release | Estrogen inhibits LH and FSH release

Luteinizing hormone (LH)

Follicle-stimulating hormone (FSH)

(B) Events in ovary (ovarian cycle)

Oocyte maturation | Developing follicle | Ovulation (day 14) | Developing oocyte | Corpus luteum

(C) Ovarian hormones and the uterine cycle

(D) Endometrium of uterus

Bleeding and sloughing (menstruation)

Highly proliferated and vascularized endometrium

Thickness of endometrium

0 7 14 21 28

Day of uterine cycle

PRINCIPLES OF LIFE, **Figure 32.10**

17. The steroid hormone levels needed to support a pregnant uterus are high. In the first trimester (first three months of pregnancy) these hormones are synthesized in one anatomical location, and during the second and third trimesters (last 6 months of pregnancy) they are synthesized in another location. Describe and explain this sequence.

18. Nursing babies evoke a rapid and positive feedback system that greatly enhances milk release from their mother's mammary glands via an oxytocin hormone reflex. Speculate on the evolutionarily adaptive values of using rapid positive feedback in this manner.

____

____

____

____

____

Concept 32.5 describes various ways that humans can alter function in their reproductive systems, either to enhance fertility or to suppress it.

19. Describe the differences between THREE categories of contraception methods.

____

____

____

____

____

20. Describe the primary risk involved in letting parents select the sex of their babies.

____

____

____

____

____

21. Males have very low levels of estrogen hormones in their blood, yet certain cells express estrogen-receptor genes that are required for normal activity. Describe one other gene that must be expressed in those cells if they are going to be able to respond to estrogens even if the only steroid entering the cells is testosterone.

____

____

____

____

____

____

Science Practices & Inquiry

In the AP Biology Curriculum Framework, there is a set of 7 Science Practices. In this chapter, we will focus on **Science Practice 7:** The student is able to connect and relate knowledge across various scales, concepts and representations in and across domains, with emphasis on 7.2: The student can connect concepts in and across domain(s) to generalize or extrapolate in and/or across enduring understandings and/or big ideas.

Question 21 asks you to connect how organisms use negative feedback to maintain their internal environments. (LO 2.16)

21. The receptors for steroid hormones are similar enough to each other that they sometimes bind to the "wrong" steroids. Prolonged use of prednisone, a cortisol-like drug used to treat inflammation, might possibly reduce fertility. On the partial model shown for a section of the three-part hormone axis regulating the female reproductive tract, show where the effects of prolonged exposure to cortisol would be likely to become apparent, leading to a lack of menstrual cycles (amenorrhea), and describe what hormonal changes an endocrinologist would see in the blood of an affected person.

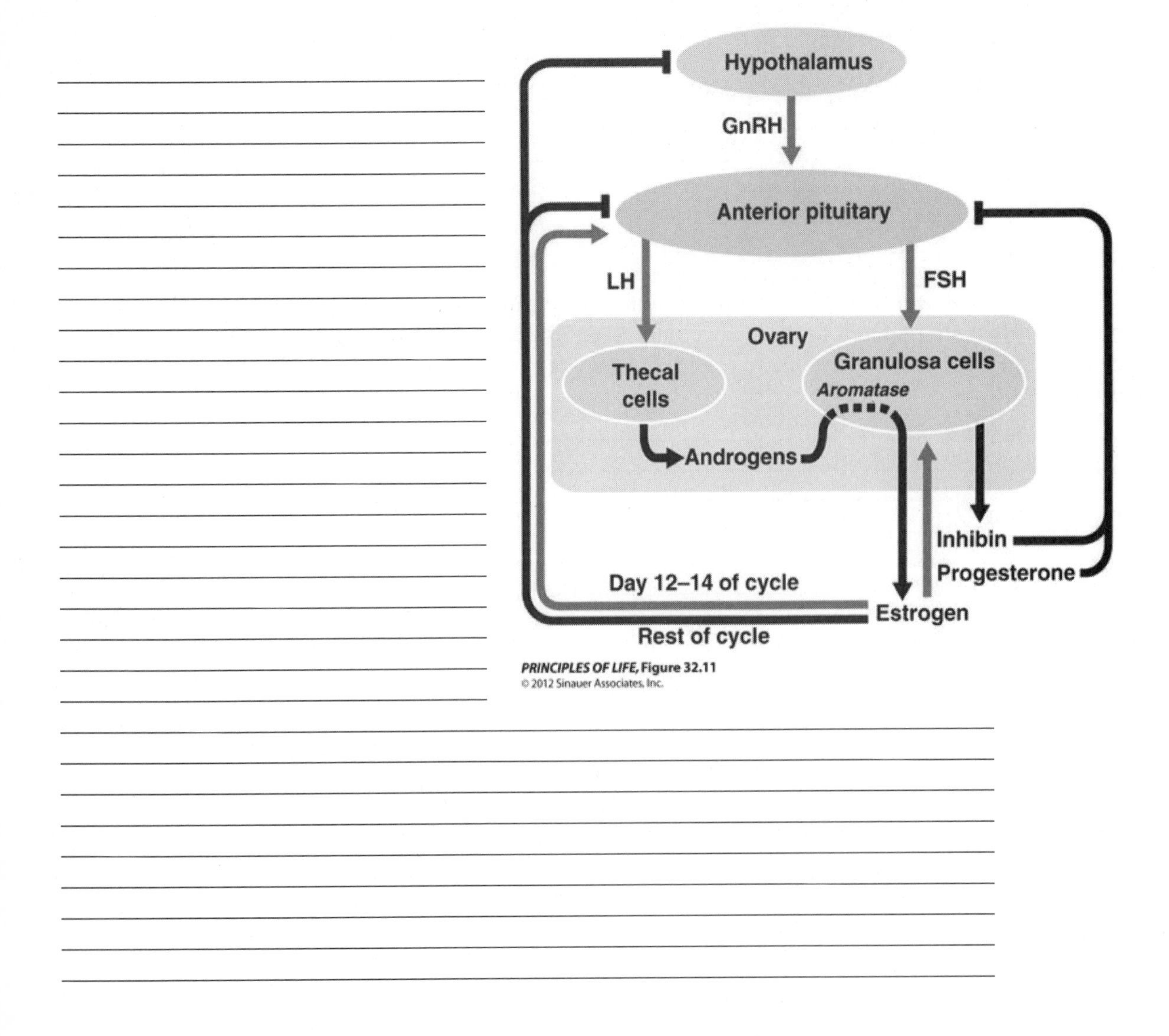

PRINCIPLES OF LIFE, Figure 32.11
© 2012 Sinauer Associates, Inc.

Chapter 33: Animal Development

Chapter Outline

33.1 - Fertilization Activates Development
33.2 - Cleavage Repackages the Cytoplasm of the Zygote
33.3 - Gastrulation Creates Three Tissue Layers
33.4 - Neurulation Creates the Nervous System
33.5 - Extraembryonic Membranes Nourish the Growing Embryo

Earlier in Chapter 14, we saw how differential gene expression acts during early development and how cytoplasmic factors in the egg play important roles in setting up the signaling cascades that orchestrate the major processes of development: determination, differentiation, morphogenesis, and growth. Throughout this chapter, we will see how these processes underlie the development of a multicellular individual from a single cell. In particular, Chapters 14 and 32 examine the cellular and molecular interactions between the egg and sperm that result in launching the development of a new individual.3

The concepts presented in Chapter 33 include three of the four Big Ideas. The Big Ideas are a means for organizing the vast amount of information we know as biology. It is vitally important that you continually work to understand these **Big Ideas** and across different **Enduring Understandings**.

Big Idea 2 states that the utilization of free energy and use of molecular building blocks are characteristic fundamental of life processes. Specifically, Chapter 33 includes:

2.d.3: Biological systems are affected by disruptions to their dynamic homeostasis.

Big Idea 3 states that living systems store, retrieve and transmit information essential to life processes. Specifically, Chapter 33 shows how cell communication is fundamental to growth and development:

3.b.2: A variety of intercellular and intracellular signal transmissions mediate gene expression.

Big Idea 4 examines the idea that biological systems interact in complex ways. Included in Chapter 33:

4.a.3: Interactions between external stimuli and regulated gene expression result in specialization of cells, tissues and organs.

Chapter Review

Concept 33.1 introduces how fertilization activates the development of a new embryo. The tiny sperm from the male parent and the considerably larger egg from the female parent do not contribute equally to all aspects of the newly formed zygote. For example, the egg is typically the source of mitochondria, so all of your mitochondria are of maternal origin. Immediately following fertilization, i.e., the union of maternal and paternal haploid DNA, a complex cascade of events occurs within the cytoplasm. Only a limited set of genes is transcribed during the early cell divisions, with signals controlling gene expression emanating from the cytoplasm.

1. Describe the maturation of an egg, up to but not including fertilization.

____
____
____
____
____
____

2. Draw and describe an unfertilized frog egg and include the cortical cytoplasm, animal pole, and vegetal pole.

____
____
____
____
____
____

3. Describe the contributions made by the sperm at fertilization, including brief discussion of the sperm's function in launching early development.

____
____
____
____
____
____

Concept 33.2 describes the events immediately following fertilization. This is a period of intense cell division with little to no increase in size of the mass of cells, i.e, the cells are smaller and smaller. As the cells divide, a ball of cells forms, which becomes the hollow ball of cells called the blastula.

4. Distinguish between the following:

zygote and blastula

____
____
____

blastula and blastocyst

____
____
____

incomplete and complete cleavage

____
____
____

5. On the blastula to the right, identify the ectoderm, endoderm, and mesoderm.

6. Identify the cells (using the letters) in the blastula to the right that will become the

a. nervous system
b. muscular system
c. digestive system
d. epidermis
e. kidneys and connective tissues
f. lungs

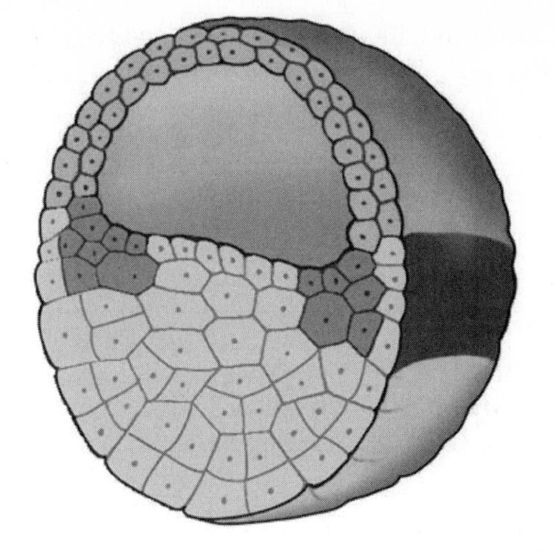

PRINCIPLES OF LIFE, Figure 33.5
© 2012 Sinauer Associates, Inc.

7. Explain why the earliest cells of the blastocyst are pluripotent and discuss their value in biomedical research.

____
____
____
____
____
____

Concept 33.3 explains how gastrulation forms three separate layers forming a body plan. The details of which cells and which layers form and fold is well-studied and complex, but are unlikely to be considered as memorized detail on the AP-Biology exam. During gastrulation, three germ layers or tissue layers form.

The ectoderm is the outer germ layer, formed from those cells remaining on the outside of the embryo. The ectoderm gives rise to the nervous system, including the eyes and ears, and forms the outer or epidermal layer of the skin. The innermost germ layer, the endoderm, forms from cells that migrate to the inside of the embryo during gastrulation. The endoderm produces the linings of internal spaces such as the digestive and respiratory tracts and the urinary bladder. It also contributes to the structure of some internal organs, including the endocrine glands, pancreas, and liver. The middle layer, the mesoderm, is made up of cells that migrate between the endoderm and the ectoderm. The mesoderm gives rise to the heart, blood vessels, muscles, bones, and several other organs.

8. Draw a simple diagram of a zygote, blastula, and gastrula of a sea urchin.

Zygote	Blastula	Gastrula

9. Much research has been done on the signaling proteins involved in development. To show that a protein is truly involved in a process it must be shown that the protein is necessary and sufficient. Explain.

10. While there are many differences between gastrulation of the urchin, frog, and chicken, there are also many similarities. Identify three broad similarities.

11. Discuss how the formation of mesoderm and endoderm differs between sea urchins and chickens. Discuss which is more similar to human tissue formation.

Concept 33.4 describes how the major organ systems form through inductive tissue interactions. Formation of the nervous system is used as an example. After the notochord has formed from mesodermal tissue, it plays a critical role in inducing neurulation, the formation of the neural tube. There are many details in this section, but memorizing details in this concept is not likely to be of great use on the AP-Biology exam

12. Explain how the nervous system arises from ectodermal tissue.

13. Your nerve cells are buried deep underneath muscle tissue, the latter being of endoderm origin. Explain how nerves are formed.

14. Segmentation can be easily seen in many animals such as the annelids (segmented worms) and arthropods (crabs and insects). Are humans segmented, and if so, how?

____

____

____

Concept 33.5 examines how the growing embryos of birds and mammals gain nutrition, with comparison of the membranes and exchange surfaces included.

15. In the diagram of a chick embryo below, label the 4 membranes (circled), and the embryo, allantois, amniotic cavity, and the gut.

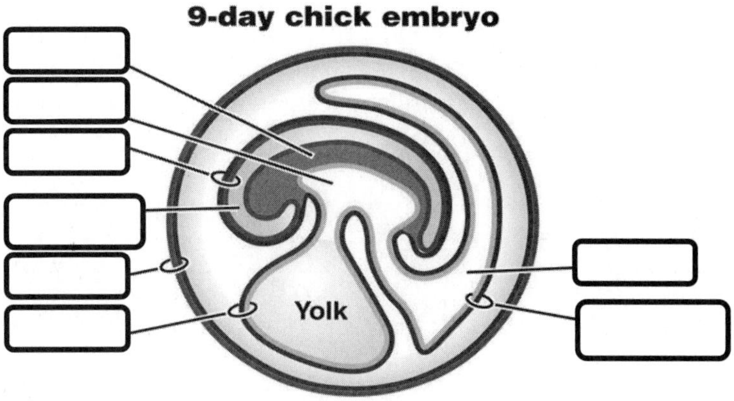

9-day chick embryo

Yolk

PRINCIPLES OF LIFE, Figure 33.15 (Part 2)

16. Describe the similarities in the mammalian placenta and the chicken egg.

____

____

____

____

____

____

____

____

____

Science Practices & Inquiry

In the AP Biology Curriculum Framework, there is a set of 7 Science Practices. In this chapter, we will focus on **Science Practice 1:** The student can use representations and models to communicate scientific phenomena and solve scientific problems. More specifically, Practice 1.3: The student can refine representations and models of natural or man-made phenomena and systems in the domain.

Question 16 asks you to refine representations to illustrate how interactions between external stimuli and gene expression result in specialization of cells, tissues and organs (LO 4.7).

16. Below is a set of early embryological drawings from several different organisms. A novice embryologist has placed them into an incorrect order. In the blanks below, place them into their correct order, and label the primary feature shown in each diagram.

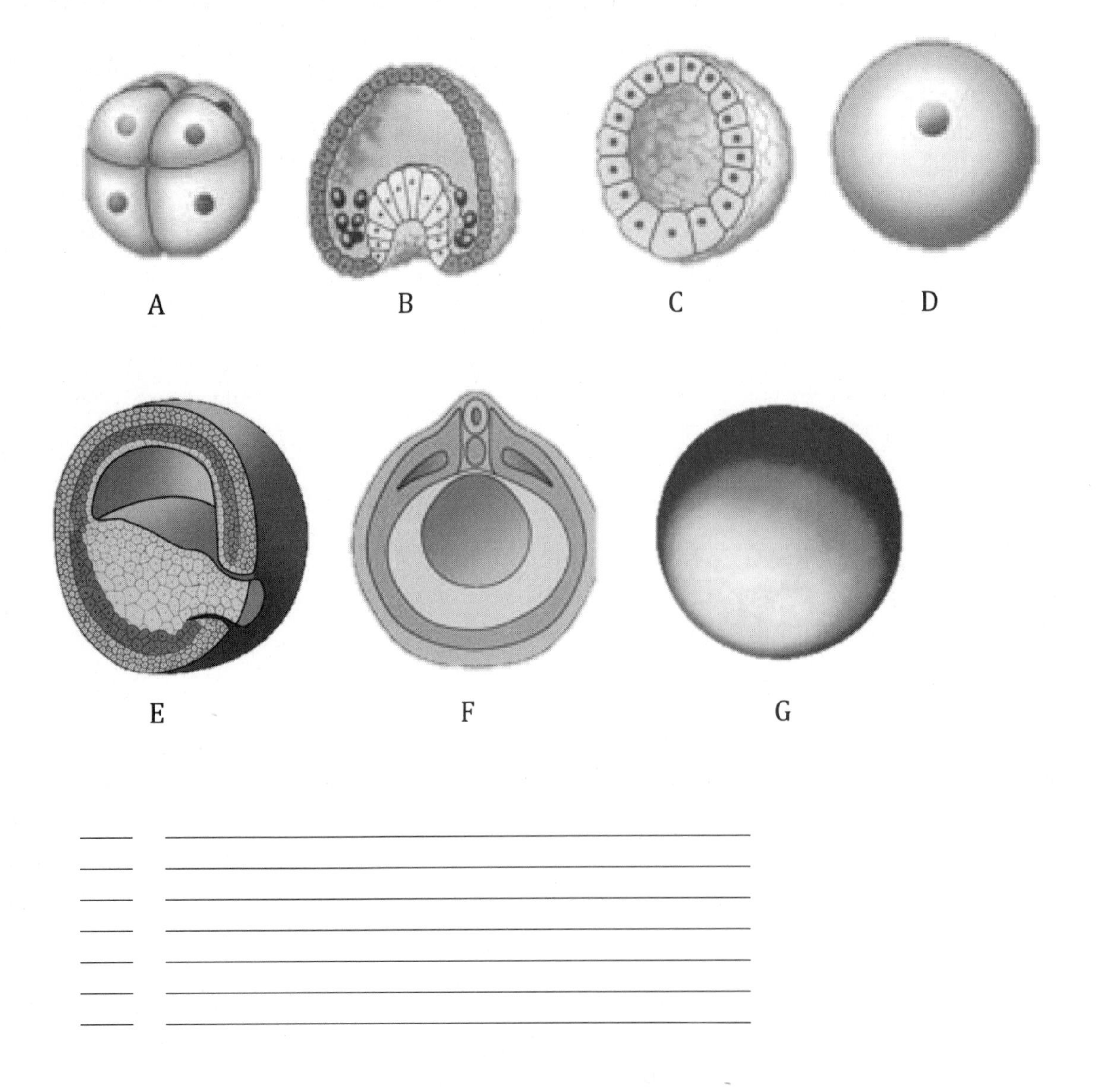

____ ____

____ ____

____ ____

____ ____

____ ____

____ ____

____ ____

Chapter 34: Neurons and Nervous Systems

Chapter Outline

34.1 - Nervous Systems Consist of Neurons and Glia
34.2 - Neurons Generate and Transmit Electrical Signals
34.3 - Neurons Communicate with Other Cells at Synapses
34.4 - The Vertebrate Nervous System Has Many Interacting Components
34.5 - Specific Brain Areas Underlie the Complex Abilities of Humans

Homeostasis, the dynamic maintenance of optimal conditions, requires sensors and effectors that regulate controlled variables, such as blood pressure. Chemical signaling via hormones is one major signaling route in homeostasis, and the other primary signaling route is the nervous system, the topic of Chapter 34.

While exploring the operations of the parts of the brain is a worthy and fun activity, the understanding of individual cells, the neurons, along with the glial cells that support neuronal metabolism are also described, aiding your comprehension of the whole system. You should find that much of this material is a review, and it is an application of material you have previously learned. You will once again see cellular organelles, membrane structure and function, the sodium-potassium pump, and membrane potentials serving important biological functions. This material is challenging to learn, but spending some time to meet this challenge will open many doors of biological understanding for you.

Chapter 34 includes components of Big Ideas Two, Three, and Four. The Big Ideas are a means for organizing biological knowledge. Refresh your familiarity of the **Big Ideas,** their **Enduring Understandings,** and **Essential Knowledge** (below).

Big Idea 2 states that the utilization of free energy and use of molecular building blocks are characteristic fundamental of life processes. Chapter 34 shows that neuronal:

2.b.1: Cell membranes are selectively permeable due to their structure.

Big Idea 3 states that living systems store, retrieve and transmit information essential to life processes. Specifically, Chapter 34 shows how neuronal communication is accomplished as:

3.d.2: Cells communicate with each other through direct contact with other cells or from a distance via chemical signaling.

3.e.2: Animals have nervous systems that detect external and internal signals, transmit and integrate information, and produce responses.

Big Idea 4 examines the idea that biological systems interact in complex ways. Implied in Chapter 34 is the concept that:

4.b.2: Cooperative interactions within organisms promote efficiency in the use of energy and matter.

Chapter Review

Concept 34.1 details the cellular components of the nervous system, starting with neurons. An individual neuron has two different functional regions. The highly branched, dendritic region is the information-receiving region, and it has trans-membrane proteins that serve as receptors to bind chemical signals called neurotransmitters. In response to a signal, the dendritic region of the cell undergoes a small change in membrane potential, called a graded potential. The second functional region of a neuron is called the axon or axonal region, and from its terminus are released neurotransmitter signals, in response to action potentials.

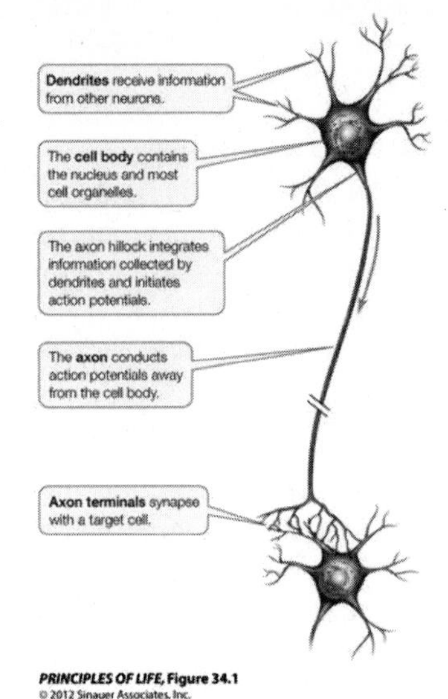

PRINCIPLES OF LIFE, **Figure 34.1**

Neurons are similar to one-way roads in that they only carry signals in one direction. Neurons that carry information into the central nervous system (spinal cord and brain) are called afferent neurons, as they "affect" what happens next in the organism. Neurons carrying commands out of the central nervous system are called efferent neurons, because they "effect" change by controlling effectors in the body. The neurons between afferent and efferent neurons are called interneurons.

1. Explain the difference between a nerve and a neuron.

____

____

____

____

2. Describe the anatomical differences between afferent and efferent neurons.

____

____

____

____

3. Describe the functional differences between afferent and efferent neurons. Give an example of the stimulus or organ controlled by each.

____

____

____

____

4. Explain why a neuron's axon hillock is regarded as the decision point for whether or not the neuron will undergo an action potential to communicate with its target.

____

____

____

____

5. Discuss the relative abundance of glia and neurons in the brain, and describe TWO functions of glia.

____

____

____

____

6. Explain why a peripheral neuron that is damaged by disease or trauma might be able to restore its function as long as its cell body remains intact.

____
____
____
____

7. Describe the blood-brain barrier and discuss its function.

____
____
____
____

Concept 34.2 reveals that neuronal function, whether it is as simple as, "Me want cookies!" or as sublime as "I finally understand the Nernst equation!" is a matter of understanding that membrane potential is all about ion movements across neuronal membrane altering the membrane potential. Millivolts (mV) describe the separation of charges from the cytosol (inside the neuron) to the extracellular fluid (outside the neuron).

The interior of a neuron, its cytosol, has many protein molecules, most of which are in the anionic (negatively charged) state, so the inside of the "resting" cell is electronegative relative to the outside of the cell, typically between -50 and -70mV. The "resting" membrane potential of a "quiet" neuron is anything but resting: it depends on a high rate of ATP hydrolysis, driving the sodium-potassium pump. (Na^+-K^+-ATPase. The unequal distribution of charge across the membrane means that it is polarized. Changes that reduce the charge difference across the membrane are called depolarizing changes, and changes that increase the charge difference are said to be hyperpolarizing changes.

The functions of neurons are based on ion movements, especially sodium and potassium ions, across the cell membrane through trans-membrane proteins called ion channels. The channels have two major characteristics: selectivity and gating. Selectivity means that a particular ion-channel protein, when it is open, has a single "best" ion that passes through it. For example, skeletal muscles move toward contraction by first opening sodium (ion) channels. Gating means that the opening or closing of the ion channel is like the operation of a gate that opens or closes in response to particular events. The sodium channels on skeletal muscles open in response to binding the excitatory neurotransmitter acetylcholine, following its release from motor neurons. Overall then, the sequence leading to skeletal-muscle contraction begins with the opening of "acetylcholine-gated sodium channels." Some of the other ion channels, including those involved in action potentials, are gated by changes in membrane voltage. Of the latter type, the voltage-gated sodium channels and the voltage-gated potassium channels are the ones most widely distributed.

Knowing that sodium ions are more abundant outside the neuron than inside, and that the inside of the cell has more negative charges than outside, one can predict that the opening of sodium channels will result in the inward movement of sodium ions, which are positively charged, thus depolarizing the neuron. Correspondingly, the opening of potassium channels will allow potassium ions, more abundant inside the cell than outside, to depart from the cell, thus hyperpolarizing the neuron. The movement of small amounts of these ions causes small changes in membrane potential, called graded potentials, such changes are most important in the dendritic region of the neuron. Graded potentials spread quickly to nearby areas but decay as they spread. Graded potentials are also summable.

Larger-scale ion movements, due to the operation of voltage-gated ion channels, are more important in the progression of changes in membrane potential called the action potential; such cycles of changes are more functionally important in the long, thin axonal region of neurons. Action potentials are "all-or-none" events, do not decay and are not summable.

8. Explain how neurons come to have more sodium ions outside the cell than in the cytosol, and more potassium ions inside the cell than in the extracellular fluid.

____

____

____

____

9. Explain what is meant by the term "threshold" in describing neuronal function.

____

____

____

____

10. The ions considered in studying excitable membranes are said to have an "electrochemical gradient." Discuss the meaning of "electro-" and "-chemical" gradients.

____

____

____

____

11. Explain why two or more graded potentials can be added together.

____

____

____

____

12. Explain why two or more action potentials cannot be added together.

____

____

____

____

13. Explain why the concept of "threshold does not apply to graded potentials.

____

____

____

____

14. Explain how the concept of "threshold" applies to action potentials.

____

____

____

____

15. Add the following letters to the graph, showing:

a. peak of increased sodium permeability.

b. threshold for an action potential.

c. resting membrane potential.

d. hyperpolarization.

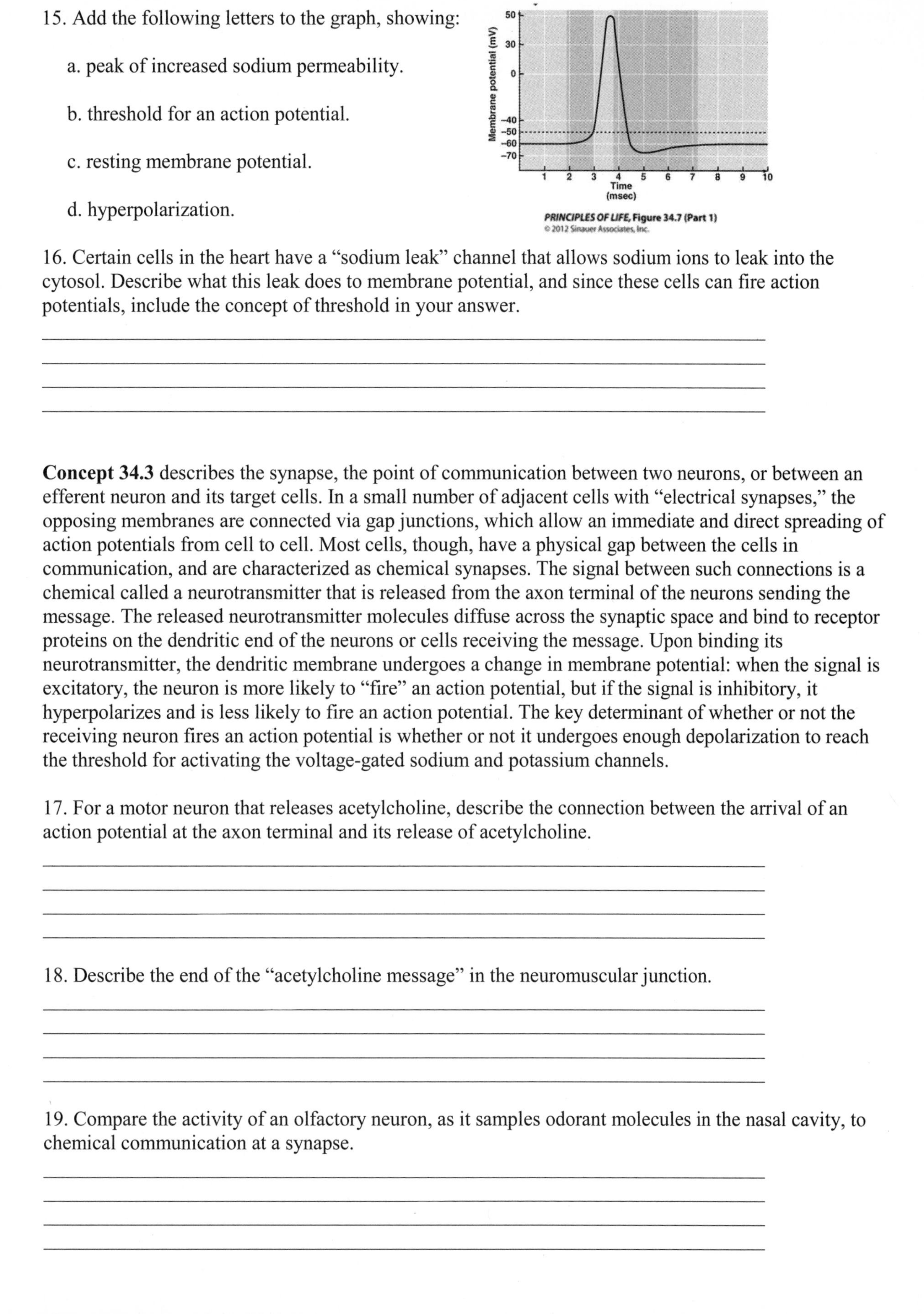

PRINCIPLES OF LIFE, **Figure 34.7 (Part 1)**

16. Certain cells in the heart have a "sodium leak" channel that allows sodium ions to leak into the cytosol. Describe what this leak does to membrane potential, and since these cells can fire action potentials, include the concept of threshold in your answer.

Concept 34.3 describes the synapse, the point of communication between two neurons, or between an efferent neuron and its target cells. In a small number of adjacent cells with "electrical synapses," the opposing membranes are connected via gap junctions, which allow an immediate and direct spreading of action potentials from cell to cell. Most cells, though, have a physical gap between the cells in communication, and are characterized as chemical synapses. The signal between such connections is a chemical called a neurotransmitter that is released from the axon terminal of the neurons sending the message. The released neurotransmitter molecules diffuse across the synaptic space and bind to receptor proteins on the dendritic end of the neurons or cells receiving the message. Upon binding its neurotransmitter, the dendritic membrane undergoes a change in membrane potential: when the signal is excitatory, the neuron is more likely to "fire" an action potential, but if the signal is inhibitory, it hyperpolarizes and is less likely to fire an action potential. The key determinant of whether or not the receiving neuron fires an action potential is whether or not it undergoes enough depolarization to reach the threshold for activating the voltage-gated sodium and potassium channels.

17. For a motor neuron that releases acetylcholine, describe the connection between the arrival of an action potential at the axon terminal and its release of acetylcholine.

18. Describe the end of the "acetylcholine message" in the neuromuscular junction.

19. Compare the activity of an olfactory neuron, as it samples odorant molecules in the nasal cavity, to chemical communication at a synapse.

20. Assuming that learning involves changes in the effectiveness of synaptic communication, explain why you might propose that learning corresponds more closely to changes in metabotropic receptors than ionotropic receptors. Define the activity of each in your answer.

Concept 34.4 samples the structural components of the vertebrate nervous system. The brain and the spinal cord make up the central nervous system. The spinal cord includes neuronal tracts to and from the brain, as well as synaptic contacts and interneurons required for reflexes, such as the "knee-jerk reflex," where sensory afferent inputs about muscle and tendon stretch modifies the amount of efferent motor outputs on associated muscles.

The autonomic nervous system controls most involuntary physiological functions in the body, and is subdivided into two contrasting sections: the sympathetic (exercise and emergency) and the parasympathetic (resting and fed) systems. In many cases, sympathetic and parasympathetic neurons influence the same organ, but in opposite directions. For example, heart rate is under autonomic influence, with sympathetic signals (epinephrine from the adrenal gland and norepinephrine from sympathetic terminals) accelerating heart rate, as appropriate during exercise or emergency, whereas parasympathetic signals (acetylcholine from parasympathetic neurons) slow your heart rate during other times.

21. Describe how the following changes in the activity of the autonomic nervous system affect heart rate. Include as many anatomical and neurochemical details as you can.

a. Increased activity in the sympathetic division.

b. Increased activity in the parasympathetic division.

22. Graphical "cartoon" representations of the motor and sensory processing areas of the brain show that particular brain regions have specific functions. Discuss how this discovery was important in moving forward the study of human consciousness.

Concept 34.5 examines several different regions and describes their functions in the CNS, demonstrating that the brain and spinal cord are organized in a predictable manner. This observation gives credence to the idea that human experience can be studied in terms of anatomy and physiology, but the huge diversity of human experiences makes consciousness impossible to fully specify in complete detail.

23. Discuss the general principles showing that production of words, as in speaking, uses a different part of the brain than does understanding and responding to words.

24. For the following, provide the requested information.

a. Add labels to the *y*-axis.

b. How many instances of being awake are shown?

c. How many REM episodes are shown?

d. How were these data obtained?

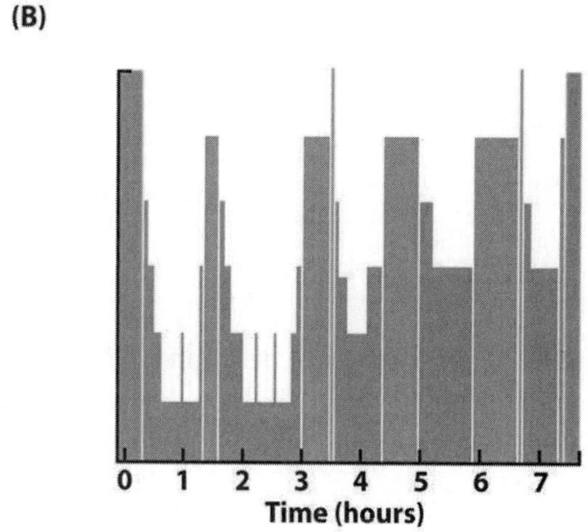

(B)

0 1 2 3 4 5 6 7
Time (hours)

PRINCIPLES OF LIFE, Figure 34.19 (Part 2)

Science Practices & Inquiry

In the AP Biology Curriculum Framework, there is a set of 7 Science Practices. In this chapter, we will focus on **Science Practice 1:** The student can use representations and models to communicate scientific phenomena and solve scientific problems. More specifically, practice 1.2: The student can use representations and models to analyze situations or solve problems qualitatively and quantitatively.

In answering Question 25, you evaluate a model of a synapse and describe how nervous systems transmit information. (LO 3.45)

25. Respond to each of the following prompts, using the following terms: graded potential, threshold, and action potential. Assume that the four axon terminals shown in the drawing represent excitatory synapses.

a. Explain how graph A represents spatial summation.

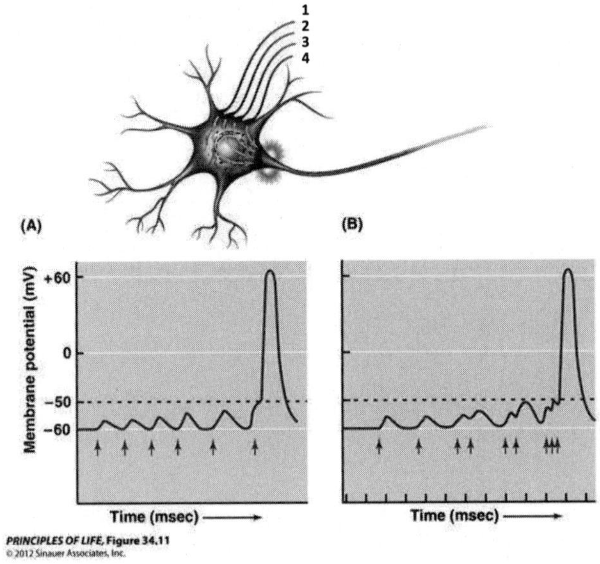
1
2
3
4
(A)
(B)
Membrane potential (mV)
+60
0
–50
–60
Time (msec)
Time (msec)
PRINCIPLES OF LIFE, Figure 34.11
© 2012 Sinauer Associates, Inc.

b. Explain how graph B represents temporal summation.

c. Add a fifth <u>inhibitory</u> axon terminal to the diagram, close to the axon hillock, and discuss the effect of its simultaneous activity with attainment of threshold.

Chapter 35: Sensors

Chapter Outline

35.1 - Sensory Systems Convert Stimuli into Action Potentials
35.2 - Chemoreceptors Detect Specific Molecules or Ions
35.3 - Mechanoreceptors Detect Physical Forces
35.4 - Photoreceptors Detect Light

Chapter 35 explores the sensory abilities of animals. In the process of sensory transduction, sense organs convert environmental stimuli to changes in the membrane potentials of neurons, leading to information processing in the nervous system. In other words, information about the outside world and the internal milieu is converted to electrochemical signals in cells that can lead to changes in an organism's behavior and physiology. The neural signals from different sensory systems arrive at, and are processed in, different locations in the brain.

Sensory organs such as the ears, eyes, tongue, nose allow the sense cells to collect, filter, and magnify the many stimuli encountered from moment to moment. In addition to sensing the outside world, sensory cells monitor the internal environment in key locations, including oxygen concentration in the blood, body temperature, etc. The three main types of sensory cells addressed in detail are chemoreceptors, mechanoreceptors, and photoreceptors.

Knowing the specific details of neuroanatomical pathways is beyond the scope of the AP-Biology curriculum. Organize your studies on the many similarities of sensory receptors in detecting stimuli via changes in membrane potentials and then transmitting information with action potentials and neurotransmitters.

The information in Chapter 35 is especially relevant for Big Ideas Two and Three. Continue to reflect upon these **Big Ideas** across their many **Enduring Understandings**.

Big Idea 2 states that the utilization of free energy and use of molecular building blocks are characteristic fundamental of life processes. Specifically, Chapter 35 includes:

2.c.2: Organisms respond to changes in their external environments.

Big Idea 3 states that living systems store, retrieve and transmit information essential to life processes. Specifically, Chapter 35 explains how animals use :

3.e.2: Animals have nervous systems that detect external and internal signals, transmit and integrate information, and produce responses.

Chapter Review

Concept 35.1 explains how the arrival of stimuli at sensory cells leads to changes in their membrane potentials, a process called sensory transduction. As in Chapter 34, changes in membrane potential are due to changes in the movement of ions across the membranes of the sensory cells/neurons, resulting in graded potentials. To reiterate the focus on sensory reception, these graded potentials are usually called receptor potentials. Depending on the sensory modality involved, the receptor potentials can lead to action potentials, e.g., olfaction, or changes in neurotransmitter release, e.g., vision and audition. In both cases, the next neuron in the pathway is altered in its activity, and so on along the pathway to the central nervous system, leading to organismal responses and contributing to homeostatic regulation.

1. The sensation of warmth is due to changes in membrane potential that occur in response to "temperature-gated ion channels." Discuss the phrase in quotes.

____

____

____

____

2. The sensation of touch is due to changes in membrane potential that occur in response to "mechanically-gated ion channels." Discuss the phrase in quotes.

____

____

____

____

3. We appear to be able to "ignore" some stimuli, e.g., the pressure of a seat on one's buttocks while seated, but not others, e.g., holding some ice. Describe these phenomena, at the level of sensory transduction and central neural processing, and comment on the evolution of sensory adaptation, based on stimulus quality.

____

____

____

____

4. Explain how the brain's anatomical organization allows an organism to respond appropriately to touching a warm surface *versus* a rough surface such as sand paper.

____

____

____

____

Concept 35.2 details how environmental chemicals, including odors, pheromones, and tastes can initiate chemosensory transduction. Animals can respond to thousands of different chemicals (odorants and pheromones) in unique and appropriate fashion, suggesting a high level of selectivity at the level of the more than 1,000 receptor proteins (products of >1,000 genes) and subsequent neural processing. Olfactory and pheromonal perceptions begin with the arrival of a chemical in the nasal cavity and the binding of that chemical to its receptor proteins. Tasting begins with the appearance of chemicals on the tongue. The five basic tastes are sweet, salty, sour, bitter, and *umami* (a savory or meaty taste). With the exception of salty taste, tastant molecules bind to receptor proteins to initiate chemosensory transduction.

5. An individual olfactory neuron in a dog's nasal cavity will likely express only one of more than 1,000 odorant-receptor genes, yet it seems dogs are capable of distinguish tens of thousands of different odorants. Discuss some possible mechanisms by which this diverse sampling capability might occur.

____

____

____

____

6. "The world is one gigantic synapse," wrote one olfactory biologist. Describe the meaning of this statement using principles of synaptic communication (Chapter 34).

____

____

____

____

7. Describe four of the five taste modalities, and discuss how receptor proteins might be involved in each category.

8. Inside blood vessels, chemoreceptors sensitive to glucose are the sensors for glucose homeostasis in the body. Describe the first chemosensory pathway activated upon the ingestion of environmental glucose and explain whether or not the internal chemoreceptors might work in the same manner.

9. A segment of the digestive system modifies secretion of bicarbonate ions depending on the pH of the fluids passing through. Describe the possible chemosensory transduction involved, stating whether it is more like olfaction or gustation.

Concept 35.3 examines how mechanoreceptor neurons respond to stretching and deformation of their membranes by altering trans-membrane ion traffic, thus generating receptor potentials and initiating transduction. Among their many roles, some mechanoreceptor mechanisms are part of a system that adjusts the strength of muscle contraction so that it can more accurately match load requirements. Mechanoreceptor mechanisms also underlie hearing (audition) and vestibular functions (balance and eye movements).

10. Describe the anatomical and functional relationships between the tympanic membrane and the incus, malleus and stapes.

12. Explain how loud noises can temporarily AND permanently impair hearing.

13. "Ringing" noises in the ears are not uncommon after a blow to the head, even when it is quiet. Explain how physical jostling of the hearing apparatus can lead to the perception of sound that does not really exist.

14. As you walk down the street with your head frequently changing its physical position, you are able to keep your eyes fixed on a target ahead of you. Describe the system that helps to guide the contraction of eye muscles to achieve this result.

____

____

____

____

15. As you fill a pail with water, you can manage to hold it steady under the faucet even as the mass of the filling pail becomes greater and greater. Describe the sensory feedback system that adjusts the strength of contraction to match the load of the filling bucket.

____

____

____

____

Concept 35.4 examines vision, starting with photosensory transduction. The transducing cells, found in the retina, are the rods and cones, the latter for color vision. In the case of rods, the absorption of photons of light by pigments, e.g., rhodopsin, is linked to changes in the membrane potential of the photoreceptor cell. In turn, a dreaded depolarization of the rod changes its secretion of neurotransmitters, altering the activity of the neural pathway leading to visual perception.

16. Describe how the exposure of a rod to a brief flash of light affects its membrane potential and release of neurotransmitter.

____

____

____

____

17. Explain how and why colors are vivid in daylight and dull and grayish at night.

____

____

____

____

18. Describe what you would expect to see in examining the relative populations of rods and cones in nocturnal owls compared to birds that are diurnal in behavioral activity.

____

____

____

19. Macular degeneration is a disease that is accompanied by proliferation of blood vessels in the eyes. Describe the cells in the visual pathway that would most immediately affected by this disease.

____

____

____

____

____

Science Practices & Inquiry

In the AP Biology Curriculum Framework, there is a set of 7 Science Practices. In this chapter, we will focus on **Science Practice 6:** The student can work with scientific explanations and theories. More specifically, practice 6.2: The student can construct explanations of phenomena based on evidence produced through scientific practices.

Question 20 asks you to analyze a hearing problem experienced by some patients every year. In this case it is hearing loss caused by a claustiatoma tumor in the middle ear. From a science practice perspective, you need to be able to construct an explanation, based on scientific theories and models, about how nervous systems detect external and internal signals, transmit and integrate information, and produce a response (LO 3.43).

20. A tumor in the middle ear damaged the bones in a patient's middle ear on the left side of her head, impairing her sense of hearing. After the tumor was removed, the audiology doctor tested the patient's ability to hear by placing a tuning fork near the left pinna, and then again while touching vibrating tuning fork to the patient's left forehead. The patient could hear the tuning fork better when it was touching the left forehead. When the unaffected right ear was tested, the tuning fork was heard more clearly when it was placed near the right pinna than when it was placed on the right forehead. Refer to the diagram to the right in explaining these observations.

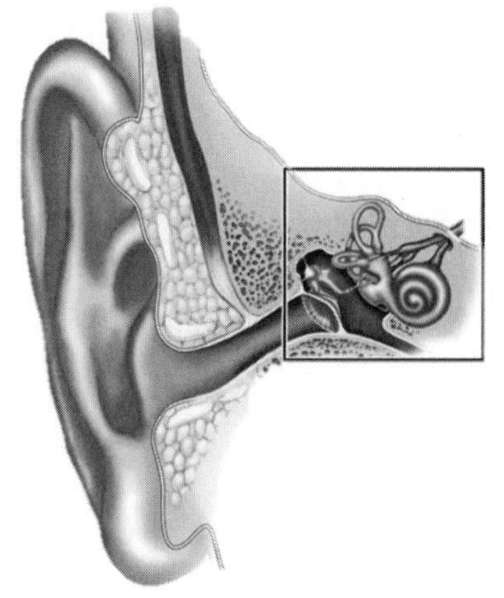

Chapter 36: Musculoskeletal Systems

Chapter Outline

36.1 - Cycles of Protein-Protein Interactions Cause Muscles to Contract
36.2 - The Characteristics of Muscle Cells Determine Muscle Performance
36.3 - Muscles Pull on Skeletal Elements to Generate Force and Cause Movement

Understanding skeletal muscle contraction has long been a thrill for humankind, as this knowledge might provide relief to those who are paralyzed, and it could enhance human performance to new heights. Understanding smooth muscle and cardiac muscle contraction are also of considerable interest, as these underlie homeostasis of blood pressure, impacting all physiological systems. Chapter 36 provides the opportunity for you to synthesize your knowledge of excitable cells, the regulation of muscle contraction, and some very basic physics principles of movement.

Chapter 36 includes **Big Ideas 3 and 4**. The Big Ideas are the organizing principles of the AP Biology Framework. You should often review the Big Ideas, the Enduring Understandings and the Essential Knowledge points as you prepare for the AP-Biology exam. Note that the exam will not expect you to have memorized many details specific to muscle contraction, but this chapter allows you to see how scientific knowledge is based on reliable theories and principles.

Big Idea 3 centers on the ways that information is coded for living organisms. In the case of musculoskeletal systems, you will see that:

3.e.2: Animals have nervous systems that detect external and internal signals, transmit and integrate information, and produce responses.

Big Idea 4 examines the complex interactions with a living organism and between organisms. In Chapter 36, we see that:

4.a.4: Organisms exhibit complex properties due to interactions between their constituent parts, and that

4.b.2: Cooperative interactions within organisms promote efficiency in the use of energy and matter:

Chapter Review

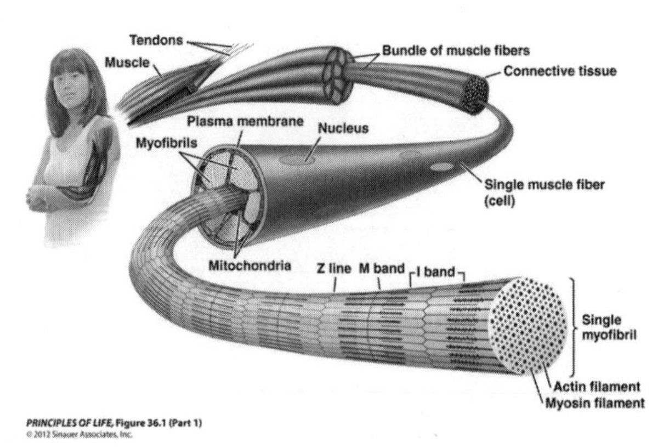

PRINCIPLES OF LIFE, Figure 36.1 (Part 1)

Concept 36.1 provides many anatomical details of skeletal muscle organization, far more than you are expected to memorize. Skeletal muscles, like the ones you are using right now to hold your skeleton in an appropriate posture, are built of bundles of muscle fibers. Anatomically, these fibers are groups of cells with membranes that merged during fetal development to function together, and as such, they are not replaceable during adulthood. Each muscle fiber has many myofibrils, which viewed more closely, at the level of the sarcomere, contain the actin and myosin filaments whose interactions are responsible for the shortening of the muscle during contraction. Both types of filaments are long, repeating polymers of actin and myosin proteins. The myosin molecules have a hinged, cross-bridge section that can bind to actin, and then flex, pulling the actin, causing it to slide. The cross-bridge alternates binding and pulling, using ATP as it does so, in a type of ratcheting pattern to cause actin to slide.

The junction of the nervous system and the muscle fibers is a synapse called the neuromuscular junction (NMJ). Neuronal activity releases acetylcholine into the NMJ, activating a sequence of changes in ion permeability that leads to an action potential on the muscle fibers. The action potential causes a brief "spike" of the concentration of calcium ions inside the muscle cell. Calcium ions bind to troponin, a protein associated with actin, changing its shape. Calcium ions bind to proteins to activate the "sliding" of actin filaments, shortening the muscle fiber. Relaxation occurs quickly, as the active transport of calcium ions out of the cytosol and back into the smooth endoplasmic reticulum (in muscles this is called the sarcoplasmic reticulum) terminates contraction. Contraction in smooth muscles and cardiac (heart) muscle is also regulated by calcium ions in the cytosol.

1. Describe the three proteins that can associate with actin in skeletal muscle, and how each is altered by a contraction sequence.

____
____
____
____

2. Describe the two binding sites on the myosin cross-bridge.

____
____
____
____

3. Explain how *rigor mortis* occurs. Include in your answer a discussion of how actin and myosin dissociate after muscle contraction.

____
____
____
____

4. Describe how action potentials spread from one cardiac muscle cell to the next.

____
____
____
____

5. Discuss "excitation-contraction coupling" relative to muscle contraction.

____
____
____
____

6. Describe and compare the proteins in skeletal and smooth muscle that can bind calcium ions.

____
____
____
____

Concept 36.2 explores graded force production by muscles. Holding a pencil steady in your hand requires less force than does holding your text book, and you understand that greater numbers of skeletal muscle fibers are undergoing excitation and contraction in the latter case. At a certain load, all of the muscle fibers are undergoing excitation and contraction, and the maximum tension possible is produced: this is called tetanus.

There are different types of muscle that vary in response to their loads and energy metabolism. Postural muscles that hold you upright are active most of the time, and hydrolyze lots of ATP in accomplishing their work. This "aerobic" type of muscle is resistant to fatigue and has lots of blood supply and mitochondria for ATP production. In contrast, the bulging muscles of a weightlifter can rapidly generate a great deal of tension, but such muscles are more prone to fatigue as energy sources can be rapidly depleted in these muscles.

7. Describe the specific meaning of "twitch," the minimum unit of contraction in skeletal muscles.

8. In the three curves on the accompanying graph, you can see that curve 1 is labeled.

Add the labels "glycolytic" and "aerobic" to the other two curves and explain what the curves tell us about ATP supplies during a 5-minute sustained period of muscle activity.

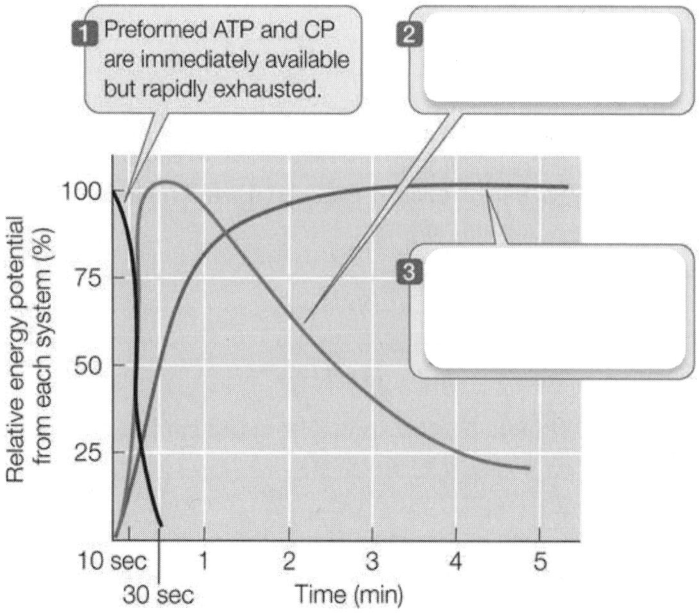

1 Preformed ATP and CP are immediately available but rapidly exhausted.

2

3

Relative energy potential from each system (%)

100, 75, 50, 25

10 sec, 30 sec, 1, 2, 3, 4, 5

Time (min)

PRINCIPLES OF LIFE, Figure 36.11

Concept 36.3 describes the parts of the body that are moved as a result of muscle contraction. For vertebrate animals, it is usually the bony skeleton that moves, but in other animals, for example earthworms, there is a "hydrostatic skeleton" that is a fluid-filled cavity that changes it shape as fluids are moved by muscle contraction.

9. Explain what is meant by the observation that our muscles are arranged in antagonistic pairs.

10. Explain whether this statement is true or false: "While resting in his seat, Lenny's skeletal muscles were completely relaxed."

11. Describe both points of attachment of skeletal muscles to bones.

12. Describe the attachments between certain bones and other bones for cases where there is not a muscle present. [Hint: patella, kneecap]

13. Compare the functions of osteoclasts (catalyze calcium release) and osteoblasts (build bone).

Science Practices & Inquiry

In the AP Biology Curriculum Framework, there is a set of 7 Science Practices. In this chapter, we will focus on **Science Practice 1:** The student can use representations and models to communicate scientific phenomena and solve scientific problems. More specifically, practice 1.3: The student can refine representations and models of natural or man-made phenomena and systems in the domain.

In answering Question 14, you will evaluate the effects of muscle stretch on tension development in muscle and then extrapolate this information to another type of muscle during which will need to refine representations and models to illustrate biocomplexity due to interactions of the constituent parts (LO 4.10).

14. These data are presented in the "Apply The Concept" Box in Chapter 36, p. 716. The normal "resting length" of the relaxed skeletal muscle in its appropriate anatomical position was measured, and then stimulated with an electrode to contract, while measuring force generation with electronic detectors. The same measurements were made after the muscle was rearranged to varying lengths, presented here as percentages of resting length, while measuring force, as before.

LENGTH OF FIBERS AT BEGINNING OF CONTRACTION AS A PERCENTAGE OF RESTING LENGTH	FORCE GENERATED AS A PERCENTAGE OF MAXIMUM FORCE DURING TETANUS
35%	10%
75%	50%
100%	100%
115%	100%
130%	80%
150%	50%
175%	10%

A. Graph the data, placing "length" on the *x*-axis and "force" on the *y*-axis, and connect the points with a curve. Label all parts of your graph.

B. Write a caption that explains the meaning of the graph.

C. The picture on the right shows two possible arrangements of actin and myosin. Assume that one represents muscle that is stretched beyond its resting length and the other is at its resting length. Draw, on your graph, similar representations showing actin and myosin overlap at THREE data points (left of middle, middle and right of middle.

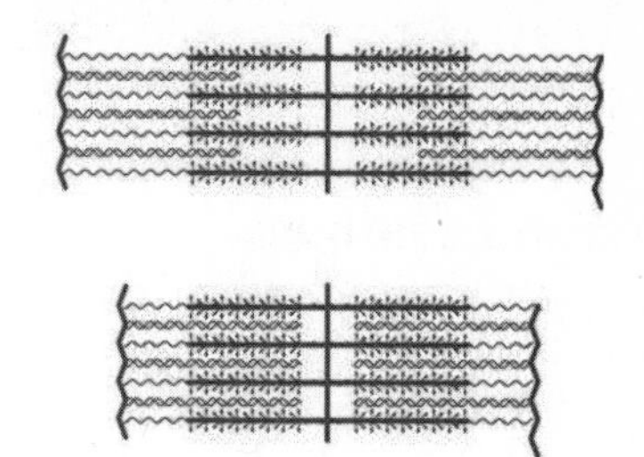

D. Assuming that smooth muscle can generate as much force as skeletal muscle, but that it can do so over a much wider range of stretch than can skeletal muscle, add a second curve to the graph to represent your estimate of what the data for smooth muscle would look like if measured similarly. Write a caption below summarizing this curve.

____
____
____
____
____
____

Chapter 37: Gas Exchange In Animals

Chapter Outline

Chapter 37 considers gas exchange in the context of homeostasis, as oxygen delivery is required for aerobic respiration, and carbon dioxide is a byproduct of glucose catabolism. The exchange of oxygen (O_2) and carbon dioxide (CO_2) between the organism and its environment occurs via simple diffusion, as these gases move across cell membranes in the directions predicted by their concentration gradients. Several other factors determine how fast small nonpolar molecules like O_2 and CO_2 diffuse, including temperature, the partial pressure of a gas, and the distance involved, but the concentration gradient is the most relevant and most variable determinant of the rate and direction of gas diffusion.

As water is the most abundant molecule in animals, we must consider gas movements in the context of the gases being dissolved in water. Although our atmosphere is slightly more than 20% oxygen (> 200,000 ppm; parts per million), in seawater, oxygen has a maximum concentration of only 20 ppm. The primary reason that cells are small and that many animals are small and thin is that O_2 is of low solubility in water, and once dissolved, it diffuses very slowly. On the time scale that oxygen is used in animals, each cell must be within 100 μm of an oxygen supply. Thus, there is a physiological limit on cell size and strong evolutionary pressure for having an effective circulatory system for oxygen delivery. In contrast, CO_2 is readily soluble in water, its concentration gradient is of great magnitude, so preventing excess CO_2 buildup is easier than acquiring oxygen.

Some organisms utilize direct exchange of gases between the cells and the nearby environment, e.g., jellyfish, while others, larger and more complex, e.g., fish, need both respiratory and circulatory systems. The respiratory organs of organisms have evolved to maximize gas diffusion, not only between the organism and its environment, but also between the circulatory system and the cells of the organism. Gas diffusion is maximized by large surface areas, short distances, and large concentration gradients.

Mechanisms that maximize gas diffusion include frequent ventilation (breathing), keeping the blood underneath the respiratory tissue moving, and counter-current circulatory exchange systems. Most vertebrate lungs are tidal in nature, i.e., air moves back and forth in the same stet of tubes, but birds maximize gas exchange using unidirectional air flow in their lungs. For organism with hemoglobin in their red blood cells, this protein takes O_2 out of the plasma, also maximizing the concentration gradient.

As with many systems, homeostatic control of gas exchange is controlled by negative feedback. In mammals, the medulla oblongata in the brainstem senses pH levels. As the level of CO_2 increases in the blood, the concentration of hydrogen ions also increases; the resulting decrease in pH is a potent chemical stimulus in a neural reflex that causes vertebrate animals to breathe faster. Should breathing is too fast, too much CO_2 is lost, and blood's pH increases. Chemoreceptors on the aorta and other large blood vessels monitor pH and oxygen levels in the blood. The affinity for hemoglobin to bind O_2 is reduced at increased temperatures and reduced pH, so these local factors near active and hot, lactic-acid producing tissues facilitate O_2 delivery. Hemoglobin is under allosteric control, such that increased acidity and higher temperatures reduce its affinity for O_2; this facilitates "unloading" of O_2 at more active, warmer, acidic tissues.

Chapter 37 spans Big Ideas 2 & 4. The Big Ideas are a means for organizing the vast amount of information we know as biology. It is vitally important that you continually work to understand these **Big Ideas** and across different **Enduring Understandings**. Chapter 37's emphasis is primarily in Big Idea 2, but also includes some of Big Idea 4 as well focusing on the nature of interrelated systems within organisms.

Big Idea 2 states that the utilization of free energy and use of molecular building blocks are characteristic fundamental of life processes. Specifically, Chapter 37 includes:

2.a.3: Organisms must exchange matter with the environment to grow, reproduce, and maintain organization.

2.d.1: All biological systems from cells and organisms to populations, communities, and ecosystems are affected by complex biotic and abiotic interactions involving exchange of matter and free energy.

2.d.2: Homeostatic mechanisms reflect both common ancestry and divergence due to adaptation in different environments.

Big Idea 4 examines the idea that biological systems interact in complex ways. Included in Chapter 37:

4.b.2: Cooperative interactions within organisms promote efficiency in the use of energy and matter.

Chapter Review

Concept 37.1 describes the physical processes governing diffusion including partial pressures, temperature, and distance to diffuse. Air has a much higher concentration of oxygen than water, as much as a factor of 10, thus making access to O_2 in water a greater challenge that extracting it from the atmosphere.

1. Fisk's law of diffusion is not an equation that is likely to appear on the AP exam, but it does serve as a nice tool to review the variables that influence the rate of gas exchange between two locations. The equation for rate of exchange (Q) is:

$$Q = DA\frac{P_1 - P_2}{L}$$

Explain how each component of this equation effects the diffusion of gases:

a. D (diffusion coefficient): ______________________________

____

b. A (cross sectional area across which the gas is diffusing): ______________

____

c. P_1 and P_2 (partial pressures of the gases at the two locations): __________

____

d. L (distance between the two locations): ______________________________

____

2. Another way of looking at the above equation is to say that $\frac{P_1 - P_2}{L}$ is a partial pressure gradient. Explain why using air rather than water is more effective for respiration.

____

____

____

____

3. Discuss which variable in the Fick's equation is most likely to vary in your lungs within the time span of a single hour, going from rest to exercise. Explain your answer.

4. Describe TWO characteristics of molecular oxygen that allow it to readily diffuse across cellular and mitochondrial membranes.

5. Assume that you are an oxygen-starved fish with a choice between cold water and warm water. Explain which temperature of water would benefit you the most.

6. Explain the effect of temperature on the metabolic rate of ectotherms, e.g., fish.

7. Based on your answers to the two questions above, describe the effect on fish of a power plant releasing warm water into a large lake.

Concept 37.2 examines the different adaptations that have evolved for respiratory exchange. These systems' primary function is to maximize the partial pressure gradients. Examples of different systems include external gills, internal gills, lungs, and tracheae. Additionally, countercurrent exchange systems, with the exchange media moving past each other in opposite directions, allow increased efficiency for the diffusion of gases across membranes.

8. Discuss why fish utilize a countercurrent exchange system in their gills to acquire dissolved oxygen from the environment.

9. Describe how the structure of lungs in a terrestrial animal maximizes the ability to acquire oxygen from the environment.

10. Countercurrent flow aids gas exchange and temperature regulation. Explain how countercurrent flow in a fish's gills is similar to what you would expect to find in the leg circulation of a penguin.

____

____

____

____

11. Birds have been sighted flying above Mt. Everest in the Himalayan Mountains, well above the altitude where humans can breathe unaided. Describe an anatomical feature of bird lungs that allows such high altitude gas exchange.

____

____

____

____

Concept 37.3 discusses the anatomy and function of mammalian lungs. While it is not one of the major required systems in the new AP® biology curriculum, it is often used as an illustrative example for obtaining nutrients and eliminating wastes.

12. Explain the forces that pull air into the human lungs.

____

____

____

____

13. Explain the forces that move air out of the human lungs.

____

____

____

____

14. Describe the path of an O_2 molecule as it travels from the atmosphere to the mitochondrion inside a muscle cell of a human. Include in your answer an account of each membrane the O_2 molecule passes through, and describe what happens to the atoms of oxygen that are "used" in the mitochondrion.

____

____

____

____

____

____

Concept 37.4 examines the regulation of breathing in mammals. As in all homeostatic systems, your goal should include knowing the controlled variable, the sensors, the signaling mechanisms and the effectors.

15. An untreated diabetic who lacks insulin will metabolize fats for energy and produce an excess of acidic end products, lowering the blood pH. How would you expect this condition to affect the individual's breathing?

____

____

____

____

16. Hyperventilating, or rapid deep breathing, reduces the amount of CO_2 in your lungs. Explain why it is dangerous to hyperventilate before trying to swim underwater.

Concept 37.5 describes how gases are transported by hemoglobin in the blood. The Bohr effect states that hemoglobin, an allosteric protein, reduces its affinity for O_2 when pH decreases. Carbon dioxide is very soluble in blood, and readily forms carbonic acid, a weak acid. Carbonic acid readily disassociates to H^+ ions and bicarbonate ions, the latter serving as a buffer against acidity in the blood.

17. Hemoglobin is described as a protein with four subunits. Identify how this protein shows the 4 levels of protein structure.

a.

b.

c.

d.

18. Explain why CO_2 is more soluble in blood plasma than is O_2. Write the equilibrium equation for carbonic acid as part of your answer.

19. Myoglobin has a higher affinity for O_2 than hemoglobin does, so it picks up and holds O_2 as an oxygen reserve in muscle tissue. Human fetal hemoglobin has a higher affinity for O_2 than does maternal hemoglobin, facilitating O_2 transfer in the placenta. Knowing this, identify which lines in the diagram to the right are

a. human hemoglobin
b. fetal hemoglobin
c. human maternal hemoglobin
d. myoglobin

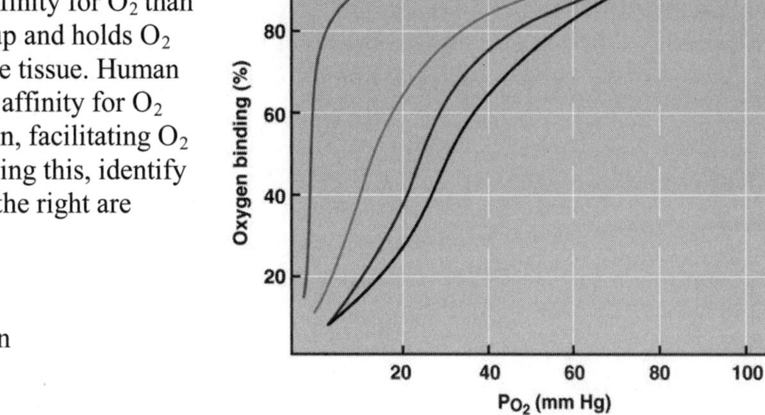

20. Explain the effect of high metabolic activity, e.g., exercising, on pH of the blood. Compare the pH in the lungs to that of exercising muscle tissue, explaining how the altered pH affects the "loading and unloading" of oxygen on hemoglobin molecules.

____

____

____

____

____

____

____

____

Science Practices & Inquiry

In the AP Biology Curriculum Framework, there is a set of 7 Science Practices. In this chapter, we will focus on **Science Practice 2: The student can use mathematics appropriately.** More specifically, practice 2.2: The student can *apply mathematical routines* to quantities that describe natural phenomena.

Question 21 asks you to use calculated surface area-to-volume ratios to predict which cell(s) might eliminate wastes or procure nutrients faster by diffusion. (LO 2.6)

Cylinder equations where h= height or length of the cylinder:
Volume = $\pi r^2 h$
SA = $2\ \pi rh + 2\ \pi r^2$

21. Calculate the surface area to volume ratio for each of the two cells below:

(a) A cylindrical shaped neuron which is 10 microns in diameter and a meter long.

(b) A typical cubic cell measuring 20 microns on a side.

(c) Neurons need to be able to conduct impulses over long distances. Explain why they are shaped the way they are.

Chapter 38: Circulatory Systems

Chapter Outline

Diffusion is so slow that if you filled your stomach with glucose, then it would take years, perhaps decades, for glucose to reach every place in your body if diffusion from the stomach was the only route of delivery. The main benefit of having a circulatory system is that it speeds delivery of diffusible materials within 100 μm of nearly every cell in the body; over that short distance, diffusion of glucose is at equilibrium in less than 5 seconds.

At first blush, building a circulatory system is a straightforward problem. You need plumbing (blood vessels) and a pump (heart). You connect them together, fill the system with blood, turn on the pump, and watch the blood circulate. In reality, a great number of subtle factors, like the size of the plumbing, the vigor of the pump, matching of flow in cases where there are two circuits, present interesting and understandable challenges. Given the essential importance of good circulation, evolutionary adaptations have shaped this system to a very high level of performance.

Chapter 38 includes **Big Ideas 1 and 2**. The Big Ideas are the organizing principles of the AP-Biology Framework. Review the Big Ideas, the Enduring Understandings and the Essential Knowledge points as you prepare for the AP Biology exam.

Big Idea 1 is that evolution binds together all living organisms and is the major component in the framework of biological knowledge. Considering circulatory systems, you will see:

- 1.b.1: Organisms share many conserved core processes and features that evolved and are widely distributed among organisms today.

Big Idea 2 establishes the premise that molecular building blocks explain the systems and functions of living organisms. Chapter 38 shows how circulation operates as:

- 2.a.3: Organisms must exchange matter with the environment to grow, reproduce and maintain organization.
- 2.c.1: Organisms use feedback mechanisms to maintain their internal environments and respond to external environmental changes.
- 2.d.2: Homeostatic mechanisms reflect both common ancestry and divergence due to adaptation in different environments.
- 2.d.3: Biological systems are affected by disruptions to their dynamic homeostasis.

Chapter Review

Concept 38.1 introduces transport of "good stuff" (nutrients, etc.) and "bad stuff" (wastes, etc.) as the immense benefit provided by the development of a circulatory system. The general idea is that a fluid containing solutes needed for cellular function is delivered to active cells, and the fluid picks up waste products as it passes near the active cells. Other sites in the body, e.g., lungs, permit for exchange between the fluid and the organism's external environment. Distinction is made between open systems, where the exchange fluid leaves blood vessels during its journey, and closed systems, in which the exchange medium is always contained within blood vessels. The fluid in the closed vessels moves more slowly through the tiniest vessels (capillaries), allowing adequate time for exchange to occur.

1. Identify the following parts of the circulatory system in the earthworm picture and describe the function of each.

a. ______________________________

b. ______________________________

2. Is the circulatory system shown in the question above is an "open" or a "closed" circulatory system? Explain why.

3. Explain which type of circulatory system, "open" or "closed," supports a higher rate of metabolism.

4. Arrange the terms below in the correct anatomical sequence as blood flows from the heart to the body and returns; the starting place (1) is completed for you.

_____ capillaries
__1__ heart
_____ arterioles
_____ venules
_____ arteries
_____ veins

Concept 38.2 delves deeper into specialized functions in circulation, leading to a description of a "dual" circulation that includes a circuit for environmental exchange with the organism, and a second circuit that delivers blood through the rest of the body for gas and nutrient exchange at the active tissues.

Among vertebrates, the simplest circulatory pump is the two-chambered heart of fish. In this arrangement, all of the blood pumped out of the heart is delivered to the gills to "load" it with oxygen and "unload" its carbon dioxide. The blood vessels in the gills are large enough that there is adequate hydrostatic pressure from the heart's contraction cycles to deliver the blood to the systemic circuit after passage through the gill circulation. Lungfish have a slighter different plan, as the development of a small lung to enhance gas exchange has taken place. Associated with the lung is a shunt circulation that moves the circulating blood away from the gills and to the lung. The lungfish's heart and the blood vessels associated with the lung have also undergone evolutionary changes, with a septum separating the atrium so that one side receives blood from the lung and the other side receives the venous return from the body.

Reptiles and adult amphibians have a three-chambered heart, along with two distinct circuits, pulmonary and systemic, which are kept relatively separated by the development of a partial septum in the ventricles. There is some mixing of relatively deoxygenated and fully oxygenated blood, however, and the system does not support a very high metabolic rate.

Birds and mammals, along with crocodilians, have the four-chambered hearts you are probably most familiar with, along with fully distinct pulmonary and systemic circulations.

5. Explain why the blood in the fish gill flows in the direction opposite to the direction of blood flow in the nearby blood vessels.

____
____
____
____

6. Explain why the volume of blood moving out of your left ventricle per unit of time exactly matches that moving out of your right ventricle.

____
____
____
____

7. Describe a possible evolutionary sequence for a large blood vessel becoming altered over evolutionary time to function as a chamber of the heart.

____
____
____
____

Concept 38.3 concentrates attention on the human heart, describing the right heart and its circuit as the pulmonary circuit, and the left heart and its circuit as the systemic circuit. The contraction cycle of the heart is explained as being the result of pacemaker cells periodically initiating action potentials, independently of the nervous system. The interval between action potentials can be shortened by signals from the sympathetic division of the autonomic nervous system, to increase cardiac output for exercise or emergencies. Similarly, the period of the cardiac cycle can be lengthened by signals from the parasympathetic division, when the person is resting and fed. Action potentials spread from cell to cell in the heart via gap junctions.

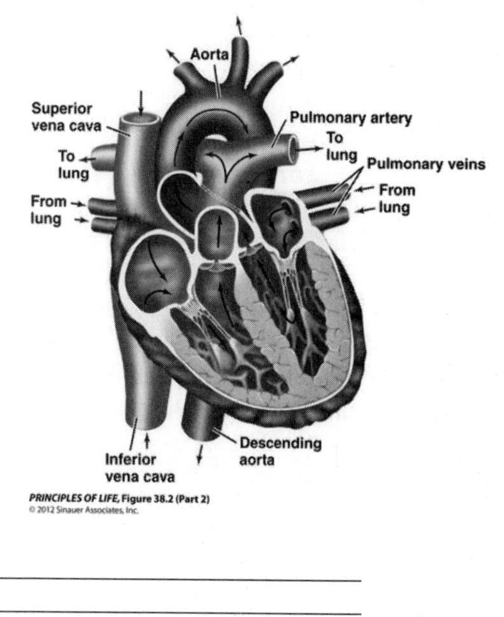

PRINCIPLES OF LIFE, Figure 38.2 (Part 2)

8. Describe systole and diastole.

____
____
____
____

9. Early investigators measuring membrane potentials of the pacemaker cells in the heart found that putting drops of acetylcholine onto a beating heart caused it to become more hyperpolarized, and the approach to threshold for an action potential was also slower. Describe the effects of these two changes on heart rate, and identify which division of the autonomic nervous system was modeled in the experiment.

____
____
____
____

10. Epinephrine, also called adrenaline, is a neurotransmitter and hormone. Predict the effects of drops of epinephrine on membrane potential in the setting described in question 9, and identify the division of the autonomic nervous system that is modeled by adding epinephrine.

____
____
____
____

11. Arrange the terms below in the correct sequence of firing action potentials during one cardiac cycle, using numbers; the starting place (1) is completed for you.

_____ Purkinje fibers
__1__ sinoatrial node
_____ bundle of His
_____ ventricles
_____ atrioventricular node
_____ atria

12. A red blood cell moving into the heart from the venous return, to the lungs and back, and then into the kidneys would pass a number of distinct anatomical landmarks on its journey. Arrange the terms below in the correct anatomical sequence, using numbers; the starting place (1) is completed for you.

_____ pulmonary vein
_____ pulmonary artery
__1__ vena cava
_____ right atrium
_____ left ventricle
_____ right ventricle
_____ left atrioventricular valve (bicuspid)
_____ right atrioventricular valve (tricuspid)
_____ left atrium

13. In studying skeletal muscle, you learned that the length of the muscle when it is stimulated can be a determinant of its force of contraction. This is also true of the heart. Explain how a stretched heart might contract more forcefully than a heart less stretched. Describe circumstances related to both conditions as your cardiac output varies throughout the day.

____
____
____
____
____

14. In the twitch contraction of skeletal muscle, every troponin binds a calcium ion in the affected muscle fiber. Explain whether or not every heart cycle would expect to be accompanied by a similar "saturation" of troponin with calcium ions.

____

____

____

____

Concept 38.4 describes the fluid that is pumped around the circulatory system. This fluid is blood, and it is composed of watery plasma and suspended cells, primarily erythrocytes (red blood cells). Erythrocytes are often described as small bags containing 1,000,000 molecules of hemoglobin, the protein that greatly increases the blood capacity to transport oxygen. Erythrocytes, lacking a nucleus and the genetic code, "live" for only four months and replacements of these cells are always needed. If the number of erythrocytes drops low enough to activate oxygen sensors in the kidneys, these organs secrete the hormone erythropoietin, which travels to the bone marrow and accelerates the formation of new erythrocytes, helping to restore the oxygen-carrying capacity of the blood to its physiological optimum. Blood also contains platelets, cell fragments that are stimulated to clump together on exposed collagen by chemical signals that are released as a result of an injury (a cut). A longer lasting "plug" on the injured vessel comes as clotting proceeds to form a clot made of a fibrin meshwork.

15. Erythrocyte shape can be altered by the osmolarity of the plasma. Explain what would happen to erythrocyte volume if it were placed in a beaker of water. Make another prediction of changes that would happen if it were placed into very salty water.

A. Placed in water ____

____

____

____

B. Placed in hyperosmolar solution ___________________________________

____

____

____

16. Describe the possibly fatal sequence of events that might occur if an inexperienced person injects pure water into the veins of a severely dehydrated hiker after being rescued.

____

____

____

____

17. Erythropoietin is an important hormone, and a drug form is now available. Explain why a person who has undergone a fairly toxic round of chemotherapy and radiation treatment for cancer would benefit from this drug.

____

____

____

____

Concept 38.5 tours the blood vessels traversed by the blood after it leaves the heart. Each day, about two liters of plasma are lost from the capillaries, and is carried as lymph through lymphatic vessels back to the blood. Blood vessels, with the exception of the capillaries, are surrounded by circular smooth muscle, the contraction of which reduces vessel diameter, thus increasing resistance to blood flow, and forcing the blood to flow elsewhere if there is an open, low resistance vessel to accept the flow. The regulation of blood vessel diameter, through vasodilation and vasoconstriction, dynamically determines how much blood flow has access to each part of the body. For example, if you are frightened by a scary noise in the night, sympathetic signals cause vasoconstriction in your kidneys and digestive tract, simultaneously causing vasodilation in large blood vessels in your legs, in case you must flee the danger.

18. Starting in the aorta, name, in correct anatomical sequence, the categories of blood vessels that will carry the blood away from and then back to the heart.

____
____
____
____

19. Hydrostatic pressure on plasma forces some of it to "leak" out of the capillaries and enter the extracellular space. Explain the force that helps return some of this fluid into the capillary, and predict what would happen in a situation where a patient has abnormally high levels of protein in the blood.

____
____
____
____

20. Explain how skeletal muscle contraction and one-way valves assist the flow of venous blood.

____
____
____
____

Concept 38.6 provides an overview of the reflexes that help regulate blood pressure. The two biggest determinants of overall blood pressure are the amount of blood pumped per unit time from the heart, and the relative amount of vasodilation and vasoconstriction in the body. The cardiovascular reflexes have a primary function of always maintaining adequate blood supplies to the heart and brain, as oxygen deprivation to either can quickly lead to permanent impairment and death of the individual. The cardiovascular control center in the medulla oblongata of the brainstem is the location where the autonomic efferent regulation takes place, based on sensory information from sensory afferent neurons whose activity levels are related to blood pressure in critical locations, i.e., in the aorta, and in the blood supply to the brain.

21. The heart and state of smooth muscle contraction in blood vessels have direct effects on blood pressure, but the kidneys also have an indirect effect; explain.

____
____
____
____
____

22. Discuss the meaning of the term "baroreceptors," include the role of afferent neurons in your answer. Be sure to describe the impact of baroreceptors on the regulation of blood pressure.

____
____
____
____
____

23. Discuss the effects on the autonomic nervous system of ingesting and absorbing excessive amounts of water.

____
____
____
____
____

Science Practices & Inquiry

In the AP Biology Curriculum Framework, there is a set of 7 Science Practices. In this chapter, we will focus on **Science Practice 1:** The student can use representations and models to communicate scientific phenomena and solve scientific problems. More specifically, practice 1.4: The student can re-express key elements of natural phenomena across multiple representations in the domain.

In answering Question 23, you will evaluate the effects of stretching the heart and seeing how much volume of blood can be pumped per contraction cycle. You will use representations or models (data and graphing) to analyze quantitatively and qualitatively the effects of disruptions to dynamic homeostasis in biological systems (LO 2.28).

24. These data are similar to those presented in the "Science Practices" question from Chapter 36. The length of the heart was varied as shown while the amount of blood pushed out of the ventricle was measured, as shown in the table.

HEART LENGTH AT BEGINNING OF CONTRACTION AS A PERCENTAGE OF MAXIMUM LENGTH	VOLUME OF BLOOD PER CONTRACTION (mL)
35%	3,800
45%	4,200
55%	4,700
65%	5,100
75%	5,700
85%	8,800
100%	10,000

A. Graph the data, placing the independent variable on the *x*-axis and the dependent variable on the *y*-axis, and connect the points with a smooth curve. Label all parts of your graph.

B. Summarize the results.

C. Compare the data you have graphed here with the data you graphed in the "Science Practices" question for Chapter 36. Explain whether or not the comparison shows the heart muscle to have the same length-tension relationship you saw in skeletal muscle.

____

____

____

____

____

D. A resting athlete might pump 5L of blood per minute from each ventricle. At maximum effort, that value could jump to 25L of blood per minute. Based on that information, discuss whether or not data from stretching the heart are relevant to real life.

____

____

____

____

____

Chapter 39: Nutrition, Digestion, and Absorption

Chapter Outline

The three physiological systems tested on the AP exam are the nervous, immune, and endocrine systems, but the digestive system is frequently used as an illustrative example to examine enzyme action, regulation, structure and function, obtaining nutrients and eliminating waste, cooperative behavior, specialization of organs, and the evolution of eukaryotic structures or processes.

Animals are heterotrophs, meaning they acquire their nutrition from eating other organisms. The catabolism of many of the molecules in ingested and absorbed food releases energy needed for metabolic needs, whereas other absorbed molecules are used as specific nutrients as cells synthesize materials needed in the body. The reactions that hydrolyze and break apart food molecules for energy transfer are catabolic reactions. Anabolic reactions, in contrast, are those that use nutrients to form the proteins, lipids, and structural carbohydrates that the animal uses to build its body.

There are many micronutrients and macronutrients listed in textbooks, but memorizing such lists is not required by the AP-Biology exam. Rather, develop your general understanding of how nutrients are utilized in organisms. Vitamins are coenzymes or parts of coenzymes important in our diet. Minerals may be required in large quantities, e.g., calcium needed for bone health and for muscle and neural functioning. Other minerals like molybdenum are only found in a few enzymes.

Digestive systems break apart large molecules (macromolecules) so that smaller products can be absorbed into the organism. Complex eukaryotic digestive systems with two openings evolved from simple digestive systems with one opening. The simplest systems are called gastrovascular cavities and are found in organisms such as jellyfish and flatworms, like *Planaria*. A major evolutionary advancement was the development of the one-way digestive system. Simple organisms with a one-way system are the annelids or segmented worms. As you examine the digestive system, be sure to pay careful attention to how the structure of an organ complements its function. An example is that the folded lining of the stomach and three distinct overlapping muscle layers are specialized for squeezing and mixing the stomach, and provide surface area for secretion.

The pancreas is a major exocrine gland, with its many digestive enzymes secreted into the intestine, but it is also an important endocrine gland, secreting hormones, including insulin and glucagon, that regulate glucose concentration in the blood. The secretion of these hormones is a classic example of negative feedback, with high levels of glucose initiating the release of insulin, thereby increasing the rate of glucose absorption into many cells. If glucose levels in the blood are low, the heart and the brain are at risk of damage, so the pancreatic hormone glucagon is released. Glucagon stimulates liver cells to break down glycogen and release glucose.

Chapter 39 has ideas that span Big Ideas 2 & 4. The Big Ideas are a means for organizing the vast amount of information we know as biology. It is vitally important that you continually work to understand these **Big Ideas** and examine your knowledge across different **Enduring Understandings**.

Big Idea 2 states that the utilization of free energy and use of molecular building blocks are characteristic fundamental of life processes. Specifically, Chapter 39 includes:

- 2.a.1: All living systems require constant input of free energy.
- 2.a.2: Organisms capture and store free energy for use in biological processes.
- 2.a.3: Organisms must exchange matter with the environment to grow, reproduce, and maintain organization.
- 2.c.1: Organisms use negative feedback mechanisms to maintain their internal environments and respond to external environmental changes.
- 2.d.2: Homeostatic mechanisms reflect both common ancestry and divergence due to adaptation in different environments.
- 2.d.3: Biological systems are affected by disruptions to their dynamic homeostasis.

Big Idea 4 examines the idea that biological systems interact in complex ways. Included in Chapter 39:

- 4.a.4: Organisms exhibit complex properties due to interactions between their constituent parts.
- 4.b.2: Cooperative interactions within organisms promote efficiency in the use of energy and matter.
- 4.c.2: Environmental factors influence the expression of the genotype in an organism.

Chapter Review

Concept 39.1 introduces the idea of heterotrophs and their need of external nutrition. Food provides both nutrition and essential minerals and vitamins for heterotrophs.

1. Explain the difference between an autotroph and a heterotroph. Provide an example of each.

____
____
____
____

2. Identify each process below as either anabolic (A) or catabolic (C).

____ a. enzymatic digestion of a protein into amino acids
____ b. producing starch from glucose
____ c. conversion of hydrogen peroxide into water and oxygen gas
____ d. hydrolysis of a carbohydrate
____ e. building muscle tissue

3. The average basal metabolic rate is 1,600 Cal/day. Walking burns 5X as many calories as at rest, and jogging burns 8X as many calories. Calculate the time required to burn off the calories from ingesting each of the following:

	Calories	Resting	Walking	Jogging
Starbucks venti Frappuccino, Peppermint Mocha	435			
Chipotle Barbacoa Burrito with cheese, sour cream and guacamole	1180			
Dominos, 1 slice, Brooklyn Deluxe pizza	300			

4. Explain why a 100% fat free diet is virtually impossible to attain, and why it might be an unhealthy choice.

Concept 39.2 examines the digestive systems of different animals, beginning with the gastrovascular cavity of jellyfish. One way or tubular guts have an opening at each end improving efficiency and allowing for specialization of various organs.

5. Describe TWO evolutionary advantages that a one-way (complete) digestive system has over a gastrovascular cavity.

6. Most organisms utilizing gastrovascular cavities also lack a circulatory system. Explain why they do not need one.

7. Explain how a person standing on his or head can drink a glass of water.

8. Very few birds are herbivores, and many of the ones that are specialize in seeds that are high in oils. Explain why.

9. Explain how you would expect the digestive tract of a carnivore to differ from that of an herbivore.

10. Explain the function of bacteria living in the gut of termites.

11. It has been said that if the lining of the small intestine were laid out flat it would cover the area of a tennis court. Explain how such a large area can be squeezed into your abdomen.

Concept 39.3 discusses how food is digested in the vertebrate gut.

12. Frequently toxins can enter with food ingested. Explain how vertebrates prevent toxins that are ingested with their food from getting into their body.

13. For each major nutrient, identify what happens to it in each part of the digestive system below:

	Carbohydrates	Proteins	Lipids
Mouth			
Stomach			
Small intestine			
Large intestine			

14. Explain the role of each of the following in digestion

bile

saliva

teeth ____

____

____

water ____

____

____

15. Explain how a person should modify his/her diet if the gallbladder had to be removed.

____

____

____

____

Concept 39.4 reviews the regulation of the distribution and use of nutrients in the body.

16. Complete the diagram below by placing the letter of each choice in the correct blank:

A. Breakdown of glycogen in liver
B. Metabolic energy production, glycogen synthesis
C. Increase in circulating insulin
D. Increases circulating glucagon
E. Release of glucose to blood
F. Stimulates pancreas to secrete glucagon
G. Stimulates pancreas to increase insulin
H. Uptake of glucose by cells

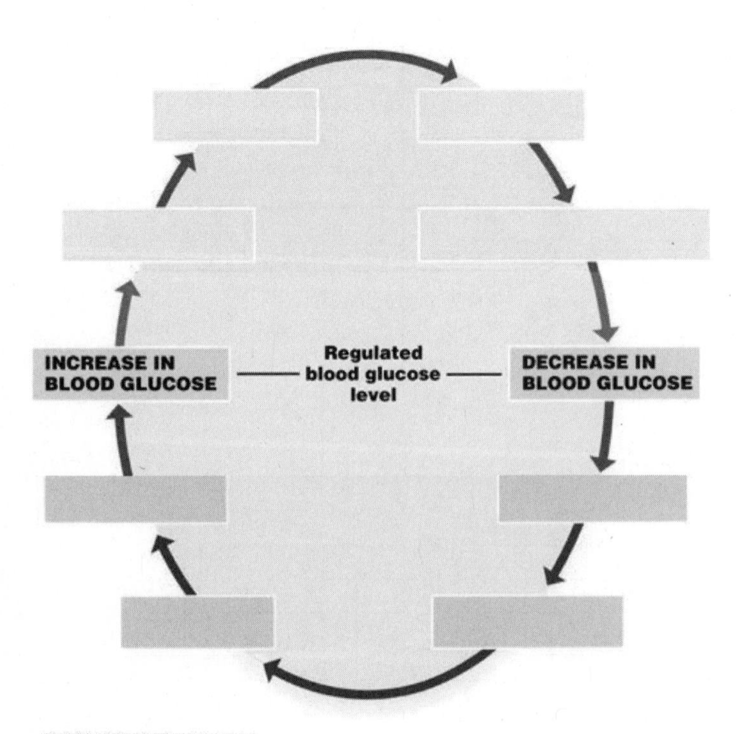

PRINCIPLES OF LIFE, Figure 39.13

17. Explain why insulin is considered an anabolic hormone.

18. Predict which of these circumstances would allow you to measure glucagon at its peak concentration the blood: during rest, after a high carbohydrate meal, after an overnight fast, or during times of stress. Explain your selection.

19. For each setting below situations, identify whether it demonstrates positive feedback or negative feedback and explain why.

a. Pancreatic enzymes are produced and secreted as inactive forms called zymogens. The zymogen trypsinogen is converted to its active form trypsin by enterokinase. Trypsin can then activate other trypsinogen molecules.

b. Acidic chyme arriving in the duodenum stimulates cells in the duodenal epithelium to release secretin, which slows the release of chime from the stomach.

c. Peptides in the stomach cause acid and pepsinogen to be released and this in turn creates more peptides in the stomach, which causes more acid and pepsinogen to be released.

Science Practices & Inquiry

In the AP Biology Curriculum Framework, there is a set of 7 Science Practices. In this chapter, we will focus on **Science Practices 5 and 6.**

Science Practice 5: The student can perform data analysis and evaluation of evidence.

Science Practice 6: The student can work with scientific explanations and theories.

More specifically, practices

5.3 The student can evaluate the evidence provided by data sets in relation to a particular scientific question.

6.4 The student can make claims and predictions about natural phenomena based on scientific theories and models.

Parts a & b of question 20 ask you to evaluate data that show the effect(s) of changes in concentrations of key molecules on negative feedback mechanisms (LO 2.17.) Parts c & d of this question take a slightly different angle and ask you to predict the effects of a change in an environmental factor on the genotypic expression of the phenotype (LO 4.24.)

20. In mice the Ob gene codes for the protein leptin, a satiety factor secreted by fats cells. Leptin appears to signal the brain to stop eating when adequate food has been consumed to assure adequate adipose tissue. The recessive *ob* allele is a mutant allele, resulting in the absence of leptin. Mice that are *ob/ob* do not experience satiety and become obese.

In an experiment, leptin was purified and injected into *ob/ob* and *Ob/Ob* (wild-type) mice daily. The data in the table were collected before the injections began (baseline) and 10 days later.

	Baseline		Day 10	
Parameter	***ob/ob***	***Ob/Ob***	***ob/ob***	***Ob/Ob***
Food intake (g/day)	12.0	5.5	5.0	6.0
Body mass (g)	64	35	50	38
Metabolic rate (ml O_2/kg/hr)	900	1150	1100	1150
Body temperature (°C)	34.8	37.0	37.0	37.0

a. Summarize the data from this experiment.
b. Do the data support the claim that leptin is a satiety signal?
c. Explain why the amount of nutrients absorbed is considerably more than the increase in body mass.
d. Predict what an injection of leptin would do for an *OB/ob* mouse.
e. A mutation resulting in non-functional leptin receptors has been discovered. Select the group of mice above that you would expect to represent the phenotype of mice without functional leptin receptors. Explain your answer

Chapter 40: Salt and Water Balance and Nitrogen Excretion

Chapter Outline

Chapter 40 focuses on an essential homeostatic process: the maintenance of the ionic composition and volume of the extracellular fluid, with emphasis on the vertebrate renal system. Additional attention is concentrated on the removal of nitrogenous wastes, the inevitable byproduct of our protein-based existence, but the importance of maintaining the extracellular fluid cannot be overstated.

The optimal conditions for life rarely match what is present in the non-living environment, and cell survival depends on optimal concentrations of ions and nutrients in the cytosol. Organisms take in nutrients, gases and minerals from the environment, in order to "condition" the blood with the "good stuff" needed by the cells in the body. In turn, the blood "conditions" the extracellular fluid around the cells, delivering nutrients and oxygen, and removing wastes, the "bad stuff," including carbon dioxide.

Chapter 40 reiterates many parts of the Curriculum Framework, especially **Big Idea 2**.
Big Idea 2 establishes the premise that molecular building blocks explain the systems and functions of living organisms. Chapter 40 shows how water and salt balance are based on the principles that:

2.b.1: Cell membranes are selectively permeable due to their structure.
2.d.2: Homeostatic mechanisms reflect both common ancestry and divergence due to adaptation in different environments.

Chapter Review

Concept 40.1 establishes the homeostatic emphasis of studying salt and water balance. Osmotic equilibrium, the balancing of dissolved solutes inside the cell with the concentration in the extracellular fluid, provides a starting point for this exploration. Note that osmotic equilibrium or balancing refers only to the number of dissolved solutes, not the chemical identities of the solutes. For example, the higher concentration of potassium ions inside cells and the higher concentration of sodium ions outside cells are partially offsetting each other in the consideration of the osmotic balance between inside and outside the cell. Osmotic imbalance, depending on whether solutes are more abundant inside or outside the cell, can lead to water loss, or gain, respectively, and impairment or even death of the cell.

Animals are classified as osmoregulators or osmoconformers, depending on whether they actively regulate or not, respectively, the osmolarity of the extracellular fluid. Freshwater and terrestrial environments compel organisms to be osmoregulators, and marine invertebrates are the most common examples of osmoconformers. At extreme concentrations of salts, for example, in "salt" lakes of greater salinity than is present in the ocean, failure to osmoregulate will lead to shrinking of cell volume and eventually death of the cells.

1. Describe the sequence of osmotic changes that might happen to a marine invertebrate, of limited osmoregulatory capacity, trapped in a tide pool of ocean water after high tide. Assume that the pool changes over the course of the next 6 hours, as described below.

a. Salinity in the pool is equal in salinity to ocean water:

b. Salinity in the pool decreases below the salinity of ocean water after rain falls:

c. Salinity in the pool greatly exceeds salinity of ocean water after evaporation:

2. Protein structure can be altered by changes in pH, possibly impairing cellular functions. Describe the role of carbonic anhydrase and bicarbonate ions in the homeostasis of acid-base balance. Include the role of the kidneys in reducing the amount of bicarbonate ions lost in the urine. Include in your answer the bicarbonate/carbonic acid disassociation equation and explain its significance with equilibrium.

3. When climbing to high altitude, mountain climbers can suffer from "respiratory alkalosis" as they hyperventilate. Explain the relationship between the breathing rate and acid-base balance in the tissues.

Concept 40.2 describes nitrogenous waste products associated with amino acid metabolism. The hydrolysis of amino acids releases ammonia, a fairly toxic waste material. For animals that have extensive contact with water, e.g., bony fishes, diffusion of ammonia from the blood to the water moving over the gills is adequate to remove the ammonia from the body. Other animals, e.g., mammals, convert ammonia to less toxic materials, including urea, while others, e.g., birds convert nitrogenous wastes to uric acid.

4. Describe the number of nitrogen atoms in one molecule of ammonia, one molecule of urea, and one molecule of uric acid. For the complete catabolism of a protein with 32 amino acids, how many molecules each of ammonia, urea, and uric acid are required, assuming the organism only utilizes one type of nitrogenous waste.

5. Compare the three types of nitrogenous waste in the amount of water required as a solute for the elimination of the waste.

____
____
____
____

Concept 40.3 demonstrates that urine is a filtrate formed from the extracellular fluid. In animals with blood vessels, the filtrate is typically formed from the blood.

Among annelids, e.g., earthworms, the filtrate is formed near a tubular structure called the metanephridium. Filtrate moves into the tubes, and solutes needed by the annelid are reabsorbed, with the excess fluid exiting the body through a pore to the outside.

Among insects, the excretory system is located in the midgut section. Excess ions and nitrogenous wastes in the open circulatory system of insects are collected in tubules called Malpighian tubules, which drain into the digestive tract, leading to the accumulation of these materials in the rectum. Concentration of the wastes results from active ion transport in the rectum and the final excretory product is concentrated to such a high degree that insects are well adapted to surviving in arid conditions.

Among animals with kidneys, e.g., humans, a filtrate is formed from the blood passing through the glomerular capillaries in the kidneys. The solutes in the filtrate are actively and passively reabsorbed along the tubule though which the filtrate moves. The rate of filtration, called the glomerular filtration rate (GFR), is an important indicator of renal (kidney) health and function. The complexity of the tubule, called the nephron, determines how the filtrate changes in composition as it moves closer to excretion from the body.

6. Clara and Josie are having an argument. Clara claims that the metanephridia of earthworms operate as a one-stage (filtration) process, but Josie claims they operate as a two-stage (filtration and reabsorption) process. Discuss both claims and explain whose claim is closer to the best description.

____
____
____
____

7. Describe the location and operation of Malpighian tubules:

a. Including the type of organisms that utilize these structures for processing ions and nitrogenous waste.

____
____
____
____

b. Describing how the rectum of these organisms further processes ions and nitrogenous wastes, including a description of how these processes affect organismal water balance.

____
____
____
____

9. Name and describe the functional unit that helps frogs maintain water and salt balance in spite of spending most of their lives associated with freshwater (low salinity) waters.

___

___

___

___

10. Describe a key functional difference in water handling between the urinary bladder of a toad that estivates and the urinary bladder of humans.

___

___

___

___

Concept 40.4 details the processing of the filtrate along the mammalian nephron, including descriptions and mechanisms of secretion and reabsorption. The geometry of the nephron and the blood vessels associated with it offer a helpful reiteration of the importance of counter-current exchange in living organisms; recall the descriptions of gas exchange and thermoregulation.

11. Add the following labels to the diagram.

A. Collecting Duct

B. Filtration occurs here

C. Proximal Convoluted Tubule

D. Distal Convoluted Tubule

E. Ascending Limb of Loop of Henle

F. Descending Limb of Loop of Henle

G. Vasa Recta

H. Glucose reabsorption occurs here

I. ADH receptors are found here

J. Water permeability is <u>always</u> low here

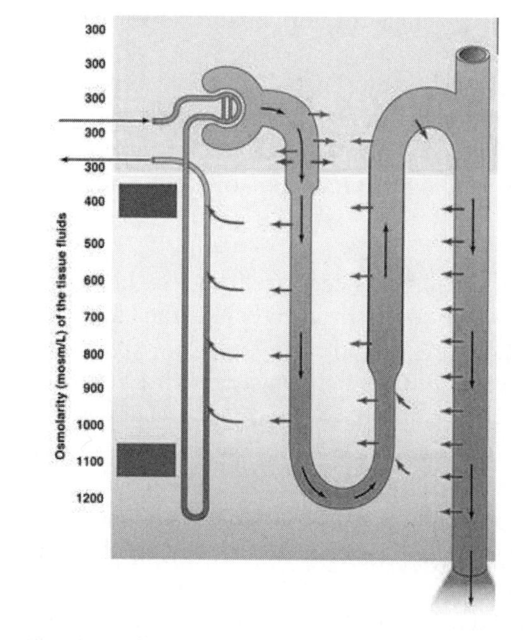

PRINCIPLES OF LIFE, Figure 40.8
© 2012 Sinauer Associates, Inc.

12. For humans, filtrate is formed at the rate of about 180 L/day, yet urine excretion is only 2 L/day. Describe the part of the nephron where most water reabsorption (~75%) takes place, and describe the process that drives water reabsorption in that location.

___

___

___

___

13. Blood in the capillaries of the kidneys passes through an area with high osmolarity and must osmotically equilibrate, yet by the time that blood departs the kidney in veins, its osmolarity has returned to plasma-typical levels. Describe how the shape of the vasa recta makes this possible.

____
____
____
____
____
____

14. Consider the role of the kidneys in the homeostasis of H^+ levels in the body. Describe the causes of acidosis and alkalosis and, for each, discuss renal responses that will help restore optimal conditions.

____
____
____
____
____
____

Concept 40.5 details the hormonal mechanisms serving the regulation of kidney function in mammals.

Following excess water ingestion and absorption, blood volume becomes abnormally high, so blood pressure increases. In such conditions, the kidney functions are altered by atrial natriuretic peptide, (ANP) secreted from an abnormally stretched heart. In response, GFR increases, and reabsorption of water decreases, leading to a reduction of blood volume and restoring blood pressure lower, toward the optimal pressure.

When blood volume is decreased, e.g., by blood loss and/or dehydration, the resulting decrease in blood pressure is sensed by baroreceptors in the kidney and by baroreceptor (pressure-sensitive) neurons in blood vessels. As a result, two hormone-systems release signals that increase the reabsorption of fluid from the renal filtrate, thus conserving the body's water by reducing the volume of urine excreted. One such hormone, anti-diuretic hormone (ADH) is secreted from the posterior pituitary gland in response to neural reflexes activated by decreased blood pressure and by increased osmolarity of body fluids, typical of what occurs during dehydration. The hormone's name describes its action: anti-diuresis, or more plainly, reduction in urine excretion. The other hormone system "protecting" against hypotension (low blood pressure) is the renin-angiotensin-aldosterone system, also called RAAS. The signals of the RAAS, especially angiotensin II and aldosterone, increase reabsorption of filtrate to the body, thus reducing urine formation. The RAAS is a target of intervention in patients with hypertension, high blood pressure, which can damage the blood vessels. Treatments using antagonists of the RAAS signals can reduce blood pressure, as can the ingestion of an inexpensive diuretic drug, which by increasing urine excretion, reduces blood volume, thereby lowering blood pressure.

15. Explain how ADH responses and RAAS responses are complementary and yet each is activated, at least partially, by distinct stimuli.

____
____
____
____
____
____

16. Dehydration causes many of the symptoms of overindulging in the consumption of ethanol, even when the ethanol is present at relatively low concentrations, e.g., beer. Explain how the consumption of so much water, in the context of beer, can lead to dehydration.

____
____
____
____
____
____

17. Explain the variable permeability of the collecting duct to water. Describe the cellular/molecular mechanism that brings about the altered permeability, and describe the signals that cause these mechanisms to be activated.

____
____
____
____
____
____

Science Practices & Inquiry

In the AP Biology Curriculum Framework, there is a set of 7 Science Practices. In this chapter, we will focus on **Science Practice 7:** The student is able to connect and relate knowledge across various scales, concepts, and representations in and across domains.
In answering Question 18, you will be asked to connect what you know about osmotic gradients, water movements, glucose reabsorption from renal filtrate and neurobiology.

18. As he prepared for his AP test, Tommy found himself frequently visiting the bathroom, thirsty, and hungry. Distracted for a moment, he found the following data set in a lab notebook his mom wrote in the 1980s.

Plasma glucose concentration (mg/100 mL plasma)	Reabsorption rate of glucose from renal filtrate (mg/min)
0	0
100	100
200	200
300	300
400	400
500	400
600	400

A. Graph the data, placing "glucose concentration" on the *x*-axis and "reabsorption rate" on the *y*-axis, and connect the points with a curve. Label all parts of your graph.

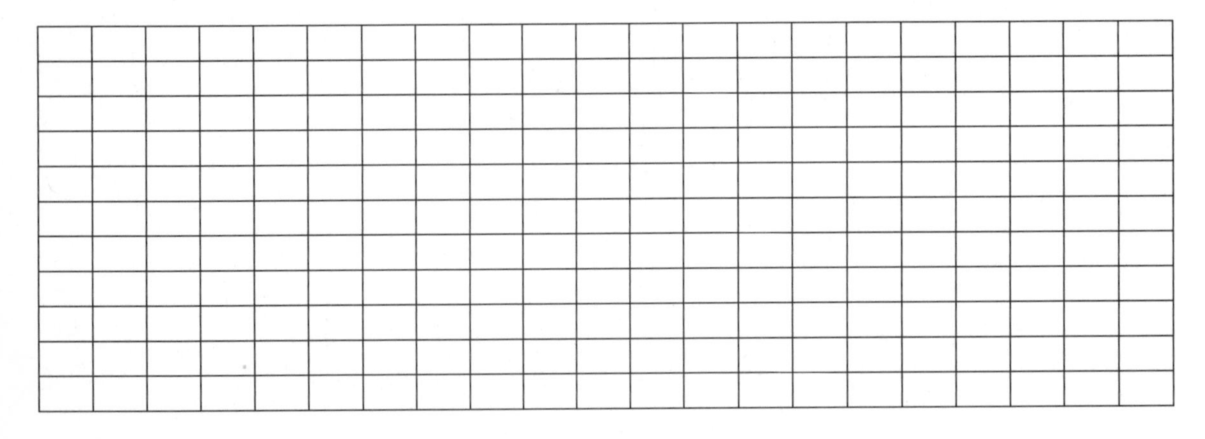

B. Write a caption that explains the meaning of the graph.

____
____
____
____
____
____

C. Examine the data to determine whether or not the phenomenon of "saturation" can be demonstrated in the data, and state where on the graph you are focusing. Explain what is being saturated, and where in the nephron saturation might occur.

____
____
____
____
____
____
____

D. Discuss the possible evidence that Tommy might be suffering from *diabetes mellitus*, and relate his symptoms to excessive loss of sodium ions from his body.

____
____
____
____
____
____

Chapter 41: Animal Behavior

Chapter Outline

Behaviors are fittingly one of the last topics we examine as they incorporate information from most of the chapters in the book. Behaviors have genetic underpinnings, have evolved in response to selective pressures, and are learned from previous generations. This chapter examines the proximate and ultimate causes of behavior. Proximate explanations often answer "how" questions. How does an organism react to a particular stimulus? How do genetic, physiological, neurological, and developmental mechanisms influence the behavior? Ultimate explanations answer "why" questions, focusing on the evolutionary explanations for behaviors. Why do organisms react the way they do? Why does this behavior yield an evolutionary advantage for the individual?

Many behaviors are complex and have genetic associations. These ties are studied through molecular genetic approaches. Recall that DNA codes for RNA, which is spliced to make mRNA, which is then translated to guide protein synthesis. Thus, a mutation in a gene, for example, a gene for an olfactory receptor protein, can change how an organism will react to its environment, if that gene product alters its capability to detect odors that are relevant to the selection pressures acting on that individual.

The behaviors of animals develop and mature over time as the nervous system and other systems develop. Juvenile behaviors such as suckling or begging for food give way to adult behaviors, such as migratory and courtship behaviors. Hormones play a large part in the developmental changes in behaviors. Some behaviors only develop at specific times in an animal's life, e.g., imprinting, a parent-offspring bond, happens only within a few hours of birth in many species. Young songbirds learn species-specific songs best while very young.

Environmental cues that can influence behavior include circadian rhythms and migration. Circadian rhythms in organisms are based on the circadian rotation of the Earth on its axis. Most organisms kept in continuous darkness or in continuous light, will continue to exhibit a sleep/wake cycle that repeats approximately every 24 hours. Migratory behaviors are seasonally expressed, and temporally guided by visual references, homing to return to a specific location, and long distance migration using the sun, stars, or the Earth's magnetic field for reference.

Natural selection shapes animal behavior. Every behavior can be examined from the perspective of its cost-benefit ratio, including energetic cost, risk cost, and opportunity cost. How much energy is expended to carry out the behavior? What are the risks of being injured or killed while performing the behavior? What benefits are lost if the behavior cannot be performed? This approach explains why some organisms will sacrifice themselves for the greater good of a colony, or serve as lookouts for a colony.

Chapter 41 has ideas that span Big Ideas 2 & 3. The Big Ideas are a means for organizing the vast amount of information we know as biology. It is vitally important that you continually work to understand these **Big Ideas** and across different **Enduring Understandings**.

Big Idea 2 states that the utilization of free energy and use of molecular building blocks are characteristic fundamental of life processes. Specifically, Chapter 41 includes:

- 2.c.2: Organisms respond to changes in their external environments.
- 2.e.2: Timing and coordination of physiological events are regulated by multiple mechanisms.
- 2.e.3: Timing and coordination of behavior are regulated by various mechanisms and are important in natural selection.

Big Idea 3 states that living systems store, retrieve and transmit information essential to life processes. Specifically, Chapter 41 lays the groundwork of cell communication:

- 3.a.3: The chromosomal basis of inheritance provides an understanding of the pattern of passage (transmission) of genes from parent to offspring.
- 3.d.2: Cells communicate with each other through direct contact with other cells or from a distance via chemical signaling.
- 3.e.1: Individuals can act on information and communicate it to others.

Chapter Review

Concept 41.1 introduces the broad nature of behavior, considering many disciplines of behavioral analysis. Animal behavior can be broadly grouped into four categories: causation, development, function, and evolution. Causation and development are considered to be proximal causes of behavior while function and evolution are ultimate causes of behavior.

1. Explain the difference between proximate and ultimate explanations of behavior.

____
____
____
____

2. For each of the fixed action patterns below, describe and discuss the "releaser."

a. Mating dances by birds.

____
____
____
____

b. Aggression to anything red in the water during breeding season by red-bellied sticklebacks.

____
____
____
____

c. Moths folding their wings and dropping to the ground when an ultrasonic sound from a bat is encountered.

____
____
____
____

3. Gull chicks instinctively peck at the red dot on the parent's bill, a behavior that induces the parent to regurgitate food into the chick's mouth. Below is a chart showing the results of an experiment showing the percent of times a model was pecked compared to the "control" releaser-model shown on the left.

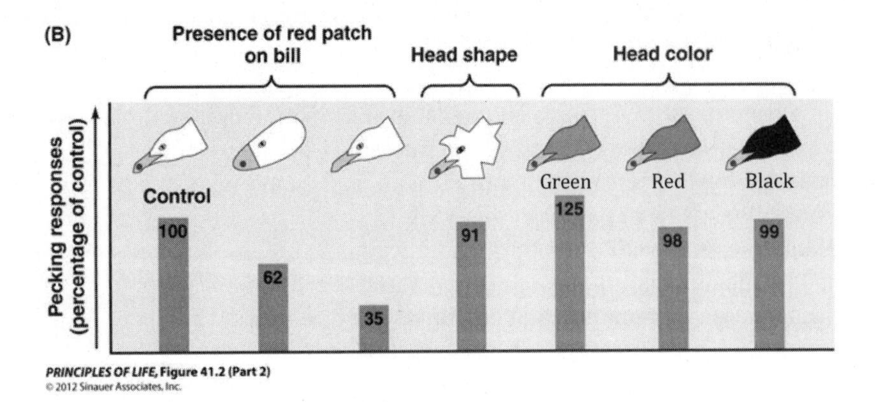

a. Summarize the results of the experiment.

____

____

____

____

____

____

b. Explain how a result of 125% is possible.

____

____

____

____

____

____

c. Provide an explanation, based on the data, as to whether the presence of a red dot on the bill, or the color/shape of the head, is a more effective stimulus.

____

____

____

____

____

____

4. Explain why fixed action patterns are highly adaptive in some circumstances.

____

____

____

____

Concept 41.2 explains how genes and behaviors are connected. Breeding experiments have revealed that a small number of genes are responsible for a behavioral difference. Induced mutations in these genes can reveal the genetic determinants of behavior.

5. In a honeybee colony, when pupae die, a normal hygienic bee (*uu rr* or ur) will uncap the cell and remove the dead pupae. Two genes control this behavior, uncapping (u) and removing (r). Recall that in honeybees, females are diploid and males are haploid. Below is a diagram depicting this.

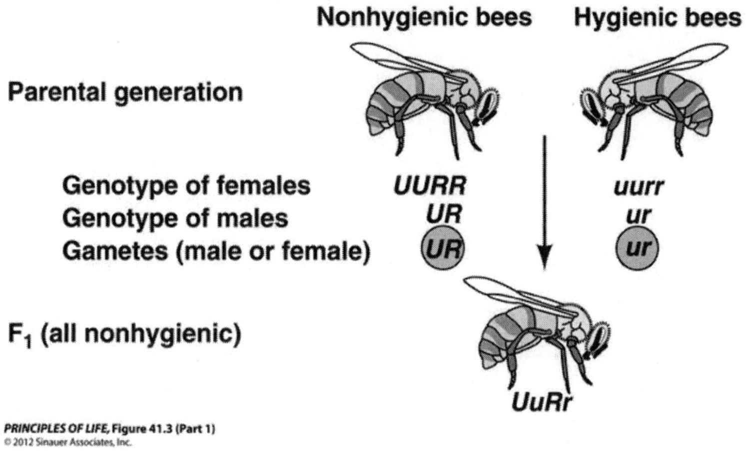

Nonhygienic bees | Hygienic bees

Parental generation

Genotype of females *UURR* *uurr*
Genotype of males *UR* *ur*
Gametes (male or female) *UR* *ur*

F_1 (all nonhygienic)

UuRr

PRINCIPLES OF LIFE, Figure 41.3 (Part 1)

a. Diagram and explain the outcome of the cross of the F1 female in the diagram above when she is crossed with a hygienic male.

b. Diagram and explain the outcome of the cross of the F1 female in the diagram above when she is crossed with a nonhygienic male.

c. State which of the two crosses is more useful in a behavioral study of bees.

6. Explain what a gene knockout is.

Concept 41.3 studies how behaviors develop over a lifetime. Not all behaviors are exhibited throughout one's lifetime. Some behaviors, such as begging for food, may be adaptive early on in development, but disappear while other behaviors such as migration develop later in life. The development of the nervous system as well as maturation of other systems influences many different behaviors.

7. Explain the relationship between imprinting and critical period.

8. A researcher frequently wears yellow boots around his research area. Explain why baby ducks will sometimes follow this person as though he is their mother, but only when he wears the yellow boots.

9. Male white-crowned sparrows frequently sing a species-specific courtship song in spring, but females do not sing this song unless injected with the hormone testosterone. Discuss the cues necessary for a white crowned sparrow to sing.

10. Discuss the adaptive value of imprinting.

Concept 41.4 examines how complex behaviors result from the interactions of genetic, physiological, and environmental factors. Circadian rhythms control the daily cycle of behavior and are normally timed to a light/dark cycle. Other environmental cues are used by organisms to navigate, such as orienting to landmarks or the celestial objects.

11. Explain why circadian rhythms are adaptive.

12. Do all circadian rhythms result in animals being awake during the day? Explain.

13. Give two examples of circadian rhythms in plants and explain how these rhythms are adaptive.

14. Martin Ralph and colleagues used artificial selection to produce two strains of hamsters: one with a short circadian period and one with a long circadian period.

Two clusters of neurons, the suprachiasmatic nuclei (SCN) in the brain, appear to control circadian rhythms of behavior. Destroying the SCNs of adult hamsters caused the animals to become arrhythmic. After several weeks of arrhythmia, scientists transplanted SCN tissue from fetal hamsters into the brains of the adult hamsters whose SCNs had been destroyed. Long-period adult hamsters received tissue from short-period fetuses, and short-period adults received tissue from long-period fetuses. The effects of these treatments on the hamsters' circadian periods are shown in the table below.

	TREATMENT			
RECIPIENT	NONE	SCN DESTROYED	SCN DESTROYED AND SHORT-PERIOD TRANSPLANT	SCN DESTROYED AND LONG-PERIOD TRANSPLANT
Short-period adult	Short-period	Arrhythmic	Not done	Long-period
Long-period adult	Long-period	Arrhythmic	Short-period	Not done

PRINCIPLES OF LIFE, Apply the Concept, Ch. 41, p. 808

a. Why did destroying the SCN make the hamsters arrhythmic?

b. How could the researchers have designed a control for this experiment? (Hint: examine the data table closely). What results would you expect from this?

15. Leatherback sea turtles frequently travel thousands of miles from the beach where they hatched, often crossing to the other side of the ocean. Yet, many years later, they unerringly return to the beach they hatched from. Discuss the types of navigation sea turtles probably use as they swim underwater.

16. Two forms of orientation behavior, **kinesis** and **taxis**, have been tested in behavioral studies. Taxis is a directed motion forward or away from a stimulus, while kinesis is random motion not resulting in any particular orientation in relationship to the stimulus. For each of the scenarios below, determine if the orientation is taxis or kinesis and provide a brief explanation about its adaptive value.

a. Cockroaches will move away from a light source.

b. With increased humidity there is an increase in the percentage time that a woodlouse will remain stationary.

c. *Euglena*, a flagellate protozoan with chloroplasts, are attracted to light.

Concept 41.5 focuses on the cost-benefit approach that behavioral ecologists utilize to study the relationship between behavior, the environment, and fitness. In other words, how does natural selection shape behavior?

17. Discuss each of the below as it relates to cost-benefit analysis of behavior:

Energetics

Risk

Opportunity

18. Using the cost-benefit approach, explain the two behaviors below. Be sure to include all three aspects of cost-benefit in your discussion.

a. Why sea birds will not defend feeding rights to parts of the ocean yet will vigorously defend nesting areas on a beach.

b. Why male elephant seals often fight to the death to stake out territories on a beach where females will soon arrive.

Concept 41.6 defines and explores social and sexual behaviors, including polygyny and polyandry, altruism, and kin selection.

19. Explain the adaptive value of polygynous and polygamous societies.

20. Define the term fitness. Distinguish between individual and inclusive fitness.

21. Pigeons fly and forage in flocks. The larger the flock, the lower the chances of a goshawk are to catch one of the pigeons. Use the cost-benefit analysis to decide if there is an optimum size of a flock.

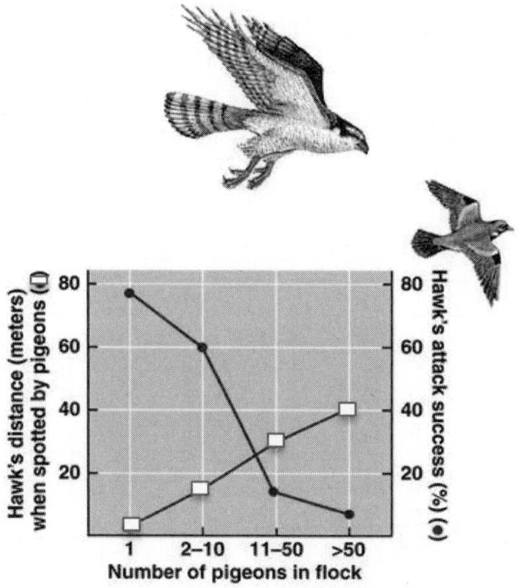

Hawk's distance (meters) when spotted by pigeons (□)
80
60
40
20
1
2–10
11–50
>50
Number of pigeons in flock
Hawk's attack success (%) (●)

22. In a honey bee colony, the queen is the only reproductive female. She mates once and stores sperm form the mating the rest of her life. Some of the eggs are not fertilized and develop into males, while most of the eggs are fertilized and develop into sterile females (worker bees). Discuss the adaptive value of the eusocial behavior of female bees.

Science Practices & Inquiry

In the AP Biology Curriculum Framework, there is a set of 7 Science Practices. In this chapter, we will focus on **Science Practice 7: The student is able to connect and relate knowledge across various scales, concepts and representations in and across domains.** More specifically, practice 7.2: The student can *connect concepts* in and across domain(s) to generalize or extrapolate in and/or across enduring understandings and/or big ideas.

Question 23 asks you to connect concepts in and across domains to predict how environmental factors affect responses to information and change behavior (LO 2.40).

23. For each of the three scenarios below, identify the type of behavior shown and discuss the environmental cues that lead to these behaviors. Include in your discussion the relationships between genetics, physiology and the outward behaviors exhibited.

a. Hand reared, endangered baby cranes are shown an ultralight aircraft when they are born, which is later used to lead the young cranes on their migratory routes.

b. When a female goose notices an egg outside the nest, it begins a repeated movement to drag the egg with its beak and neck. However, if the egg slides off, the goose continues to repeat the movements even if the egg is absent, until it reaches the nest, at which point, the goose repeats the motion over and over until the egg is returned.

c. In a group of wild turkeys, a subordinate turkey may help his dominant brother put on an impressive display that is only of direct benefit to the dominant turkey.

Chapter 42: Organisms in Their Environment

Chapter Outline

42.1 - Ecological Systems Vary in Space and over Time
42.2 - Climate and Topography Shape Earth's Physical Environments
42.3 - Physical Geography Provides the Template for Biogeography
42.4 - Geological History Has Shaped the Distributions of Organisms
42.5 - Human Activities Affect Ecological Systems on a Global Scale
42.6 - Ecological Investigation Depends on Natural History Knowledge and Modeling

An ecological system (biome) is composed of the populations of organisms living and interacting together in a particular environment. Some ecological systems are small, such as a pond, while others can be quite large, such as the boreal forest stretching across much of Canada. A particular ecosystem, such as a small pond, might appear to have well defined borders, but there are likely additional organisms from neighboring ecosystems that feed on organisms in the pond, such as herons that eat its fish and raccoons that eat its shellfish. No ecosystem is completely isolated, particularly from physical parameters, such as the abiotic factors water and sunlight (energy), which are constantly moving into and out of ecosystems.

An ecosystem has developed and exists within its long-term climate trends, primarily patterns of temperature and moisture. On any given day or week, weather describes short-term changes in atmospheric conditions. By comparison, long-term averages over many years or decades are known as climate. Latitude, elevation, and topography are the primary factors that determine an area's climate and thus the distribution of different types of terrestrial biomes.

Chapter 42 spans Big Ideas One, Two, and Three. The Big Ideas are a means for organizing the vast amount of information we know as biology. As you prepare for your exam, it is crucially important that you continually work to understand these **Big Ideas** and across different **Enduring Understandings**.

Big Idea 1 recognizes that evolution ties together all parts of biology. Chapter 42 develops the knowledge that:

1.b.2: Phylogenetic trees and cladograms are graphical representations (models) of evolutionary history that can be tested.
1.c.2: Speciation may occur when two populations become reproductively isolated from each other.

Big Idea 2 emphasizes that living organisms undergo energy transfers and use molecules as building blocks. Specifically, Chapter 42 details knowledge that:

2.c.2: Organisms respond to changes in their external environments.
2.d.1: All biological systems from cells and organisms to populations, communities, and ecosystems are affected by complex biotic and abiotic interactions involving exchange of matter and free energy

Big Idea 4 explores interactions in biological systems. Included in Chapter 42 are the knowledge foundations for understanding that:

4.b.2: Cooperative interactions within organisms promote efficiency in the use of energy and matter.
4.b.4: Distribution of local and global ecosystems changes over time.
4.c.4: The diversity of species within an ecosystem may influence the stability of the ecosystem.

Chapter Review

Concept 42.1 introduces the ecological concepts. An ecosystem comprises a biological community interacting with its environment. This concept requires consideration of the interactions within and between biotic (living organisms) and abiotic (physical environment) factors. A community description includes all of the populations in a given area interacting with each other, while a population is one group of individuals of a single species living and interbreeding in a particular location.

1. A school administrator wrote: Our school is committed to providing a welcoming environment for our diverse community. Describe TWO things that are inaccurate in this statement, considering ecological/biological perspectives.

____
____
____

2. Identify three biotic components found in most forest ecosystems.

____
____
____

3. Identify three abiotic components found in most forest ecosystems.

____
____
____

4. Interactions between the members of a population of unicellular organisms can be compared to the interactions between the cells of a multicellular organism. Discuss TWO similarities and TWO differences between the bacterial community of the human gut and the bacterial community of a forest community.

____
____
____
____
____
____

5. Interactions between them member of a population of unicellular organisms often lead to increased efficiency and utilization of energy and matter, much like that seen in a multicellular organism. Explain three benefits that humans gain from "hosting" the microbial community of the human gut.

____
____
____
____
____
____
____
____

6. Deep sea vent communities are found at the bottom of the sea floor typically near diverging plate boundaries where super-heated water issues from the sea floor. This seawater typically has a temperature ranging from 60°C to 400+°C, and is often highly acidic (<pH 3). This water also contains hydrogen sulfide that many species of bacteria can oxidize to provide energy transfers for chemosynthesis.

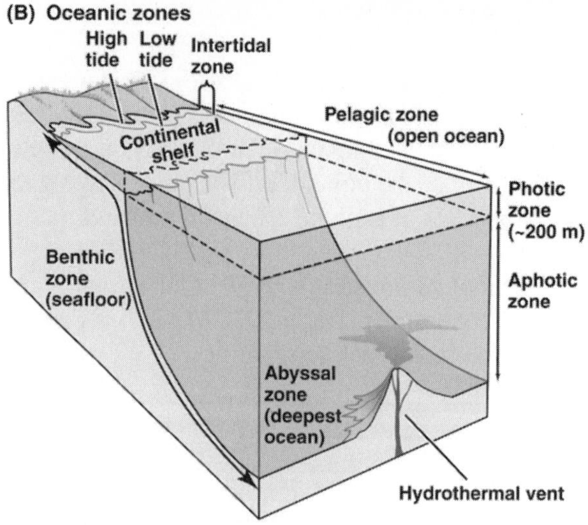

PRINCIPLES OF LIFE, Figure 42.13 (Part 2)
© 2012 Sinauer Associates, Inc.

One novel organism found in the community of organisms surrounding these deep sea vents is the giant tubeworm. These particular tubeworms have no mouth or digestive tract, but harbor inside them they have many bacteria, typically billions of bacteria per gram of tubeworm tissue.

Describe and discuss the route by which carbon sources needed by the tubeworms and bacteria become available in the deep ocean.

____

____

____

____

____

____

Concept 42.2 discusses the abiotic components of ecosystems. The uneven distribution of solar energy across the Earth's surface sets global wind patterns in motion, which in turn also drive the global ocean surface currents. Water and climate diagrams summarize the climate of ecosystems.

7. Explain the difference between a "climate change" and a "weather change." Provide an example of each.

____

____

____

____

8. Describe how the Earth's weather patterns are affected by the tilt of the Earth on its axis. Include equatorial and temperate comparisons.

____

____

____

Describe what would happen if each of the below were to occur:

a. The tilt of the Earth was changed to 5°.

____

____

b. The tilt of the Earth was reversed.

____

____

c. The tilt of the Earth was to increase to 25°.

____

____

Climate diagrams summarize the temperature and precipitation data for a localized area. Answer questions 9 – 12 using the data in the following climate diagrams.

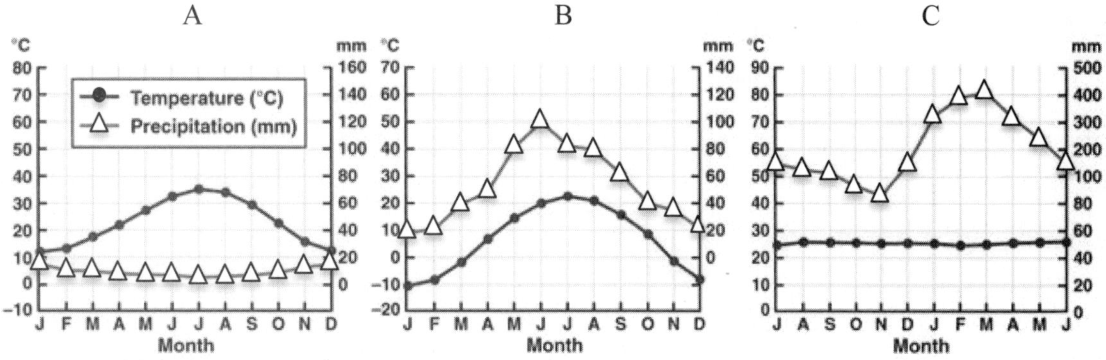

A B C

PRINCIPLES OF LIFE, **Apply the Concept, Ch. 42, p. 829**

9. Select the diagram (A, B or C) that best represents climate near the equator and explain your choice.

____

____

____

____

10. Select the diagram (A, B or C) that represents the latitude 30°, where there is typically a high rate of evaporation from the surface, and dry, warm air sinking to the surface; and explain what type of biome would be found here.

____

____

____

____

11. Select the diagram (A, B or C) that best represents a location that would have the longest growing season and explain your choice.

____

____

____

____

12. Diagram C includes varying precipitation during the year. Select the month with the least precipitation and discuss why this reduction occurs at that time. Also suggest a biome type that might be found in a region with this pattern.

____

____

____

____

13. Organisms' activities are affected by interactions with abiotic factors. Explain how each of the factors below could affect the behavior and health of the following marine organisms:

	Coral reef	Fish	Sea Otter
Nutrient availability			
Temperature change from 72° to 65°			
Salinity change from 32ppt to 25ppt			
A significant pH change from 8.18 to 8.07			

Concept 42.3 surveys the major biomes of the world and their physical conditions.

14. Explain the term biome and discuss how biomes are defined. Describe the type of biome you live in, linking your description to the definition of that biome.

____
____
____
____
____
____

15. Discuss whether or not the human gut can be considered to be a biome, and indicate your perception of how small the smallest biome can be.

____
____
____

16. Draw a climate diagram (annual rhythms of temperature and precipitation) below for a temperate seasonal forest in the southern hemisphere.

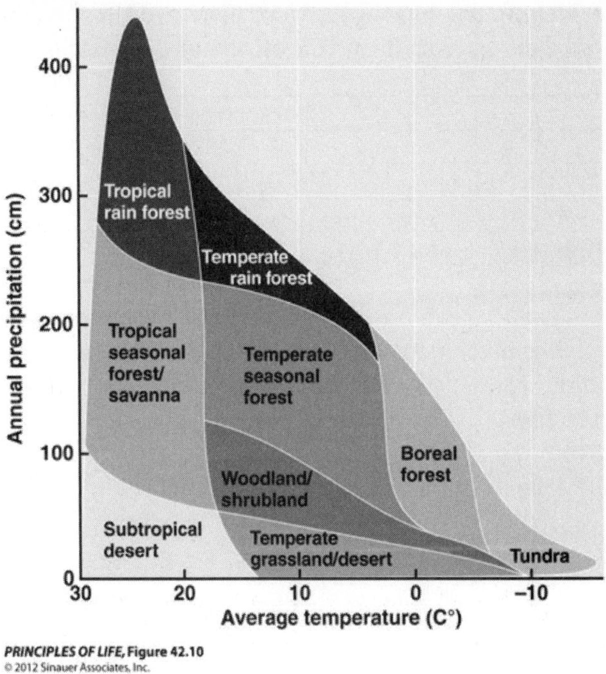

PRINCIPLES OF LIFE, Figure 42.10

Concept 42.4 examines the geologic history of Earth, including plate tectonics, in shaping the distribution of organisms we see today. The movement of the plates has shaped our world into seven major biogeographic regions.

17. Explain how major geological events such as plate tectonics impact ecosystem distribution.

____

____

____

____

____

____

____

____

18. South America and Africa split apart approximately 100 mya. Explain what evidence you would look for in the fossil record to support this.

____

____

____

____

____

____

Concept 42.5 addresses the many ways that human activity has affected the complexity and heterogeneity of the world's ecosystems. Every major period of life has been named in the past with the name generally denoting the dominant life form during that time period. Some have suggested that we are now in a new geological period, the Anthropocene, or "Age of Humans."

19. Identify two human activities that have altered ecosystems. For each, describe how humans have altered the ecosystem and the effects of the alteration on biodiversity.

____
____
____
____
____

Concept 42.6 explores how ecologists study and model the world. To do this, scientists are continually devising novel methods to explore and to observe our natural world. The better the data we have from the natural world, the better the computer models are to model our world, and we gain precision in making predictions.

20. Explain the phrase "garbage in, garbage out" as it relates to the computer modeling of the Earth's ecosystems.

____
____
____
____
____
____
____
____

Science Practices & Inquiry

In the AP Biology Curriculum Framework, there is a set of 7 Science Practices. In this chapter, we will focus on **Science Practice 6:** The student can work with scientific explanations and theories**.** More specifically, practice 6.3: The student can articulate the reasons that scientific explanations and theories are refined or replaced.

Questions 21 asks you to explain how the distribution of ecosystems changes over time by identifying large-scale events that have resulted in these changes in the past (LO 4.20).

21. Identify two large-scale events that have changed the distribution of ecosystems in the past (not human caused). For each, explain how the ecosystems have changed in the past and speculate on how they might change in the future.

Chapter 43: Populations

Chapter Outline

43.1 - Populations Are Patchy in Space and Dynamic over Time
43.2 - Births Increase and Deaths Decrease Population Size
43.3 - Life Histories Determine Population Growth Rates
43.4 - Populations Grow Multiplicatively, but Not for Long
43.5 - Extinction and Recolonization Affect Population Dynamics
43.6 - Ecology Provides Tools for Managing Populations

Chapter 43 begins with a description of populations, composed of individuals of a species that interact with one another in a given area. Population density and population size are two measures of ecological interest. There are many sampling tools for determining population size, but most involve determining the population density and then multiplying this by the area of the population's habitat.

Populations change over time in response to many factors. At a simplistic level, the change in the size of a population (dN) over time (dT), assuming no immigration or emigration, is the number of births (B) less the number of deaths (D), as shown by this equation:

$$\frac{dN}{dt} = B - D$$

Because it is usually impossible to monitor every individual in a population, ecologists calculate *per capita* birth and death rates, i.e., the average individual's number of offspring and the average individual's chance of dying. The difference between these birth and death rates is the per capita growth rate (r). By multiplying r by the population size (N), we obtain an estimate of population-growth rate over time. This is expressed by the equation:

$$\frac{\Delta N}{\Delta T} = rN$$

The life history of a species includes its growth period, development, reproduction, and death of an average individual. They are frequently represented as circles beginning with eggs or mating parents and finishing with the next generation. You probably saw life histories of animals and plants in previous chapters. To the right is the life history of the black-legged tick. Life histories can be very complex, particularly if a species relies on blood meals from other species. Other resources such as light, space, temperature also impact a species life history.

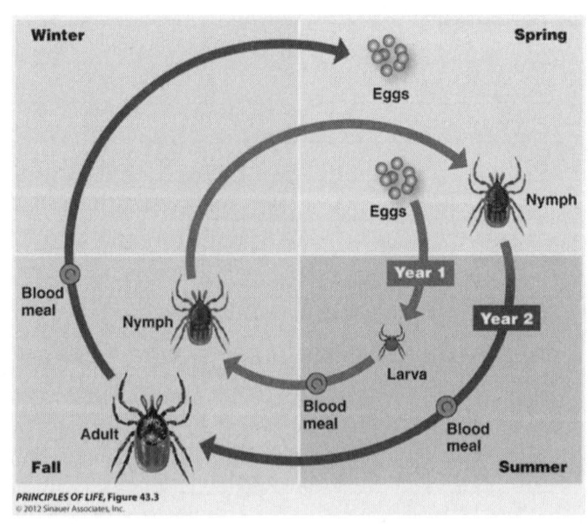

Population growth rates typically fall into two patterns, exponential (multiplicative) and logistic. The equations representing these two models are:

$\frac{dN}{dt} = r_{max}N$ represents the equation for exponential growth

$\frac{dN}{dt} = r_{max}N\left(\frac{K-N}{N}\right)$ represents the equation for logistic growth

Most species will grow multiplicatively until they reach their carrying capacity. At this point, they will fluctuate up or down in response to resource availability or changing environmental conditions. Rarely do we see a species with a consistent population size at its carrying capacity. Most species show frequent increases or decreases in population.

The Big Ideas are a means for organizing the vast amount of information in biology. It is vitally important that you continually work to understand these **Big Ideas** and their many **Enduring Understandings**. Ecology is a great opportunity to look at a topic across the Big Ideas as Chapter 43 spans three of the Big Ideas.

Big Idea 1 recognizes that evolution ties together all parts of biology. In Chapter 43 we look at how limited resources for populations can drive natural selection.

1.a.1: Natural selection is a major mechanism of evolution.
1.c.1: Speciation and extinction have occurred throughout the Earth's history.

Big Idea 2 states that the utilization of free energy and use of molecular building blocks are characteristic fundamental of life processes. Specifically, Chapter 43 includes:

2.a.1: All living systems require constant input of free energy.
2.a.3: Organisms must exchange matter with the environment to grow, reproduce, and maintain organization.
2.d.1: All biological systems from cells and organisms to populations, communities, and ecosystems are affected by complex biotic and abiotic interactions involving exchange of matter and free energy.
2.d.3: Biological systems are affected by disruptions to their dynamic homeostasis.

Big Idea 4 examines the idea that biological systems interact in complex ways. Included in Chapter 43:

4.a.5: Communities are composed of populations of organisms that interact in complex ways.
4.a.6: Interactions among living systems and with their environment result in the movement of matter and energy.
4.b.3: Interactions between and within populations influence patterns of species distribution and abundance.

Chapter Review

Concept 43.1 points out that there are many ecological, aesthetic, and ethical dimensions to the study of population dynamics. Humans have long been managing populations, including herds of cattle, fields of crops, schools of fish, and endangered populations of plants and animals.

1. Identify three populations (plants and/or animals) that are actively managed by humans. For each, identify two factors in their environment that affect its abundance.

a. ____
____
____
____

b. ____
____
____
____

c. ____

____

____

____

2. Edith's checkerspot butterfly (*Euphydryas editha*) extends from British Columbia and Alberta to Baja California. Many subpopulations of this endangered butterfly in California went extinct during a severe drought between 1975 and 1977. The only subpopulation that did not go extinct at that time was the largest one, on Morgan Hill In the San Francisco Bay area. The butterflies are divided into subpopulations, each occupying a patch of suitable habitat. Arrows indicate 9 colonization events in 1986.

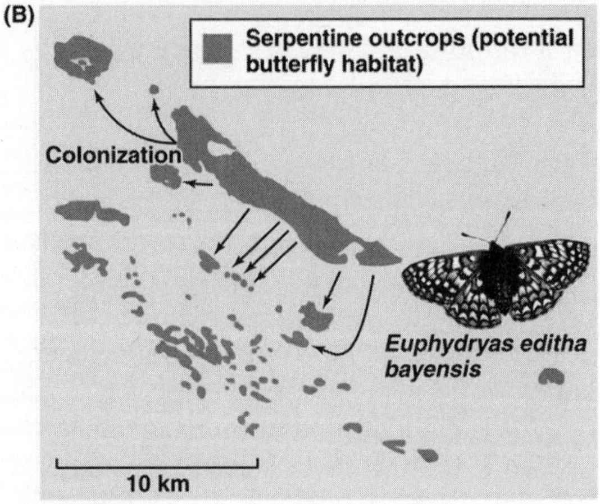

PRINCIPLES OF LIFE, Figure 43.1 (Part 2)
© 2012 Sinauer Associates, Inc.

Identify and describe three factors that would prevent Edith's checkerspot butterfly from colonizing other areas.

____

____

____

____

____

____

____

____

Concept 43.2 examines the study of demographics, how populations change over time. Many formulas are given in this section but the ones you really need to be familiar with are found on the equation sheet, which can be found in the Appendix.

3. What factors other than birth and death rates affect the size of a population?

____

____

____

____

____

____

____

4. Below is a graph showing how densities of acorns, rodents (mice and chipmunks), and the black-legged tick populations vary over time in an oak forest of the state of New York.

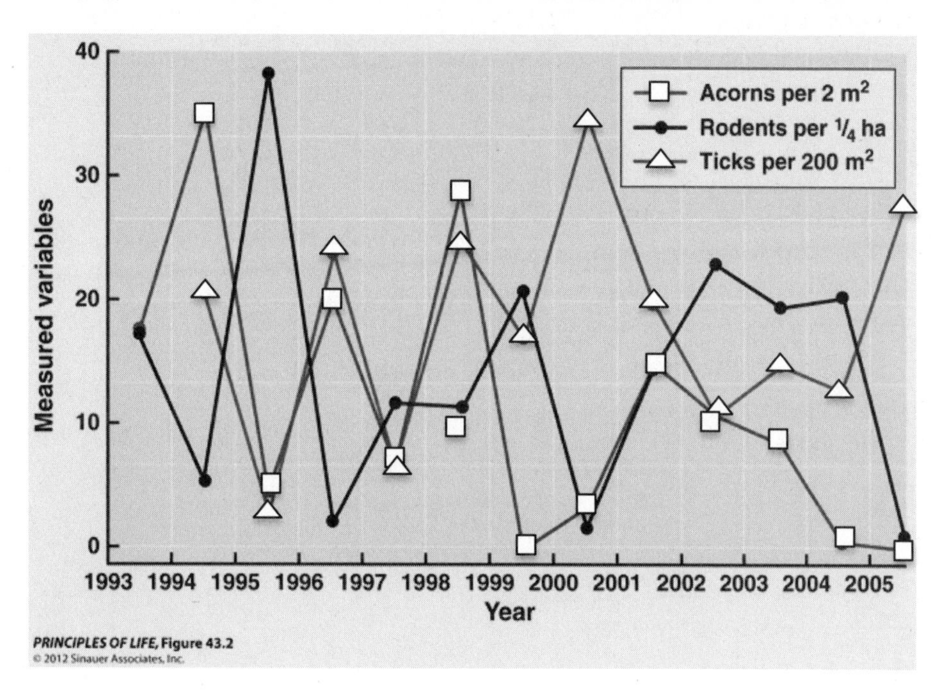

PRINCIPLES OF LIFE, Figure 43.2

a. Describe the possible relationship between acorn and rodent densities.

b. Describe the possible relationship between rodent and tick densities.

c. Ticks are a parasite relying on blood meals to reproduce. Explain how physical factors such as rainfall could influence their population.

d. For the year 2000, calculate the number of ticks in the forest if the forest is 50 square kilometers in size.

e. Identify TWO abiotic and TWO biotic factors that also could influence the population fluctuations seen in the figure.

5. The Life Table to the right shows the survivorship and fecundity for all (n=210 hatchlings) the Cactus Ground Finches born in 1978 on Isla Daphne in the Galapagos archipelago.

TABLE 43.1 Life Table for the 1978 Cohort of Cactus Ground Finch on Isla Daphne

CALENDAR YEAR	AGE OF BIRD (YEARS)	SURVIVORSHIP[a]	FECUNDITY[b]
1978	0 (hatchlings)	1.00	0.00
1979	1	0.43	0.05
1980	2	0.37	0.67
1981	3	0.33	1.50
1982	4	0.31	0.66
1983 Increased rain	5	0.30	5.50
1984	6	0.20	0.69
1985 Drought	7	0.11	0.00
1986	8	0.07	0.00
1987	9	0.07	2.20
1988	10	0.05	0.00
1989	11	0.05	0.00

[a] Survivorship = the proportion of the original cohort (here, 210 birds) surviving from fledging to age x.

[b] Fecundity = average number of young fledged per female of age x.

PRINCIPLES OF LIFE, Table 43.1

a. Identify the age range when the ground finches in this study were reproducing at their highest rate.

____

____

____

b. Survivorship at age X is the proportion of the original cohort that survived to age X. Calculate how many birds did not make it to one year of age.

____

____

____

c. Explain why increased rainfall in 1983 is associated with increased fecundity.

____

____

d. Fecundity is the average number of young per female. Assume that one-half of the population is female. Calculate the number of young born in 1983.

____

____

____

____

6. Assume that a population of 25 women and 25 men, all aged 21 years old, colonizes a previously uninhabited island. Twenty babies are born the next year. In the space below, write in possible life table values for these twenty babies.

Year	Age	Survivorship	Fecundity
1	0	1.00	0.00
5			
10			
15			
20			
25			
30			
35			
40			
45			
50			
55			
60			
65			
70			

7. Explain the survivorship and fecundity values you gave in the Life Table on the previous page.

Concept 43.4 focuses on the dynamics of population growth rates. Most populations tend to grow at either a logistic or exponential growth rate.

8. Explain why populations cannot grow multiplicatively for extended periods of time.

9. The graph to the right shows that many years include an increase in the human population.

a. Discuss whether the growth in the first 11,500 years on the graph should be classified as multiplicative or additive.

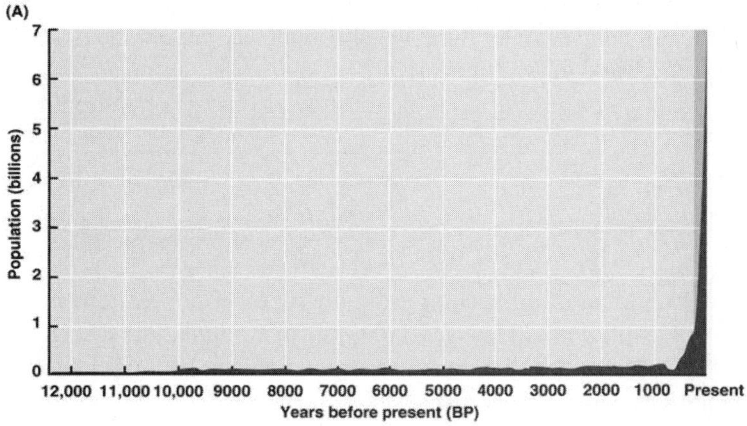

PRINCIPLES OF LIFE, Figure 43.9 (Part 1)

b. Discuss whether the growth in the last 500 years on the graph should be classified as multiplicative or additive.

c. Explain why human offspring continue to grow over a long of a period of time.

d. Discuss how the population crash of ~500 years ago might have affected human genetic diversity.

10. Yellow star-thistle (*Centaurea solstitialis*) is a spiny annual plant native to the Mediterranean region. The species is a noxious weed, unpalatable to livestock, that has invaded several regions of the United States, including an imaginary farm operated by Rancher Jane.

a. Jane carefully inspects her ranch every year. In 2001, there was no star thistle, but in 2002, she discovers that 1 hectare of the 128-hectare pasture has been invaded. In 2003, she finds the weed population has grown to cover 2 hectares. Based on this pattern, predict how many hectares would star-thistle infested in 2004, 2005, and 2006, assuming that the population is growing additively? How many hectares if the population is growing multiplicatively?

____

____

____

____

b. Imagine that Rancher Jane did not see any thistle star in 2001, but she discovers in 2002 that the star-thistle population has suddenly infested 32 hectares of her pasture. How many years does she have until the weed completely covers the pasture if its population is growing additively? Multiplicatively?

____

____

____

____

Concept 43.5 Most of the section dealing with meta-populations is beyond the scope of the AP Biology Framework. Still, it could be useful to know that population size of species does change due to more than just births or deaths. Immigration and emigration can be a big factor in population size as well as habitat fragmentation induced by humans.

11. Extinction of populations in small patches can easily occur. Explain why this is true. Identify a fragmented habitat in your local area. What animals are found in this habitat and describe the risks they face.

____

____

____

____

____

____

12. Sidewalks can easily divide an area into small meta-populations for small organisms such as snails. For each of the barriers below, pick one organism that might be limited in its distribution due to that indicated barrier, and explain why you chose it.

Parking lot

____

____

Small road

____

____

Interstate highway

____

____

Housing developments

Concept 43.6 shows how an ability to predict the dynamics of populations and metapopulations contributes to our ability to influence the fates of natural populations. On the AP exam, you will most likely not encounter a question about the black rockfish or the Edith's checkerspot butterfly that you may have read about in the text of this chapter. However, you may well see population data that you will have to graph or interpret.

13. In the study shown below, moss (dark areas) was scraped off of a rock (light areas) in the pattern shown. In the "insular" or island treatment (I), the patches are surrounded by bare rock that is inhospitable to moss-dwelling small arthropods, and thus a barrier to recolonization. In the "corridor" treatment (C), the patches are connected to the mainland by a 7 X 2 cm strip of live moss. In the "broken-corridor" treatment (B), the configuration is the same as the "corridor" treatment, except that a 2-cm strip of bare rock cuts the moss strip.

After 6 months, the number of species of small arthropods was counted in each of the four areas.

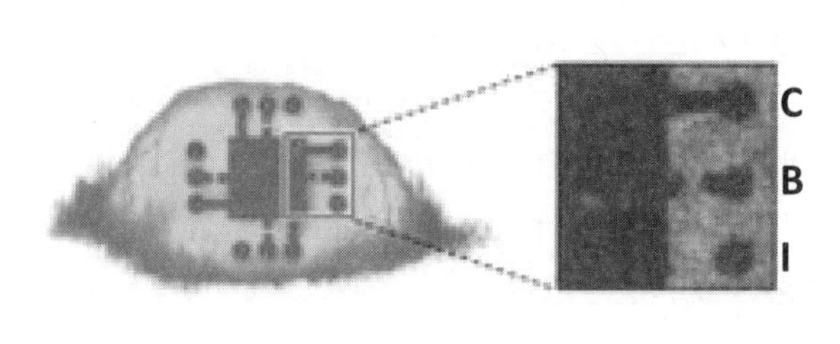

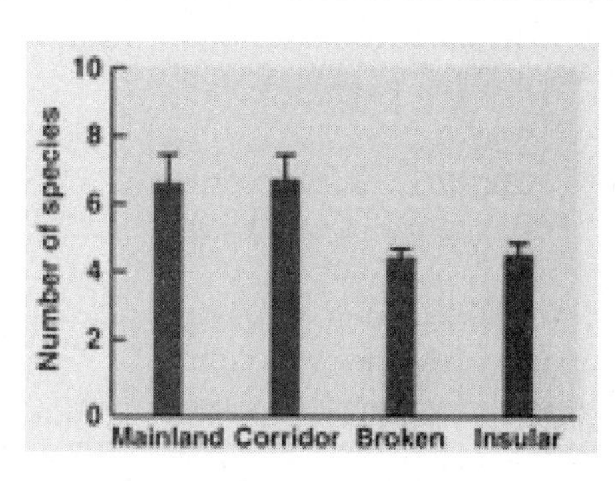

a. Explain why this experiment should be replicated multiple times before strong conclusions can be drawn.

b. Summarize the results shown in the graph.

c. Explain the purpose of the broken corridor (B) treatment.

Science Practices & Inquiry

In the AP Biology Curriculum Framework, there is a set of 7 Science Practices. In this chapter, we will focus on **Science Practice 5:** The student can perform data analysis and evaluation of evidence. More specifically, practice 5.2: The student can refine observations and measurements based on data analysis.

Question 14 asks you to use data analysis to refine observations and measurements regarding the effect of population interactions on patterns of species distribution and abundance (LO 4.19).

14. Rancher Jane (question 10) is thinking of shifting from growing cattle to growing American bison (*Bison bison*), but she needs to know how well bison will do on her ranch. She buys 50 female bison that are already inseminated and places 10 of them, picked at random, into their own pasture. These 10 females serve as a sample from which Rancher Jane collects demographic data over one year. Use her data to answer the questions below. Show all of your work.

FEMALE #	Alive at end of year?	# of offspring
1	Yes	1
2	Yes	0
3	Yes	1
4	Yes	0
5	No	0
6	Yes	1
7	Yes	1
8	No	0
9	Yes	1
10	Yes	0

a. What is the total number of births and deaths among this sample population?

____
____
____
____

b. What are the estimated birth and death rates for the entire bison herd, based on this sample?

____
____
____
____

c. Based on these estimates, what is the size of Rancher Jane's entire bison herd at the end of the year?

____
____
____
____

Chapter 44: Ecological and Evolutionary Consequences of Species Interactions

Chapter Outline

When you watch a butterfly sipping nectar from a flower or a bat swooping and diving after insects at nighttime, you are watching species interacting with each other. The butterfly is drinking the flower's nectar, but it also moves pollen from one flower to another, helping the plant achieve sexual reproduction. In turn, the butterfly is prey for many species of birds. These are but a few of the many and often intertwined relationships between different organisms.

As organisms adapt to avoid being eaten, many new novel structures and features evolve over time; thorns, camouflage, armored shells, bitter (spicy) tasting compounds, to name a few. As one organism evolves to avoid being eaten, so too does the predator adapt (evolve) to the new defensive mechanisms. For example, as seeds became larger and thicker, the bills of birds became larger to be able to crack larger seeds, like the macaw to the right. These consumer-resource interactions lead to evolutionary "arms races" with often astonishing results. Hot chili peppers, for example, synthesize a repellent compound known as capsaicin, which many organisms find too bitter or spicy to eat, yet many people enjoy, and even prefer, spicy dishes.

Picture credit:
Jen Smith/Wikipedia

Chapter 44 has ideas that span Big Ideas One, Two, and Four. The Big Ideas are a means for organizing the vast amount of information we know as biology. Continue to work to understand these **Big Ideas** and across their many **Enduring Understandings**.

Big Idea 1 emphasizes evolution across all parts of biology. Chapter 44 shows that:

1.a.2: Natural selection acts on phenotypic variations in populations
1.c.1: Speciation and extinction occurred throughout the Earth's history.
1.c.3: Populations of organisms continue to evolve.

Big Idea 2 states that the utilization of free energy and use of molecular building blocks are characteristic fundamental of life processes. Specifically, Chapter 44 includes:

2.d.1: All biological systems from cells and organisms to populations, communities, and ecosystems are affected by complex biotic and abiotic interactions involving exchange of matter and free energy
2.d.3: Biological systems are affected by disruptions to their dynamic homeostasis.

Big Idea 4 examines the idea that biological systems interact in complex ways. Included in Chapter 44 are facts and theory showing:

4.a.5: Communities are composed populations of organisms that interact in complex ways.
4.b.3: Interactions between and within populations influence patterns of species distribution and abundance.

Chapter Review

Concept 44.1 notes that symbiotic relationships between organisms can be positive, neutral, or negative. There are five broad categories of interactions between organisms: competition, consumer-resource, mutualism, commensalism, and amensalism.

1. For each of the relationships shown below, mark in the appropriate column if the first species mentioned is affected negatively (-), positively (+), or not at all (0). Do the same for the second species, then write the type of interaction in the third column. The first example is completed for you.

Example	Species 1	Species 2	Type of Interaction
American bison feeds on grasses	+	-	Consumer resource
Wrasses (small fish) clean the teeth of larger fish			
Mosquito feeds on the blood of a deer			
Bread mold secretes penicillin that kills bacteria in the local area which generally have little to no effect on the mold			
A cactus wren builds a nest in a cholla cactus without affecting the cactus.			
A hawk captures a small squirrel for food			
A rabbit rests in the shade of a small bush.			
Deer create a trail through a forest where they routinely travel.			
Plovers (a bird) remove insects from the backs of large animals.			

2. In a forest, many different types of trees can be found. Identify and discuss how three different biotic factors can affect the growth and health of the trees.

____
____
____
____
____
____

3. Explain how predation is different from parasitism.

____
____
____

Concept 44.2 examines how interspecific interactions affect growth. When two species directly compete with each other for resources, *per capita* growth rates are usually less, one species may go extinct, and/or resource partitioning may occur.

4. The intertidal zone along ocean shorelines is rarely bare rock. In most areas, a host of species compete for space to live. Twice a month, as the moon moves around the earth, the tides show increased variation and the high tides are higher than normal, submerging the shoreline which is normally only splashed by waves.

Barnacles are small crustaceans that filter seawater in the intertidal zone for small food particles. Two barnacle species, exhibit interspecific competition as they compete for open space. Rock barnacles, *Semibalanus balanoides,* are found in a narrow range between the lower intertidal zone and the average high tide line. Stellate barnacles, *Chthamalus stellatus*, are found across a broader range, between the lower intertidal zone and the highest high tide line.

a. Draw a diagram of the intertidal zone showing a region where only stellate barnacles are found. Include in your diagram the low tide line, average high tide line, and highest high tide line. Explain how these barnacles can survive above the average high tide line.
b. Explain why rock barnacles cannot survive above the average high tide line.
c. When the two barnacles are found living in the same location, the rock barnacles grow on top of the stellate barnacles, thus killing them. Draw a diagram showing what happens when these two species are found in the intertidal zone together.
d. If a researcher clears off a patch of rock in the middle of the intertidal zone, explain what will happen as both types of barnacles attempt to colonize that zone.

Diagram for (a)	Diagram for (c)

5. In a forest ecosystem, foxes prey on small mammals including rabbits and mice. Explain why foxes are unlikely to consume all of the rabbits, causing their extinction.

____

____

____

____

6. Explain what happens when two different species compete for the same resources in the same location at the same time.

____

____

____

____

Concept 44.3 explores how interactions between species can affect individual fitness, thus shaping evolutionary changes. The interests of consumer and resource species are at odds with each other, leading to an evolutionary "arms race," in which prey continually evolve better defenses and predators continually evolve better offenses, and neither gains any lasting advantage over the other.

7. Fitness in the biological sense is not how fast you can run a mile or how many push-ups an athlete can perform. Define the term fitness in its evolutionary sense and give an example of fitness.

____

____

____

____

____

8. Lake Victoria in Africa is home to many different species of fish known as cichlids. In one location, a researcher saw 5 different species of cichlids repeatedly in the same location. Using the concept of resource partitioning, explain how these different species of fish can coexist in the same location.

____

____

____

____

____

9. The Red Queen analogy is an example taken from Lewis Carroll's *Through the Looking-Glass* with the words "it takes all the running you can do, to keep in the same place." Explain how this analogy could apply to the evolutionary arms race between chili peppers and the animals that eat chili peppers.

____

____

____

____

____

10. Many mistakenly believe that plants purposefully produce fruit for other organisms to eat so that their seeds will be spread to new locations. Discuss why this is false.

____

____

____

____

Concept 44.4 considers how human-introduced species can alter species interactions. Introduced species often end up in environments that lack the regulatory controls found in their native habitat, sometimes resulting in out-of-control growth. Kudzu, purple loosestrife, fire ants, Eurasian weevil are all examples of introduced species that have grown out of control and have altered many ecological relationships in North America. Two illustrative examples of introduced species you should be familiar with follow.

Kudzu, *Pueraria lobata,* is a climbing vine native to Japan and China. It is usually classified as a weed because it climbs over trees or shrubs and grows so rapidly that it kills them due to heavy shading. Kudzu spreads primarily by vegetative propagation via runners that form new plants and by rhizomes. Kudzu was introduced from Japan into the US at the Japanese pavilion in the 1876 Centennial Exposition in Philadelphia. Many people used kudzu to control erosion as it is fast growing and stabilizes loose soil quickly. It is now common along roadsides and other areas throughout most of the southeastern United States. Kudzu can grow at a rate of approximately one foot per day under optimal conditions.

In 1935, the Soil Conservation Service began to test kudzu as a solution to the eroded lands in Alabama and Georgia. After a few years, that agency proclaimed that kudzu, with its fast growth and deep roots, could solve the erosion problems of the south. Many farmers were paid $8 per acre to plant kudzu and prevent erosion. During the Great Depression, the Civilian Conservation Corps planted over 70 million seedlings of kudzu from Maryland to Texas. In the 1950's, and scientists began to worry about kudzu spreading across the South, and kudzu fell from grace and became classified as a weed. But the damage was done, and by the 1980s, kudzu had covered an estimated 7 million acres of land in the south, and is spreading by 320,000 acres per year.

11. Discuss why kudzu continues to spread so quickly across the US.

____
____
____
____
____
____

12. Discuss the impact of kudzu on species diversity in afflicted areas.

____
____
____
____
____
____

Dutch Elm disease is due to a fungus that is spread from tree to tree by bark beetles, quickly killing susceptible elm trees. The fungus is thought to have been introduced to the US on a shipment of logs from the Netherlands in 1928. Although the name may imply otherwise, Dutch Elm disease is most likely native to Asia, and was probably introduced to Europe about 1910. Elm trees, once dominant trees across the Northeast forests of the US, known for their long-term survival of up to 400 years of age, now rarely live for more than 10 – 15 years.

13. Describe why the inadvertent introduction of introduced species has increased during the last century.

____
____
____
____

14. Identify three ways that introduced species are inadvertently introduced to new areas.

Science Practices & Inquiry

In the AP Biology Curriculum Framework, there is a set of 7 Science Practices. In this chapter, we will focus on **Science Practice 4:** The student can plan and implement data collection strategies appropriate to a particular scientific question. More specifically, practice 4.1: The student can justify the selection of the kind of data needed to answer a particular scientific question; and practice 4.2: The student can design a plan for collecting data to answer a particular scientific question.

Question 15 asks you to justify the selection of the kind of data needed to answer scientific questions about the interaction of populations within communities. (LO 4.11)

15. Some consider the black walnut tree an example of an "amensal" species, as it secretes a chemical, juglone, from its roots that harms or kills many neighboring plants. This interaction may also be considered an odd form of consumer-resource if the death of nearby plants removes competition and allows the walnut tree access to greater scarce resources. Design an experiment that would allow you to determine the correct type of ecological interaction exists between black walnut trees and its neighbors.

Chapter 45: Ecological Communities

Chapter Outline

45.1 - Communities Contain Species That Colonize and Persist
45.2 - Communities Change over Space and Time
45.3 - Trophic Interactions Determine How Energy and Materials Move Through Communities
45.4 - Species Diversity Affects Community Function
45.5 - Diversity Patterns Provide Clues to Determinants of Diversity
45.6 - Community Ecology Suggests Strategies for Conserving Community Function

Chapters 43 and 44 focused on populations of species and interactions between two different species. In chapter 45, we move on to examine how multiple groups of species interact to form communities. Communities change as the environment changes, with latitude or elevation, and with other factors such as extinction and colonization, disturbance, and climate change. When disturbances occur in ecosystems, the affected communities usually change their structure in response. Depending on the severity of the disturbance, the original community may or may not return to its native state. Catastrophic disturbances such as volcanic eruptions can create entirely new landscapes for colonization by new communities. On a smaller scale, the local extinction of one species may open a window for a small change in community structure.

Energy and materials move within and through communities. Solar energy enters communities as a readily usable and high-energy input, with one output being heat. Thus, energy flows through an ecological community in a one-way direction. Matter, on the other hand, is continuously recycled within the community as organisms dine on each other, die, and decompose, thereby releasing primary nutrients needed by others in the community. In fact, many of the molecules in your body were not too long ago part of an autotroph, maybe a corn plant or a lettuce plant. Ultimately, the molecules and atoms that comprise our bodies have been recycled millions of times and may once have been in a dinosaur millions of years ago. In brief: energy flows and matter cycles.

Estimating biodiversity in an ecological community provides an indication of its health and stability. Biodiversity is measured by both species richness and species evenness. Species richness is simply how many species are in a community. Species evenness is less intuitive and focuses on how many of each species are present and how the individual organisms are spatially distributed in the community. As biodiversity increases, the productivity and stability of a community generally increase as well.

Humans depend on many ecosystem "services" provided by different organisms or groups of organisms. For example, pollinators are critical to many forms of agriculture, and without them, many crops will fail to form their fruits or vegetables. Managing and restoring ecosystems while maintaining their health are key facets of research and practice in community ecology.

Chapter 45 spans Big Ideas One, Two, and Four. The Big Ideas are a means for organizing the vast amount of information we know as biology. It is vitally important that you continually work to understand these **Big Ideas** and across different **Enduring Understandings**. Chapter 45's emphasis is primarily within Big Idea 4, yet has many ties with Big Ideas 1 and 2.

Big Idea 1 recognizes that evolution ties together all parts of biology. In Chapter 45 we look at how evolution ties with ecology.

1.c.1: Speciation and extinction have occurred throughout the Earth's history.

Big Idea 2 states that the utilization of free energy and use of molecular building blocks are characteristic fundamental of life processes. Specifically, Chapter 45 includes:

2.a.1: All living systems require constant input of free energy.
2.a.2: Organisms capture and store free energy for use in biological processes.
2.d.1: All biological systems from cells and organisms to populations, communities, and ecosystems are affected by complex biotic and abiotic interactions involving exchange of matter and free energy
2.d.3: Biological systems are affected by disruptions to their dynamic homeostasis.

Big Idea 4 examines the idea that biological systems interact in complex ways. Included in Chapter 45:

4.a.5: Communities are composed of populations of organisms that interact in complex ways.
4.a.6: Interactions among living systems and with their environment result in the movement of matter and energy.
4.b.3: Interactions between and within populations influence patterns of species distribution and abundance.
4.b.4: Distribution of local and global ecosystems changes over time.
4.c.4: The diversity of species within an ecosystem may influence the stability of the ecosystem.

Chapter Review

Concept 45.1 states that communities are made up of groups of species that coexist and interact with one another within a defined geographic area. In some communities, the boundaries of the community are determined by the physical habitat: a pond, for example, defines a community of aquatic species that interact much more with one another than with terrestrial species outside the pond. However, the boundaries between different communities often overlap. For example, raccoons from a forest community will forage for food at a pond's edge, while deer will graze on grass in an open meadow and retreat into the forest for protection. Thus, an ecosystem's boundaries are flexible.

1. Ecotones are transitional boundaries between different ecosystems, *e. g.*, the border between a grassy area and a forest. This border area has a mixture of grasses, shrubs, small trees, and sometimes older trees all living in a thin strip. Identify another ecotone and describe the characteristics of the two ecosystems bordering it, and show how it includes some of the features of each bordering community.

____
____
____
____

2. Despite the many decades that have passed after the eruption of the volcano Krakatau, only now are stable communities of organisms becoming established on the island. Explain why these communities are not exactly identical to the communities present when the volcano erupted in 1883.

____
____
____
____

3. The new communities on Krakatau appear to be stabilizing at present. Discuss whether or not these communities will continue to be stable or change in the future.

____
____
____
____

Concept 45.2 examines how communities of organisms change over time. Succession is most dramatic in a community after a major disturbance, and is apparent as a somewhat predictable progression of species coming and going. In some rare circumstances, a new community forms where there is very little preexisting community and little to no soil, such as a lava flow, or a boulder field left by a retreating glacier. Most successional events occur as one population or community replaces another, and this called secondary succession. Climate change, invasive species, and humans are all causes of disturbances leading to succession.

4. Recall that free energy is the energy available for use in an ecosystem. Explain how changes in free energy availability can result in disruptions to an ecosystem. Give three examples of possible disruptions and describe their effects on energy availability. Be sure one disruption increases and one decreases free energy.

____
____
____
____
____
____

5. For each disruption in the question above, discuss what effect that variance will have on the number and size of trophic levels.

____
____
____
____
____
____
____

6. Explain how changes in the number of producers, *e. g.,* by disease outbreaks, can affect the number and size of other trophic levels.

____
____
____
____
____
____

Concept 45.3 describes how energy and matter moves through an ecosystem. Remember, energy flows through an ecosystem, and matter is constantly recycled. Autotrophs trap solar energy and convert it to chemical potential energy. As consumers eat autotrophs, they utilize this energy for metabolism and store some as resources for future use. A good general rule to remember is that of the 10% rule, the total biomass of each trophic level is about one-tenth that of the level it feeds on.

7. Identify two examples of free energy available in the environment autotrophs can utilize for energy.

____
____
____
____
____

8. Dung beetles utilize dung in many different ways. Some eat dung, while others lay their eggs in it, and some form balls of dung and roll the balls into their nests. The diagram shows how the species composition of dung beetles changes In a pile of dung over time.

Rollers
Dweller
Dung
Tunneler

Rollers
Dwellers
Tunnelers

Species
Copris repertus
Drepanocerus setosus
Liatongus indicus
Sisyphus neglectus
Onitis subopacus
Onthophagus cervus
Onthophagus rectecormutus
Onthophagus falsus
Sisyphus longipes
Copris davisoni
Oniticellus cinctus
Copris signatus
Onitis siva

1 3 5 7 14 21
Dung age (days)

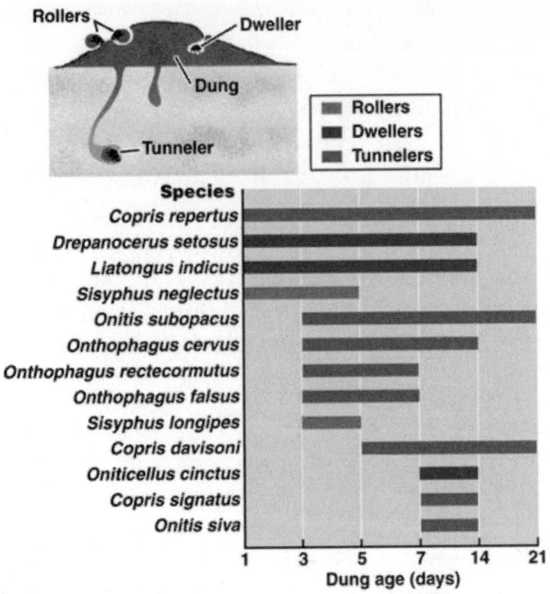

PRINCIPLES OF LIFE, Figure 45.4

Explain why the composition of dung beetles in the dung changes over time.

9. Frequently after disturbances, such as a tree falling down or a fire, an ecosystem will go through a predictable succession of changes. But the original community is not always restored. Explain why.

10. Below is a food web found in the grasslands of Yellowstone National Park. Use the food web to answer the following questions.

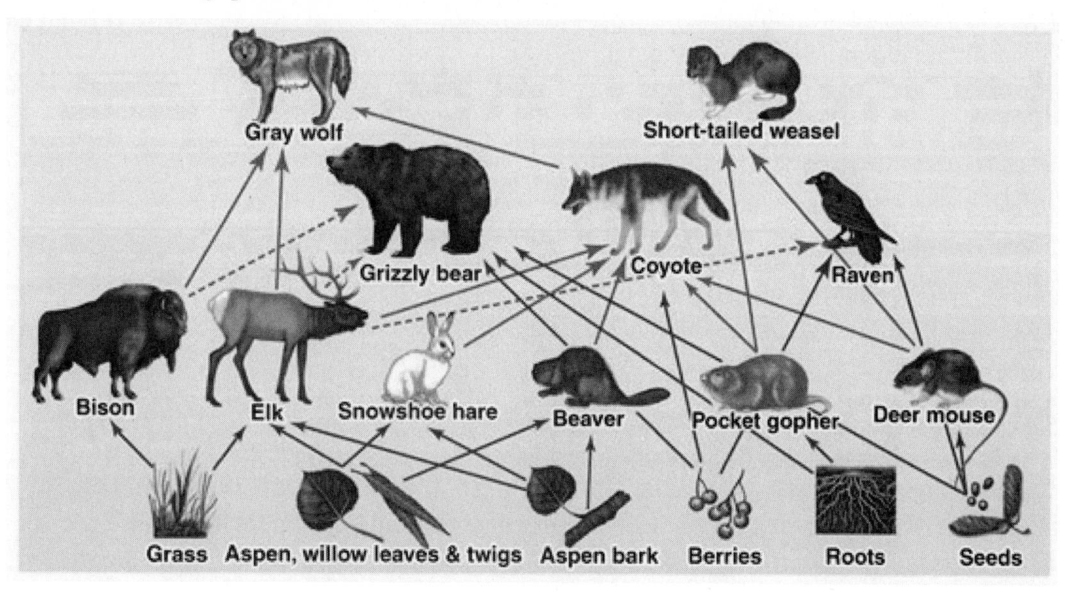

a. Briefly describe what would happen if each of the following groups were removed from the park. Limit your answer to two trophic levels.

Coyotes

____

____

All primary producers

____

____

All consumers

____

____

b. Discuss TWO possible consequences of adding a new primary consumer, one that feeds only on grasses, to the park.

____

____

____

____

11. Explain why there are fewer high-level consumers (wolves and weasels) than primary consumers in Yellowstone Park.

____

____

____

____

____

12. One theory for the development of large coal and oil deposits is that the fungi that lived prior to 300 mya lacked the ability to digest the cellulose of trees and tree ferns. Describe what the landscape must have looked like in a forest or swamp during this time period and what would have happened if fungi did not evolve the ability to digest cellulose more quickly.

Concept 45.4 examines species diversity and how it is measured through species diversity and species evenness.

13. Explain how each of the following abiotic factors can affect the stability of populations.

Water

Nutrient availability

Availability of nesting materials and sites

14. Explain how each of the following biotic factors can affect the stability of populations.

Food chains and food webs

Species diversity

Population density

Algal blooms

15. The hypothetical communities of fungi (mushrooms) pictured below are all the same size (12 individuals) but differ in species richness (3 versus 4 species) and the species' relative abundances, both of which affect diversity. Describe where you might find an example of each of the diagrams (models) in the real world. You do not need to limit yourself to just 3 or 4 species in your examples.

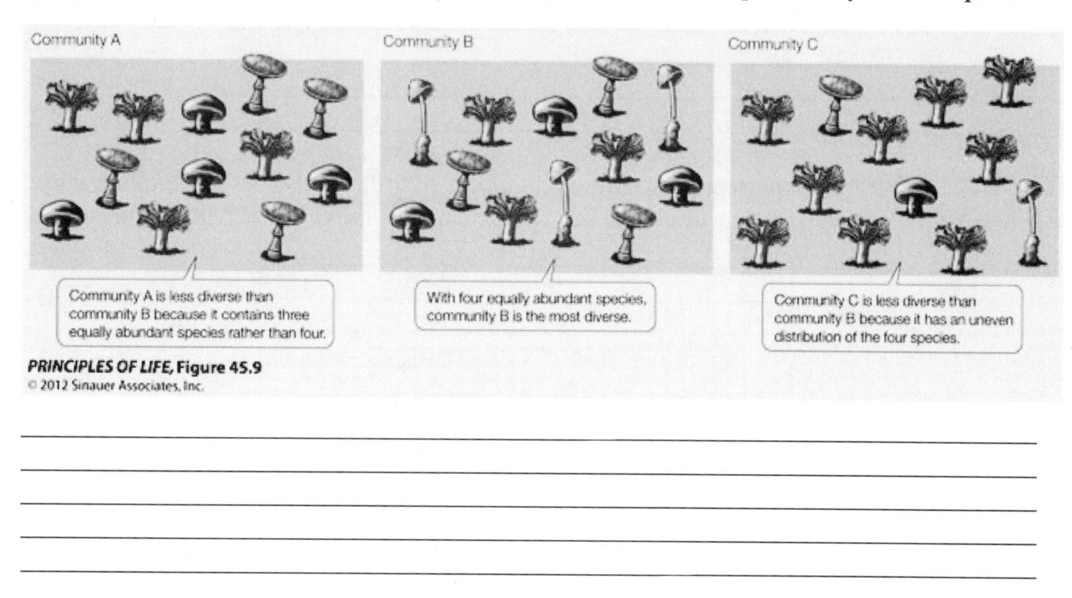

Community A
Community B
Community C
Community A is less diverse than community B because it contains three equally abundant species rather than four.
With four equally abundant species, community B is the most diverse.
Community C is less diverse than community B because it has an uneven distribution of the four species.

PRINCIPLES OF LIFE, Figure 45.9

____

____

____

____

____

Concept 45.5 studies the factors that influence biodiversity. The study of island biogeography has contributed greatly to our understanding of biodiversity and to the structure and function of ecological communities.

16. By experimentally removing all the arthropods from four small mangrove islands of equal size but different distance from the mainland, two researchers were able to observe the process of recolonization and compare the results with the predictions of island biogeography theory. Below is a graph of their data.

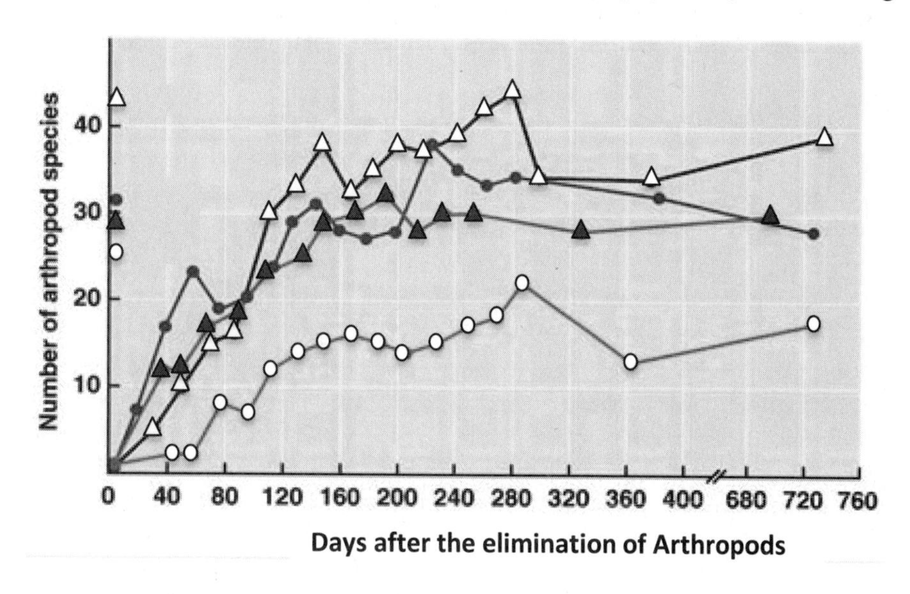

a. Which line would represent an island closest to the mainland? Explain.

____

____

____

b. Why do most of the lines level off after approximately 200 days and not continue to increase?

c. The dots at the left of the graph represent original numbers of arthropods present. Explain why the recolonization populations are similar to but not the same as the original populations.

17. Abundant biodiversity is often equated with stability in ecosystems. Explain why.

18. For TWO of the below, discuss how human impact accelerates change at local and global levels. Be sure to give specific examples or locations and species.

- Logging,
- Slash and burn agriculture,
- Urbanization,
- Monocropping,
- Infrastructure development (dams, transmission lines, roads)

19. Choose one item from the list below and discuss how introduced species can exploit new niches free of predators or competitors while devastating native species.

- Dutch elm disease
- Potato blight
- Small pox [historic example for Native Americans]

____
____
____
____
____
____
____
____

Concept 45.6 explores the value in maintaining and fostering the biodiversity of communities. The demise of many thousands of hives of bees due to colony collapse disorder has caused alarm. Colony collapse disorder is thought to be a complex interaction between multiple variables, including possible a fungus and protozoans, but why it happens and where it came form is still a puzzle

20. Some ecologists propose buying land in Central America and forming a green belt of vegetation from South America up through tropical America. Others propose saving large isolated regions as parks. Discuss which proposal would be more successful at maintaining species biodiversity throughout Central America.

____
____
____
____
____
____
____
____

Science Practices & Inquiry

In the AP Biology Curriculum Framework, there is a set of 7 Science Practices. In this chapter, we will focus on Science Practice 6: The student can work with scientific explanations and theories. More specifically, practice 6.4: The student can make claims and predictions about natural phenomena based on scientific theories and models.

Question 21 asks you to make scientific claims and predictions about how species diversity within an ecosystem influences ecosystem stability (LO 4.27).

21. Species diversity is impacted by many different phenomena.

For three of the following disruptions, explain how the disruption can impact the dynamic homeostasis or balance of an ecosystem.

- Invasive and/or eruptive species
- Human impact
- Hurricanes, floods, earthquakes, volcanoes, fires
- Water limitation
- Salination

Chapter 46: The Global Ecosystem

Chapter Outline

After looking at how organisms affect each other and how ecosystems function, we now turn our attention to looking at ecology on the global scale. All of the Earth's systems are interrelated. As the wind patterns change over the Pacific Ocean, precipitation and climate changes occur throughout the world. When a large volcano in the south Pacific erupts, injecting gases and ash into the upper atmosphere, there can be long-lasting effects for many years throughout the world.

A change in climate alters the productivity of an ecosystem. The gross primary productivity (GPP) is the total amount of energy trapped by the producers in an ecosystem, but not all of this energy is available to consumers. Much of the GPP is dissipated as heat energy as fuel molecules are metabolized. The net primary productivity (NPP) refers to the amount of energy that is captured in the tissues of primary producers. NPP is therefore a key indicator of the amount of energy available to the next trophic level, i.e., consumers. Measurements of an ecosystem's NPP are important descriptors of that ecosystem's changes.

The recycling of elements and nutrients in ecosystems can be traced in its biogeochemical cycles. The nutrients typically move from one form to another, into and out of major pools and sinks (inaccessible forms), at varying flux rates. The most important cycles to focus on are the water, carbon, and nitrogen cycles.

The biogeochemical cycle of the element carbon has direct impacts on the world's climate. The increased release of carbon dioxide and other greenhouse gases from human activities is a significant contributor to warmer winters and summers, more powerful storms, and more unpredictable weather. Much like a pane of clear glass, the atmosphere is allows the passage of most solar radiation, warming the Earth's surface. Much of this energy is reradiated as heat energy, but the greenhouse gases in the atmosphere preventing this energy from reaching space. Instead, the captured energy warms our atmosphere in the process, resulting in the global climate change we are presently observing around the world.

Climate changes cause many changes and disruptions to ecosystems: shifts in plant species, changes in migration patterns, and melting of polar ice caps, to name just a few. The challenge we face is how we as humans can minimize our influence towards climate change. Organisms have substantially changed the Earth's atmosphere in the past, most notably resulting from the evolution of photosynthetic organisms releasing oxygen gas. Other physical changes in the past have resulted in warming and cooling trends. But now, one species, *Homo sapiens,* is single handedly causing large-scale changes. With cooperation around the world, hopefully humans can mitigate our effects on the world.

Chapter 46 has ideas that span Big Ideas Two and Four. The Big Ideas are a means for organizing the vast amount of information we know as biology. It is vitally important that you continually work to understand these **Big Ideas** and across different **Enduring Understandings**.

Big Idea 2 states that the utilization of free energy and use of molecular building blocks are characteristic fundamental of life processes. Specifically, Chapter 46 includes:

2.a.2: Organisms capture and store energy for use in biological processes.

2.a.3: Organisms must exchange matter with the environment to grow, reproduce, and maintain organization.

Big Idea 4 examines the idea that biological systems interact in complex ways. Included in Chapter 46:

4.a.6: Interactions among living systems and with their environment result in the movement of matter and energy.

4.b.4: Distribution of local and global ecosystems changes over time.

4.c.4: The diversity of species within an ecosystem may influence the stability of the ecosystem.

Chapter Review

Concept 46.1 examines how net primary productivity (NPP) varies with climate change.

1. Explain why NPP does not measure the rate of exchange between organisms and all of the nutrients they need.

____

____

____

2. Use the diagram below to answer the following questions.

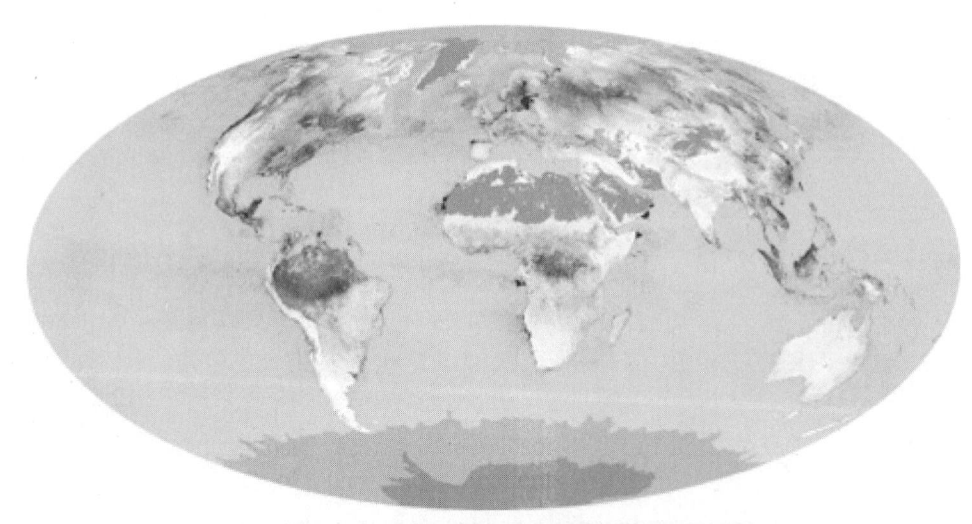

Image Credit: ASA map by Robert Simmon and Reto Stöckli, based on MODIS data.

a. Identify two reasons why NPP is higher in the equatorial regions.

____

____

____

____

b. Summarize the pattern of NPP across North America and discuss why these patterns might exist.

3. Design an experiment to determine the identity of the limiting nutrient in a grassland ecosystem.

4. Explain why the limiting nutrient in a grassland ecosystem might not be the same limiting nutrient in another ecosystem.

Concept 46.2 identifies and discusses biological, geological, and chemical processes contributing to biogeochemical cycles.

5. Describe the energy source that drives the constant recycling of matter.

6. Explain why Earth is an open system in regards to energy. Relate this to the concept of free energy and entropy.

7. Explain why Earth is a closed system in regards to matter. Relate this to the concept of free energy and entropy.

Concept 46.3 considers the three most important biogeochemical cycles in ecosystems: water, carbon, and nitrogen cycles. One result of too much nitrogen runoff from agricultural fields and animal wastes is the cultural eutrophication in bodies of water. This can result in hypoxic or anoxic zones in these ecosystems.

8. Complete the table below for each of the three major biogeochemical cycles. Sinks are the locations where the nutrient is inaccessible for long periods.

Cycle	Fluxes	Pools	Sinks
Water			
Carbon			
Nitrogen			

9. Choose either the nitrogen or carbon cycle and draw it out showing the major fluxes, pools, and sinks. Include in your diagram examples of actual living organisms.

10. Excess nutrients (nitrates) running into a body of water can create an algal bloom. Explain how the overabundance of algae can lead to anoxic zone (no dissolved oxygen) in the body of water.

____

____

____

____

____

____

11. Although nitrogen gas (N_2) is approximately 78% of the Earth's atmosphere, it is very often a limiting nutrient in many ecosystems. Explain why we need to constantly apply nitrogen as fertilizer to agricultural fields, despites its abundance in air.

____

____

____

Concept 46.4 relates biogeochemical cycles and global climate, with particular attention on the greenhouse effect and the recent increases of greenhouse gases in the atmosphere.

Use the graph below showing the concentration of carbon dioxide in the atmosphere to answer questions 12 & 13.

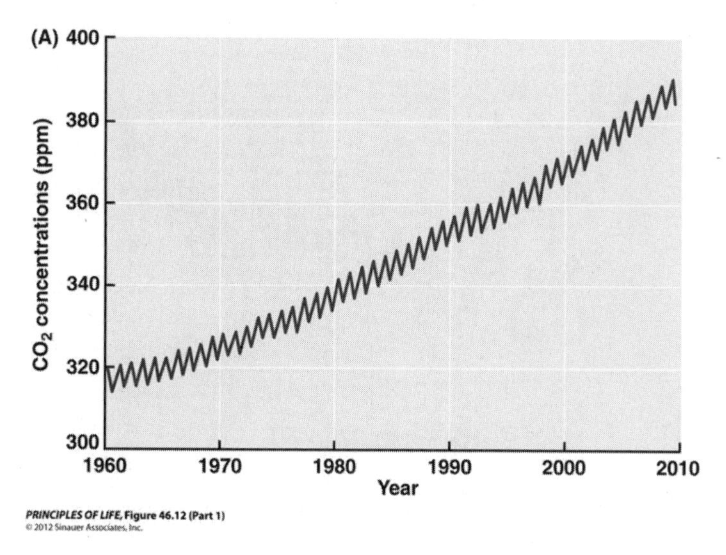

12. Draw a one-year diagram or flow chart, using living organisms, that represents the rise and fall of carbon dioxide each year in the atmosphere in the span of one year. Assume the data are from a temperate forest.

13. Calculate the percent increase of carbon dioxide in the atmosphere in 2010 compared to the amount present in 1980.

14. Carbon dioxide is not the most potent greenhouse gas, yet it is the one most often discussed. Explain why.

____
____
____
____
____

Concept 46.5 looks at the effects of global climate change on the earth's ecosystems, and discusses prediction of how ecological systems are changing as a result.

Use the graph below showing a computer's simulation of temperature in the atmosphere and the actual results (dark line) to answer questions 15 - 18.

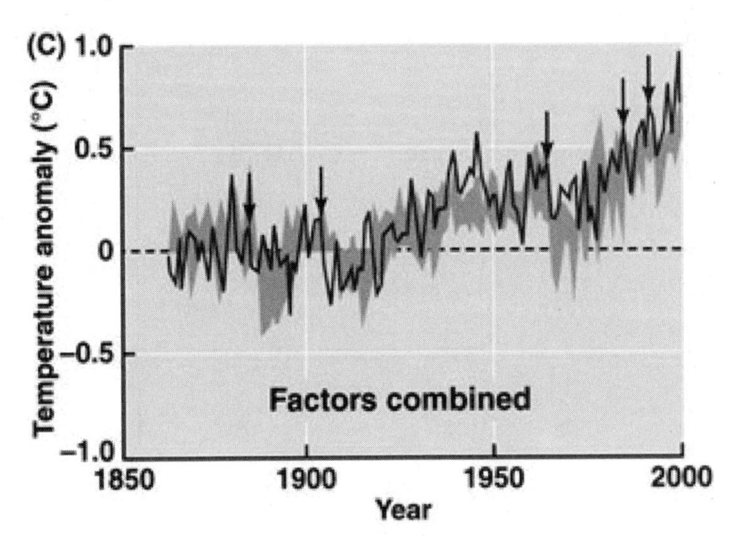

15. Explain the dashed line across the middle of the graph.

____
____
____

16. The arrows represent volcanic eruptions. Discuss what the data show for the years following volcanic eruptions and explain how this change occurred.

____
____
____

17. Discuss what likely happened to ecosystems in the years following the volcanic eruptions.

____
____
____

18. Carbon dioxide concentrations have continued to rise, and there was a major volcanic eruption in Iceland in 2011. Expand the graph to 2012 with these two factors in mind.

Concept 46.6 focuses on humans and their impact on the environment, raising the hope that humans can mitigate their effect on the environment.

19. Interview one or two older members of your local community who have spent the majority of their lives in your area. How has the local climate changed over their lifetime? Has the local vegetation or animal life changed over the years?

20. You have been elected President of the United States. What is the first thing you would do to change how Americans affect the growing issue of climate change; how would you work to influence other global leaders to reduce their greenhouse gas emissions?

Science Practices & Inquiry

In the AP Biology Curriculum Framework, there is a set of 7 Science Practices. In this chapter, we will focus on Science Practice 2: The student can use mathematics appropriately. More specifically, practice 2.2: The student can *apply mathematical routines* to quantities that describe natural phenomena.

Question 19 asks you to apply mathematical routines to quantities that describe interactions among living systems and their environment, which result in the movement of matter and energy (LO 4.14).

21. The chart below shows how NPP varies with different ecosystems. Calculate the percent NPP of temperate forests of the entire NPP for terrestrial ecosystems. Express your answer to the nearest tenth and bubble your answer into the grid provided. Show your work below.

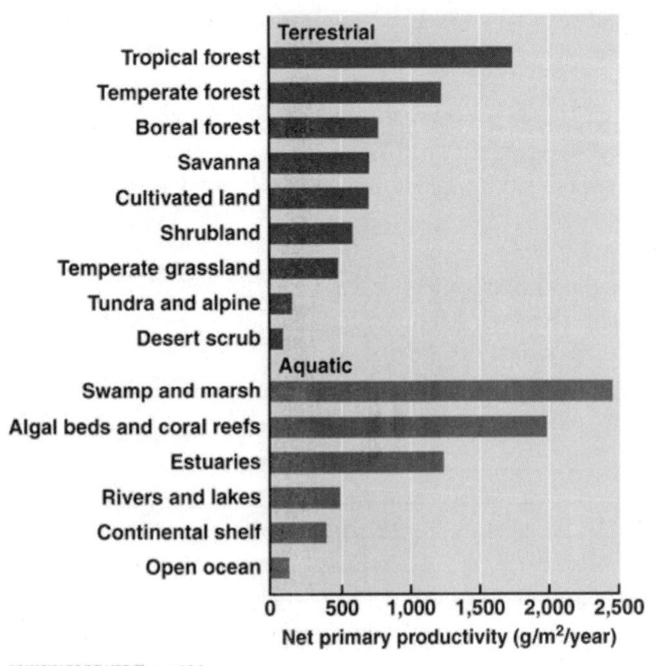

PRINCIPLES OF LIFE, Figure 46.1
© 2012 Sinauer Associates, Inc.

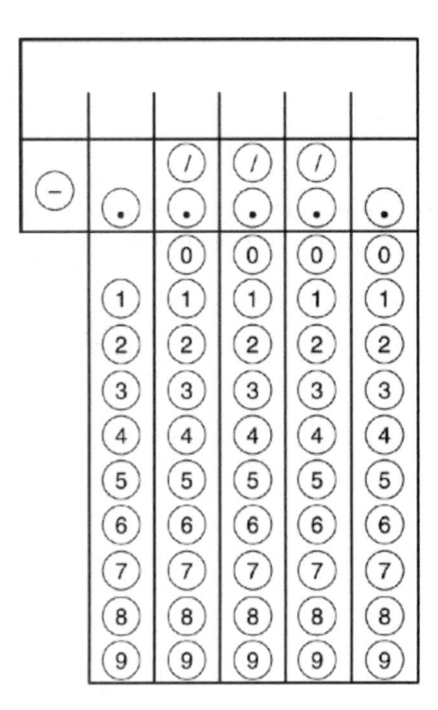

Preparing for the AP Exam

An Introduction to the New AP Biology Course

The revised AP Biology exam, to be administered for the first time in 2013, will be a very different exam from the former version in some ways and will be very familiar in other ways. Like the exams in previous years, there will be multiple choice questions and essay answers to free-response questions (FRQs). The primary difference is that the new exam has a greater focus on the conceptual understanding of biology. With some reduced content, more mathematical-based questions, and an increased emphasis on learning and inquiry, it is an exciting time to be studying AP Biology.

The revised AP Biology Science Exam is divided into four **Big Ideas**. Each of these is weighted equally on the exam. The four topics are shown below and each is described in more detail in the Course Description that can be downloaded from APCentral (apcentral.collegeboard.com).

1. **Evolution**
2. **Cellular Processes: Energy and Communication**
3. **Genetics and Information Transfer**
4. **Interactions**

There is a concise list of Learning Objectives from which test questions will be written. Gone are the days when anything from an introductory biology textbook was fair game. All of the test questions will be written as defined by the scope of the new Learning Objectives in the Curriculum Framework. Content areas including phylogeny, much of plant biology, and many of the human systems have been removed from the required material.

Many of these topics are not required, but are found in the illustrative examples. For most topics, there are three or four different illustrative examples to choose from to apply a concept. Your teacher should have used at least one illustrative example for each topic in your class. But, just because a content area is not on the AP exam, doesn't mean you cannot learn it! Pay close attention to these illustrative examples and be ready to apply them to free response questions. You should plan on reviewing at least one of these examples from each section. The examples will not show up directly on the multiple-choice section without background information, but they will most likely appear in the free response questions. As an example, if you look at Essential Knowledge 1.A.2, you will see:

c. Some phenotypic variations significantly increase or decrease fitness of the organism and the population. To foster student understanding of this concept, instructors can choose an illustrative example such as:

- Sickle cell anemia
- Peppered moth
- DDT resistance in insects

A short free response question could ask you to show how phenotypic variations can increase the fitness of a population using one of these three illustrative examples.

The format of the exam will be:

Section I (90 minutes; 50%)

Part A: 63 Multiple Choice questions

Part B: 6 Grid-In questions

Section II (10 minutes reading time + 80 minutes response time)

2 long free-response questions; 10 points each (25%)

6 short free response questions; 3-4 points each (25%)

The new AP Biology exam will have an equation sheet for students to use and you may use a calculator as well. This calculator may not be a scientific or graphing calculator. Only simple four function calculators (with square root) are allowed. The calculation questions or gridded items will be input into a grid like the one shown to the right.

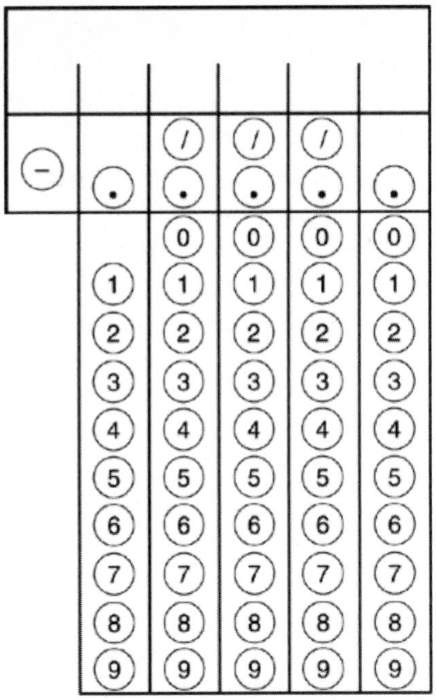

The answers for multiple-choice questions are recorded on a Scantron sheet and are graded by a computer. You will only earn points for correct responses on multiple-choice questions. There is no guessing penalty for incorrect answers, so answer **ALL** questions. Another change from previous years is that the multiple-choice items have four possible responses, a – d. Additionally, you will not find answers such as *all of the above* or *none of the above* among the answer choices and there are no negative stems such as "Which of the below is NOT…" As before, the questions are all mixed up topically. So do not expect all of Big Idea One to be first, followed by Big Idea Two.

All of the essays written by AP Biology students in response to the FRQs are hand-graded by several hundred high school teachers and college professors who come to the AP Reading each June to score the exams. To minimize the subjectivity that might come into play in assessing the answers to each FRQ, a grading rubric is developed for each question, and the graders practice grading and check each others' work for accuracy. The rubric instructs the Reader on how to allocate points and score the exam. Released free response questions and their rubrics showing point distributions can be found at AP Central. Be aware: the free response questions found from 2012 and earlier are of the long free response style. Until the summer of 2013, there were no short style questions. The College Board will release a practice exam before the summer of 2013 that will have examples of grid-in items and short free response items.

The Educational Testing Service will typically send you your scores in early July. Depending upon your choice, the scores may be sent directly to colleges and universities.

The results will fall into one of the following categories on a five-point scale:

5	Extremely well qualified (to receive college credit)
4	Well qualified
3	Qualified
2	Possibly qualified
1	No Recommendation

Many colleges and universities accept a score of 4 or 5 for credit and placement. Some colleges will give you credit for a score of 3. You may go online to the college you are interested in attending and find the score that institution accepts or contact the admissions office for more information.

Strategies for Preparing for the AP Exam

- As you study, develop your own shorthand notes and/or acronyms. For example, in studying signaling, you might realize that signaling implies a stimulus to start the signal, a sensor that detects the signal, and receptor-based effector mechanisms that lead to response. By using the underlined letters in that sentence, the acronym SSRE can help you remember some key ideas about signaling. These "codes to self" can be jotted down during the essay-reading time you are given before you start writing essay answers on the test.

- Make reading and writing an integral part of your coursework. Practice writing essays. Again: practice writing essays! Pull out some blank pieces of paper and write an essay on any topic from the course, selected by you or by your teacher, using only your memory to guide you. When you are done, open your notes and text and see what you forgot to mention, or what you stated incorrectly. Then take out some more blank paper and do it again. See how long it takes you to write your answer, and you will get an estimate of how long it will take during the actual exam. You will be better prepared to pace yourself when the real essays have to be written.
- Two to three weeks before the exam, you should start your review. Make a review schedule and stick to it! Schedule time to take the practice exam in this book, time to review each major part and time to review problematic parts in more detail.
- Re-try the *Strive* exercises, covering up your old answers with a page of blank paper.
- Take an actual full-length test in advance. One is included at the end of this book for you.
- Make sure you understand the structure of the exam and how it will be graded.
- Get plenty of rest the night before the exam. Make sure that Test Day finds you well-rested and alert, ready to focus all of your attention on the test booklet that will be placed in front of you. By this time in your school year, <u>especially</u> if you have worked faithfully on this *Strive* guide, you will know as much biology as you can. Only if you have had a decent sleep the night before the test will your brain pull that information out, instructing your arm to fill in the right bubblesand to write lucid and focused essays. Cramming late into the night before the exam will not give your brain the time it needs to digest the material and transfer it into long term memory.
- Be comfortable going into the exam – wear layers in case it is colder or warmer than you think it will be. It is important that you are comfortable while taking the exam.
- Bring a snack and some water for break time. Your brain needs energy to work at this level for sustained periods.

Strategies for Taking the AP Exam

1. The first thing that you should do is to carefully read the question. Be sure that you answer the question that is asked, and only that question, and that you answer all parts of it. As you read the question, pay particular attention to **bold** and <u>underlined</u> words; they are **<u>important</u>**.
2. Write in sentences and paragraphs. Writing in outline form will earn you NO credit!
3. A good strategy is to define terms, give an example, and then elaborate on your example – how does your example work? Say something about each of the important terms that you use. Define the simple terms and the complex terms. Often it is the easy definitions that are left out.
4. Remember, if you do not write it, you cannot get points for it!
5. Answer the question parts in the order called for. It is best not to skip around within the question. The questions do not have to be answered in any particular order. If the essay is set-up with internal parts A, B, C, then answer them in that order, clearly labeling each part.
6. Write clearly and neatly. It is foolhardy to confuse the reader with lousy penmanship. And while on this topic, please use a blue or black pen – these colors are much easier to read.
7. Go into detail that is on the subject and to the point. Be sure to include the obvious (for example, "light (photons) is necessary for photosynthesis"). Answer the question thoroughly.
8. If you cannot remember a word exactly, take a shot at it—get as close as you can. If you don't have a name for a concept, describe the concept. Remember the test is often graded conceptually. So you may well get the point for the idea you have described.
9. Remember that no detail is too small to be included as long as it is to the point. I use the narrow pattern shotgun analogy here. You want to write down everything you know about the topic, but stay on the topic. For instance, if a question asks about the structure of DNA, talk about the helix, etc. Do not waste time on RNA, expression, or Mendelian genetics. There is no such thing as something so obvious (as long as it is correct) that you should omit. By no means should you re-state the question, but make sure you fill in details that you might think are obvious, and you

should score some points for those obvious ideas! Any time you provide a nugget of truth in a written answer, you have a chance to earn some points.

10. Carefully label your diagrams (they get no points otherwise) and place them in the text at the appropriate place, not detached at the end. Be sure to refer to the diagram in your essay.
11. Widen your margins a little. This will make the essay easier for most folks to read.
12. Understand that the exam is written to be hard. The national average for the essay section will be about 50% correct, that is 5 points out of a possible 10 on each essay. It is very likely that you will not know everything. This is expected, but it is very likely that you do know something about each essay, so relax and do the best you can. Write thorough answers.
13. Do not leave a blank. Often on difficult essays the mean will be very low. If you are struggling with a question, then most likely, many others are also struggling. So if you write a little, you may well hit the mean.

DON'T

1. These are not English Essays. Don't restate the question. Don't write a rambling introduction. Don't finish with a conclusion that restates things. You will only be wasting your valuable time.
2. Don't waste time on background information or a long introduction unless the question calls for historical development or historical significance. Answer the question.
3. Don't ramble—get to the point!
4. Don't panic or get angry because you are unfamiliar with the question. You probably have read or heard something about the subject—be calm and think.
5. Don't scratch out excessively. One or two lines through the unwanted word(s) should be sufficient.
6. **Don't write in the margin**, especially on the short free response questions.
7. Don't worry about spelling every word perfectly or using exact grammar. These are not a part of the standards the graders use. It is important for you to know, however, that very poor spelling and grammar may create an impression in the sub-conscious of the grader.
8. Don't write sloppily. It is easy for a grader to miss an important word when he/she cannot read your handwriting. If a word cannot be read, it cannot earn a point.
9. Avoid writing with large loops, which extend above and below the lines. This makes it very hard to decipher your writing.

Full-Length Practice Exam

Directions:

Following is a full-length Practice Exam for you to hone your skills for the AP Biology Exam. This exam is comprised of two parts. Allocate 3 hours of time to take this exam. It is important that you practice your test taking skills by simulating real testing conditions. The answers and sample rubrics follow at the end of the test.

1. Take this test in a single three hour block.
2. Do not use headphones, or allow music, phones, or texting to interrupt you.
3. Allow 90 minutes for Section I; this includes the multiple choice and calculation questions. Then put this section away – you may not go back to it during Section II.
4. Take a 10-15 minute break, have a snack.
5. Give yourself 90 minutes to complete Section II. Watch your time carefully and stop after the time is up.
6. After you check your results, be sure to go back over and review the areas you struggled with. Each question is correlated to a specific Learning Objective in the AP Biology Curriculum Framework. Look up the ones you erred on and study these topics.

BIOLOGY: SECTION 1
Part A: Multiple-Choice Questions
Part B: Grid-In Questions
Time – 90 Minutes

Part A Directions: Each of the questions or incomplete statements below is followed by four suggested answers or completions. Select the answer that is best in each case and enter the appropriate letter in the corresponding space on the answer sheet. When you have completed part A, you should continue on to part B.

1. An example of a point mutation with a significant effect on phenotype is the one that causes sickle cell disease. Select the statement that predicts the change in phenotype that occurs if two sickle cell trait carriers have a child who is homozygous recessive for sickle cell disease.

(A) A chromosomal mutation resulting in a homozygous recessive individual for that gene will have a heterozygous advantage over the other genotypes.
(B) Point mutations in coding regions of DNA are silent and will not result in a genetic disorder.
(C) An individual who has two copies of the gene for sickle cell disease will have defective, sickle-shaped red blood cells.
(D) The sickle cell allele differs from the normal allele by one base pair resulting in a polypeptide that differs by many different amino acids from the normal protein in the red blood cells.

2. Bioluminescence in a species of bacteria requires a specific density of bacterial cells. Select the best explanation of the cell-to-cell communication that accounts for this phenomenon.

(A) One species of bacteria can release chemical substances that are sensed by another species of bacteria causing both species of bacterial cells to emit light.
(B) At low population densities, the bacteria cannot secrete the critical concentration of the chemical signal needed to evoke the bioluminescence response.
(C) Bioluminescent bacteria are able to create light when one bacterial cell comes in contact with another bacterial cell of the same species.
(D) The concept of quorum sensing is used as an explanation for the activities involved in forming biofilms.

3. Select the best explanation of a deviation from Mendel's models of the inheritance of traits.

(A) Only one copy of a rare allele is needed for expression in males while two copies of the rare allele must be present for expression in females.
(B) A male with an X-linked gene can pass it on only to his sons.
(C) Daughter carriers for an X-linked mutation can only pass the mutation on to her sons.
(D) Sex-linked disorders are "linked" to the sex of the organism.

4. Nervous tissue transmits information by electrochemical processes, generating action potentials and chemical communication at synapses. Select the statement that describes what occurs when a nerve process transmits information to a muscle cell.

(A) The hormone epinephrine (adrenalin) increases blood flow to a muscle cell which will then cause the muscle cell to secrete acetylcholine and the muscle fibers will contract.
(B) A sensory neuron's axon will stimulate the sodium-potassium pump in a muscle cell membrane to begin pumping sodium ions out of the muscle cell and the muscle cell will contract.
(C) When an action potential arrives at the neuromuscular junction, acetylcholine will diffuse from the neuron process and move across the synaptic cleft to a muscle cell membrane receptor.
(D) Acetylcholine contains calcium ions that bind to troponin in the muscle fiber allowing myosin cross bridges to form which will result in muscle contraction.

5. Select the correct model that accurately illustrates how genetic information is translated into a polypeptide.

(A) **DNA**: TTA – CCC – GUC – GAG

→ **RNA**: ATT - GGG – CAG – CTC

→ four amino acids in a polypeptide

(B) **DNA**: TAC – ACC – GTG – GGC

→ four amino acids in a polypeptide

(C) **RNA**: AUG – UGG – CUC – CCG

→ **DNA**: TAC – ACC – GTC – GGC

→ four amino acids in a polypeptide

(D) **mRNA**: AUG – ACC – CUC – CCG

→ four amino acids in a polypeptide

Questions 6 – 9

The protein leptin, a product of the obese gene (*ob*), is implicated as a hormonal signal in energy homeostasis. It is secreted by adipose (fat) cells: as they get larger, more leptin enters the blood. Leptin binds to receptor proteins on hypothalamic neurons, triggering behavioral changes that decrease food intake. In some individuals, genetic mutations that interfere with leptin signaling are associated with the development of obesity.

6. Choose the mechanism that is likely most similar to the relationship of leptin and food intake.

   (A) Increased levels of estrogens cause a gonadotropin surge in the middle of the ovarian cycle.
   (B) Clumped of platelets at the site of injury leads to a reduction in blood leakage.
   (C) In an action potential, the initial influx of sodium ions leads to more influx of sodium ions.
   (D) A person shivering from being cold stops shivering as the body heats up.

7. Several different mutations affect leptin signaling in humans. Select the category of mutations that would likely cause the portrayed defect in the mRNA of the leptin-receptor gene.

Codon (5' – 3')	131	132	133	134
Wild Type mRNA	CGA	UUA	UGA	AUA
Mutated mRNA	CGA	UUA	UAA	UA_

(A) Frameshift
(B) Nonsense
(C) Missense
(D) Point

First base	Second base: U	C	A	G	Third base
U	UUU Phe	UCU Ser	UAU Tyr	UGU Cys	U
	UUC Phe	UCC Ser	UAC Tyr	UGC Cys	C
	UUA Leu	UCA Ser	UAA Stop	UGA Stop	A
	UUG Leu	UCG Ser	UAG Stop	UGG Trp	G
C	CUU Leu	CCU Pro	CAU His	CGU Arg	U
	CUC Leu	CCC Pro	CAC His	CGC Arg	C
	CUA Leu	CCA Pro	CAA Gln	CGA Arg	A
	CUG Leu	CCG Pro	CAG Gln	CGG Arg	G
A	AUU Ile	ACU Thr	AAU Asn	AGU Ser	U
	AUC Ile	ACC Thr	AAC Asn	AGC Ser	C
	AUA Ile	ACA Thr	AAA Lys	AGA Arg	A
	AUG Met	ACG Thr	AAG Lys	AGG Arg	G
G	GUU Val	GCU Ala	GAU Asp	GGU Gly	U
	GUC Val	GCC Ala	GAC Asp	GGC Gly	C
	GUA Val	GCA Ala	GAA Glu	GGA Gly	A
	GUG Val	GCG Ala	GAG Glu	GGG Gly	G

8. Though the hormone leptin has a metabolic function in amphibians, it also appears to be a factor in mate selectivity by toads. One hypothesis is that toads with a plentiful food supply, resulting in elevated secretion of leptin, will have adequate energy to successfully reproduce.

High Water		**Low Water**	
No Leptin	**Leptin Injection**	**No Leptin**	**Leptin Injection**
Prefer own species	Prefer other species	Prefer other species	Prefer other species

   Select the research question that could be plausibly generated by the data in the table.

   (A) How does leptin activity affect mate selection in a particular species of toad?
   (B) What is the effect of water height on leptin production?
   (C) Why does decreased leptin production have no impact on mate selection?
   (D) What effect does leptin activity have on habitat selection in toads?

9. Select the organelle dysfunction that is most likely responsible for an individual in whom leptin insensitivity is due to defective insertion of leptin receptors in cell membranes.

   (A) Abnormal proton pumps in the mitochondria.
   (B) Accumulation of misfolded proteins in the endoplasmic reticulum.
   (C) Failure of the lysosomes to produce an enzyme.
   (D) Nutrient deficiencies causing disruption of the chloroplasts.

10. Signal transduction is the process by which a signal is converted to a cellular response. Signaling cascades relay signals from receptors to cell targets, often amplifying the incoming signals, with the result of appropriate responses by the cell.

    Select the statement that describes a model that expresses the key elements of signal transduction pathways by which second messengers are often essential to the function of the cascade.

    (A) G-proteins act as intermediates between a receptor and effector and can bind to three different molecules and activate the protein effector.
    (B) The binding of an epinephrine molecule leads to the production of many molecules of cAMP, thus activating many enzyme targets leading to the release of glucose molecules from glycogen.
    (C) Steroid hormones are lipid-soluble and readily pass through the membranes of target cells, beginning the cascading events when they reach receptors inside the target cells.
    (D) Protein kinase receptors on the surface of a liver cell can bind to the hormone insulin and activate a cascade of enzymatic reactions eventually catalyzing glucagon to release glucose.

11. Meiosis is a type of cell division that consists of two nuclear divisions. From the following, select the choice that is most representative of how the process of meiosis has evolved to serve its function(s).

    (A) Meiosis results in two daughter cells with the same number of chromosomes as their parent cells.
    (B) Meiosis produces four sperm cells with the haploid number of chromosomes inside each cell.
    (C) Meiosis produces a gastrula from the zygote that is genetically similar to the parent cell.
    (D) Meiosis halves the nuclear chromosomes content and generates diversity.

12. Natural selection can act on characters with quantitative variation in several different ways. The graph shows the distribution of birth weight and mortality. Which of the following statements best describes the type of selection shown in the graph?

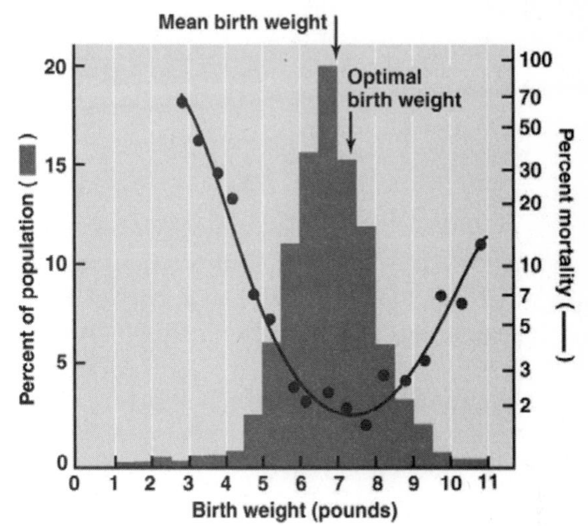

    (A) Human birth weight is influenced by disruptive selection because some babies born at either extreme for this trait may still survive.
    (B) Human birth weight is influenced by directional selection because the optimal birth weight for babies is between 7-8 pounds, shifted slightly from the middle of this graph.
    (C) Human birth weight is influenced by stabilizing selection because babies that weigh more or less than average are more likely to die soon after birth.
    (D) Human birth weight is not influenced at all by natural selection because humans are able to control almost every aspect of their environment.

Questions 13 – 15 use this pair of graphs.

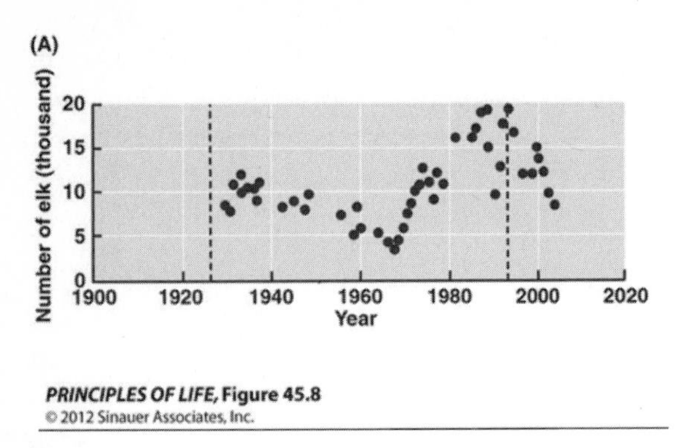

PRINCIPLES OF LIFE, Figure 45.8
© 2012 Sinauer Associates, Inc.

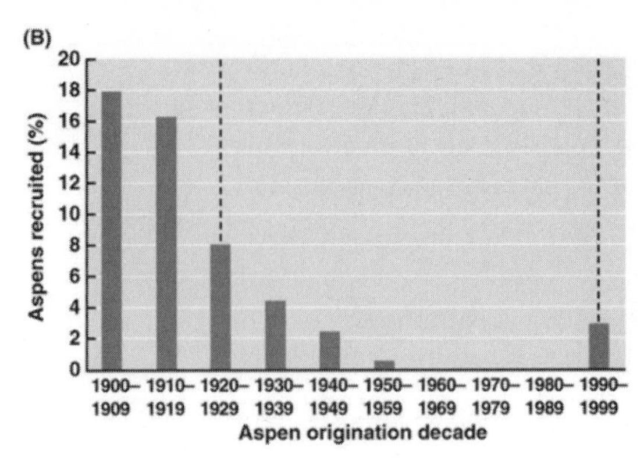

Figure A shows annual data collected in Yellowstone National Park of elk (herbivore) and aspen (tree) populations from the early 1900's to present. Prior to the 1920's there was a strong wolf presence in the park that kept the elk population in check (wolves feed on a variety of large mammals, including elk). By 1926, the wolves had been eliminated from Yellowstone; in 1995, wolves were reintroduced to the park.

13. According to the data, which of the following statements best explains what happened to the aspen population from the 1920s – 1990s?

    (A) The presence of wolves prior to 1920 made the soil incapable of supporting the aspen population.
    (B) In the absence of wolves, browsing (eating) by elk prevented the establishment of young aspen trees.
    (C) The population of aspen is dependent upon a high density of elk.
    (D) The decrease in population of big herbivores was most likely responsible for the decline of the aspen forests within this time frame.

14. According to the data, which of the following statements best describes the dynamics of these three populations in Yellowstone National Park from 1920 to present?

    (A) Predators play a key role in determining the presence and abundance of many species.
    (B) Removing predators from a community is generally the most ecologically sound way to ensure the survivorship of the herbivore species.
    (C) The reintroduction of wolves to Yellowstone National Park in 1995 was unlikely to have had an effect on the regeneration of the aspen forests.
    (D) The removal of elk from the Yellowstone community would have been an equally sound environmental practice, which would have led to the regeneration of the aspen forests.

15. Which of the below scenarios could enable scientists to determine if wolves are indeed a keystone species in Yellowstone National Park.

    (A) Count the number of aspen trees over the next 50 years after wolves have been reintroduced to the park.
    (B) Fence off half of the park from wolves and determine the ratio of wolves to aspen trees in each area to see if the wolves have an effect on aspen trees.
    (C) Fence off a large area of the park from wolves that is very similar to the area inhabited by the wolves and monitor the number of different species in each area.
    (D) Monitor the growth of the wolf population each year to determine if they are maintaining a stable population (K selected) or increasing at a logarithmic pace (r-selected).

Community A **Community B** **Community C**

***PRINCIPLES OF LIFE*, Figure 45.9**

16. The diagram above shows three hypothetical communities of fungi with varying species richness and abundance. Based on the diagram, which of the following is an accurate statement about the communities?

    (A) Community A is more diverse than community B because it contains more individuals of each species that is present.
    (B) Community B is more diverse than community C because it has four equally abundant species.
    (C) Community C exhibits the same diversity as community B because they each have four species present within the community.
    (D) Since each community is composed of the same number of individual organisms, species diversity cannot be determined.

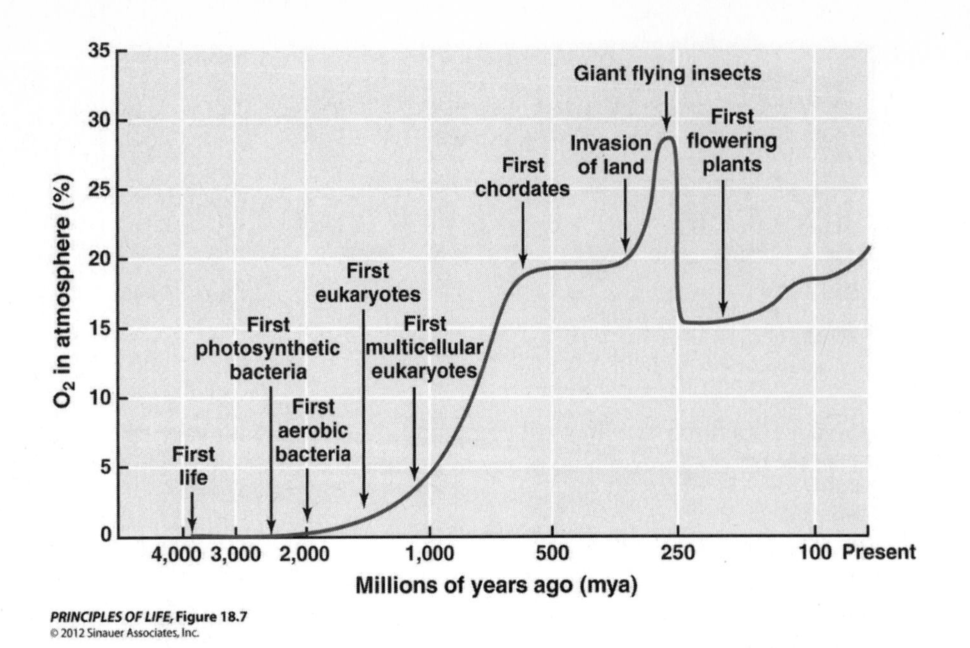

PRINCIPLES OF LIFE, Figure 18.7

17. The graph above illustrates that changes in the concentration of atmospheric oxygen over millions of years have strongly influenced, and been influenced by, living organisms. Choose the prediction that is scientifically compatible with the data.

    (A) The presence of chordates in a community tends to increase the available oxygen concentration in the atmosphere.
    (B) The presence of flowering plants in a community tends to decrease the available oxygen concentration in the atmosphere.
    (C) The removal of insects from a community increases the density of the flowering plant population.
    (D) The upper limit on body mass in poikilothermic animals is influenced by the availability of atmospheric oxygen.

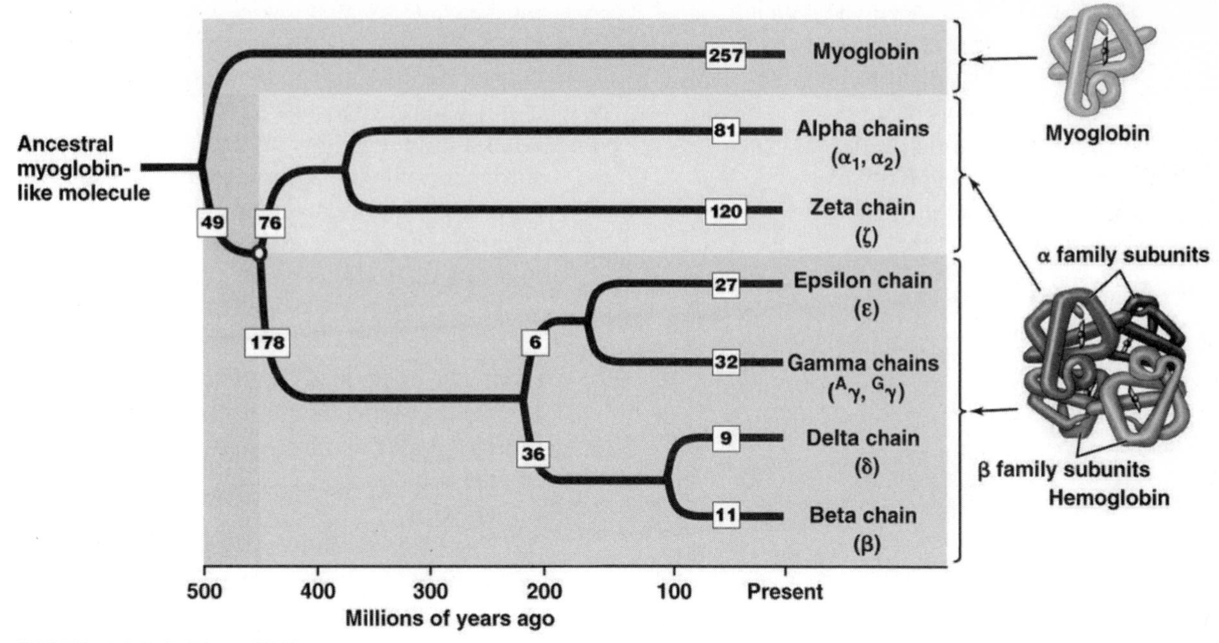

***PRINCIPLES OF LIFE*, Figure 15.22**

18. This gene tree shows the divergence of the globin molecule from an ancestral myoglobin-like molecule to present. The boxed numbers indicate the estimated number of DNA sequence changes along that branch of the tree. Choose the most correct interpretation of the data presented.

 (A) Hemoglobin and myoglobin are estimated to have diverged 450 million years ago.
 (B) The alpha chains and the zeta chain are estimated to have diverged 450 million years ago.
 (C) The α-globin and β-globin gene clusters are estimated to have diverged 450 million years ago.
 (D) The epsilon and beta chains are estimated to have diverged 450 million years ago.

19. Genome size varies tremendously among organisms with some correlation between genome size and organism complexity. The graphic above displays data for the percent of the genome encoding functional genes. Which of the following statements accurately describes the visual data display?

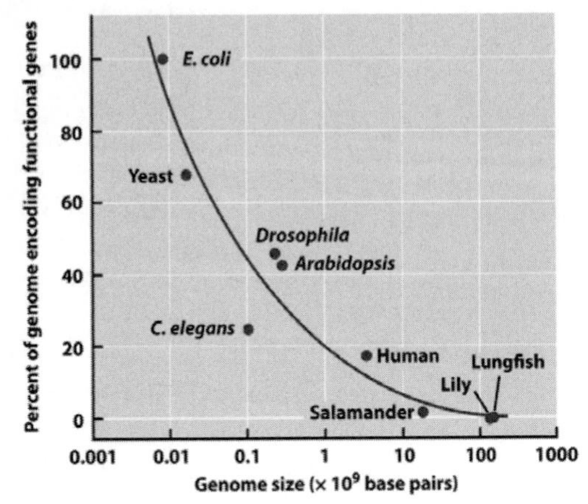

(A) Most of the DNA of bacteria and yeasts encodes RNAs or proteins, but a large percentage of the DNA of multicellular species is non-coding.
(B) Most of the DNA of multicellular species encodes RNAs or proteins, but a large percentage of the DNA of unicellular species is non-coding.
(C) Most human DNA, unlike unicellular organisms, encodes important proteins.
(D) The overall size of the genome of bacteria and yeasts is considerably larger than the genome size of multicellular species.

20. Movement of an organism is a direct attribute of two systems interacting with each other. An organism analyzes the localized environment and determines when and where to move by unleashing an action potential triggering a release of calcium ions in a cell specialized for movement, leading to a contraction. Which two systems in an organism contribute to this coordinated effort?

(A) Respiratory and Circulatory
(B) Nervous and Muscular
(C) Immune and Muscular
(D) Respiratory and Muscular

21. Gardeners know that laying down a thick layer of mulch, shredded organic material, helps to foster the growth of established plants and prevent weeds from germinating. After applying a thick layer of mulch, a gardener noticed that the established plants grew faster and very few weeds grew up through the mulch. Which of the below is a logical hypothesis as to why few weeds grow up through the mulch?

(A) Toxic compounds are released through the decomposition process.
(B) Decomposing mulch uses up many of the soil nutrients preventing weeds from growing.
(C) The decomposing mulch uses up oxygen and reduces light levels.
(D) Decomposing mulch uses up CO_2 needed by the weeds for photosynthesis.

22. Plants require many trace nutrients and minerals such as copper, nickel, and magnesium. Many enzymes and processes stabilize and make ATP biologically active with the use of magnesium. Which of the below requires magnesium?

(A) Xylem cells for transport of water in vascular bundles
(B) Aquaporins for diffusion of water in cell membranes
(C) Plasmodesmata allowing the movement of cytoplasm from cell to cell
(D) Chlorophyll for photosynthesis in chloroplasts

23. A researcher measured the amount of DNA in a group of cells immediately following mitosis and found an average of 8 picograms of DNA per nucleus. How much DNA would the cells have during the S phase of mitosis?

(A) 4 picograms
(B) 7 picograms
(C) 15 picograms
(D) 24 picograms

Questions 24 – 26

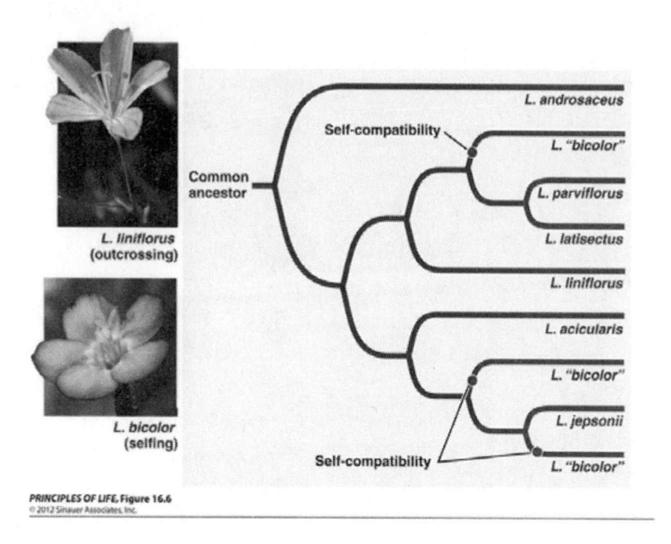

PRINCIPLES OF LIFE, Figure 16.6

The figure above shows a portion of phylogeny of the *Leptosiphon* (a flowering plant in the phlox family) genus. The red dots represent three separate occasions when the self-compatibility characteristic (the ability of a flower to regularly fertilize their ovules using their own pollen) evolved in this plant genus.

24. According to this figure, which of the following represents a monophyletic group?

    (A) *L. androsaceus, L. "bicolor", L. parviflorus*
    (B) *L. liniflorus, L. acicularis, L. "bicolor", L. jepsonii*
    (C) *L. androsaceus, L. jepsonii, L. "bicolor"*
    (D) *L. "bicolor", L. parviflorus, L. latisectus*

25. What is the best explanation for the *L. "bicolor"* species name showing up in three different locations within the phylogeny of this plant genus?

    (A) Convergent evolution of self-compatibility originally fooled taxonomists into classifying three separate species as *L. "bicolor."*
    (B) In order to eliminate confusion, taxonomists named all three of the self-compatible species in this plant genus the same scientific name.
    (C) The phylogeny suggests the exact same species evolved three different times over the course of history.
    (D) The phylogeny suggests the three species are actually varieties of the same plant species on different branches of the phylogenetic tree.

26. Which of the below techniques would be a plausible technique for scientists to determine if the three separate groups of flowers known as L. bicolor are indeed one species or are 3 separate species.

    (A) Compare the flower structure of *L. bicolor* specimens with that of other members of the *Leptosiphon* genus
    (B) Compare DNA or protein sequences of *L. bicolor* specimens with other species of the *Leptosiphon* genus.
    (C) Compare the height of different *L. bicolor* specimens with other members of the *Leptosiphon* genus.
    (D) Compare the gross productivity of different *L. bicolor* specimens with other members of the *Leptosiphon* genus

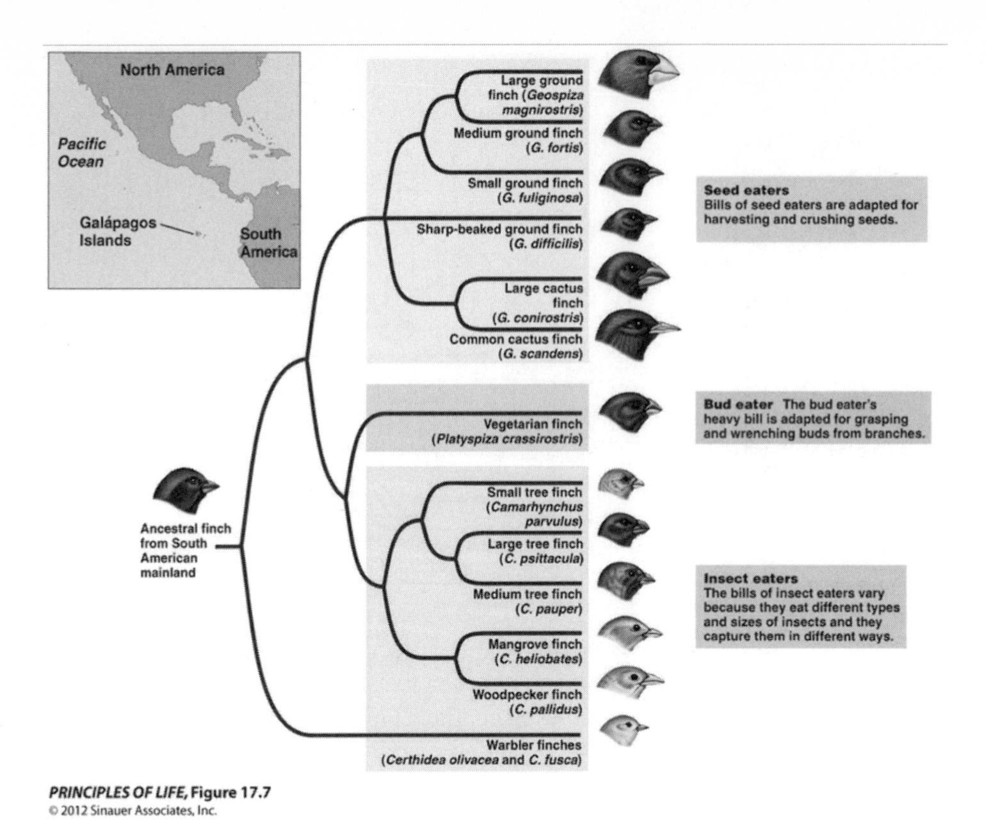

***PRINCIPLES OF LIFE*, Figure 17.7**

27. The different species of Darwin's finches shown in the phylogeny in the figure above have all evolved on islands of the Galapagos archipelago within the past 3 million years from a single South American finch species that colonized the islands. The islands are far apart with a variety of environmental conditions. Which of the following statements best represents the type of speciation that gave rise to these fourteen finch species?

(A) Sympatric speciation
(B) Allopatric speciation
(C) Post zygotic isolating mechanisms
(D) Behavioral isolation

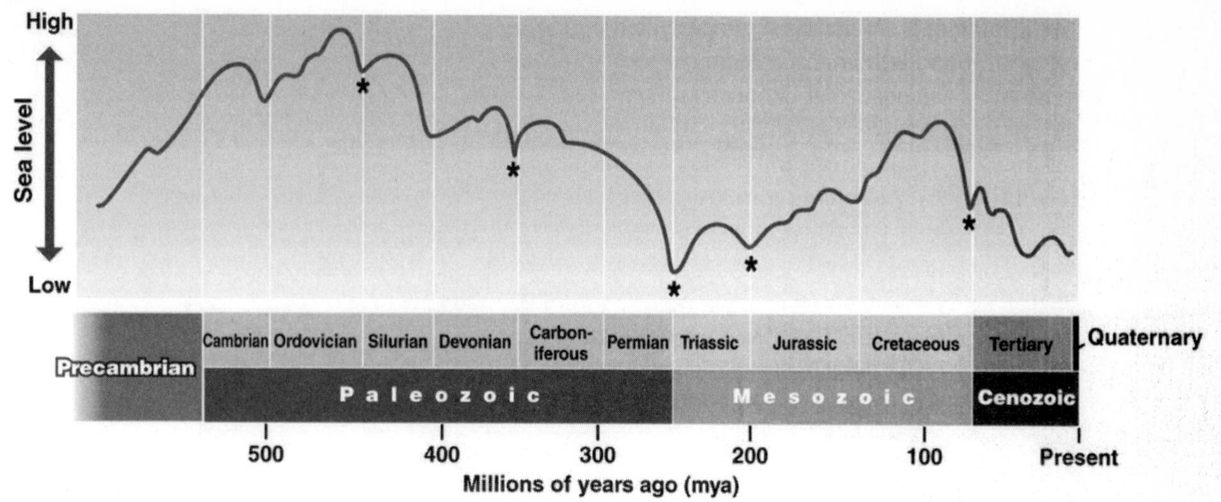

***PRINCIPLES OF LIFE*, Figure 18.3**

28. The figure above uses asterisks to identify five mass extinctions in the history of life on Earth during which a large proportion of the species living at the time disappeared. The line represents sea level rise and fall. Using information available in the figure, which of the following represents a true statement about the extinction events?

(A) The extinction at the end of the Permian period was the most devastating.
(B) There were three distinct extinction events that occurred during the Mesozoic era.
(C) Each of the mass extinctions coincides with a lowering of sea levels.
(D) Each of the eras (Precambrian, Paleozoic, Mesozoic, and Cenozoic) ended with a mass extinction event.

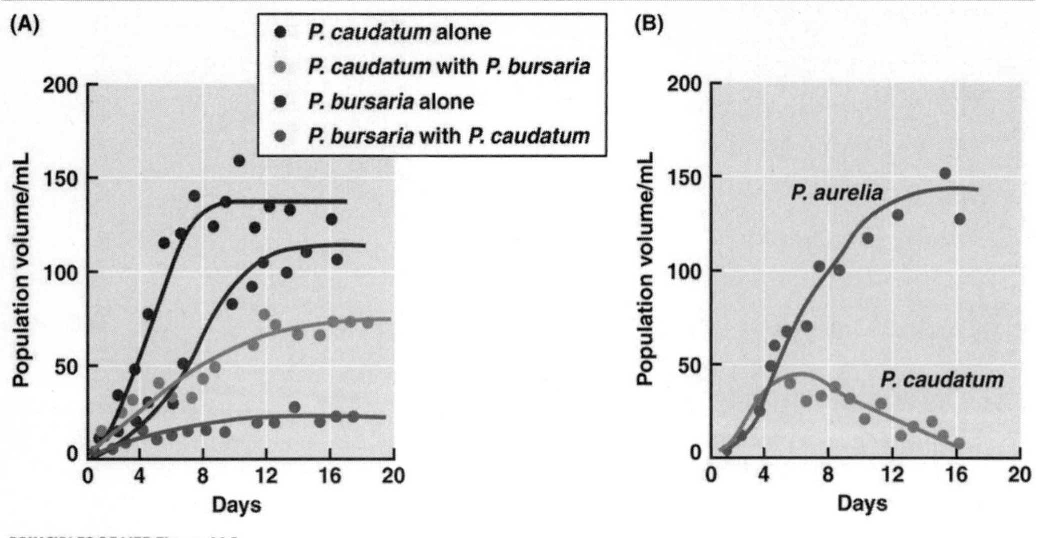

PRINCIPLES OF LIFE, **Figure 44.3**

29. The figure on the left (above) shows data collected from an experiment measuring population density of *Paramecium caudatum* and *Paramecium bursaria* (unicellular protists) when grown alone and grown together. The figure on the right shows data collected from a similar experiment using *Paramecium aurelia* and *Paramecium caudatum*. Which of the following statements is supported by the data in the figure?

(A) The population densities of both *P. caudatum* and *P. bursaria* are affected by the interspecific interaction known as competition.
(B) When *P. caudatum* and *P. bursaria* coexist, they achieve higher population densities than either would achieve alone.
(C) The presence of a competitor in this experiment increases population growth rate.
(D) *P. aurelia* population densities are unaffected by the presence of *P. caudatum*.

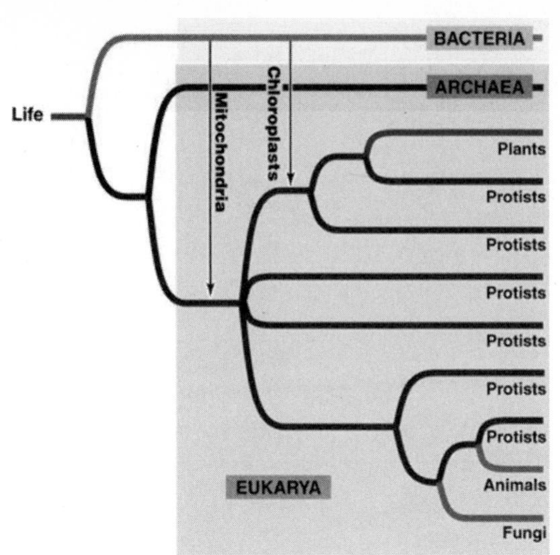

PRINCIPLES OF LIFE, Figure 1.4

30. The tree of life is a common depiction for the evolution of life from a common ancestor. Which of the below further supports the idea that life has evolved from a common ancestor?

(A) Aerobic respiration relies on oxygen gas.
(B) Aerobic respiration is common to many life forms.
(C) All cells use stomata to regulate their internal environments.
(D) All living organisms use DNA to store genetic information.

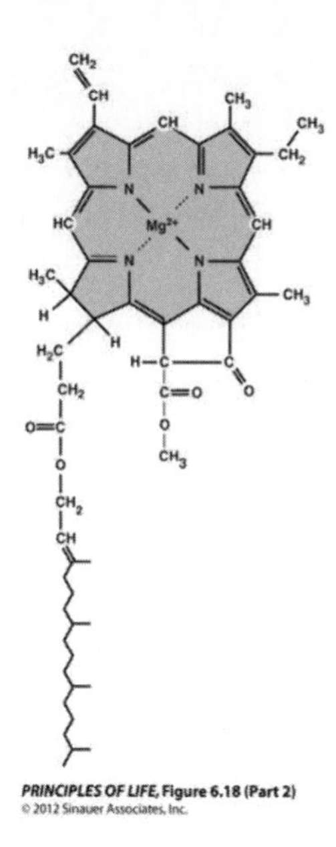

PRINCIPLES OF LIFE, Figure 6.18 (Part 2)

31. Chlorophyll is an important pigment involved in photosynthesis. Its structure is highly amphipathic with a hydrophilic ring structure and a very hydrophobic tail structure. Chlorophyll is found

(A) inside the stroma of chloroplasts
(B) in the thylakoid membranes of chloroplasts
(C) in the cytoplasm of plant cells
(D) inside the hydrophobic region of cell membranes

32. Choose the sequence that would be a "silent mutation" of an RNA strand composed of the nucleotides:

AUG AAA CAA AGC

(A) AUG AAG CAA AGC
(B) AUG CAA AGC
(C) UAG AAA CAA AGC
(D) AUG AAA CAA GAC

Second letter

First letter	U	C	A	G	Third letter
U	UUU UUC Phenyl-alanine UUA UUG Leucine	UCU UCC UCA UCG Serine	UAU UAC Tyrosine UAA Stop codon UAG Stop codon	UGU UGC Cysteine UGA Stop codon UGG Tryptophan	U C A G
C	CUU CUC CUA CUG Leucine	CCU CCC CCA CCG Proline	CAU CAC Histidine CAA CAG Glutamine	CGU CGC CGA CGG Arginine	U C A G
A	AUU AUC AUA Isoleucine AUG Methionine; start codon	ACU ACC ACA ACG Threonine	AAU AAC Asparagine AAA AAG Lysine	AGU AGC Serine AGA AGG Arginine	U C A G
G	GUU GUC GUA GUG Valine	GCU GCC GCA GCG Alanine	GAU GAC Aspartic acid GAA GAG Glutamic acid	GGU GGC GGA GGG Glycine	U C A G

PRINCIPLES OF LIFE, Figure 10.11

33. Meiosis results in the formation of gametes in vertebrates. One of the important side effects of meiosis is the incredible diversity produced as a result of this process. Which of the diagrams below depicts the formation of diversity during meiosis?

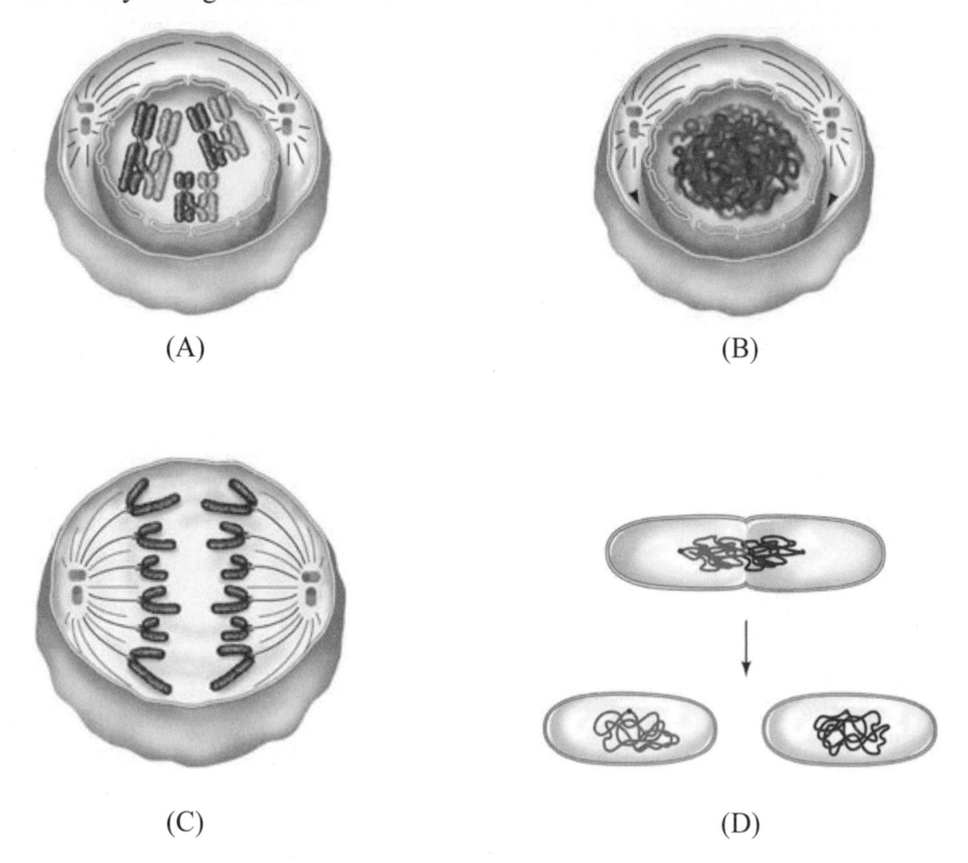

The figure below illustrates the cell cycle.

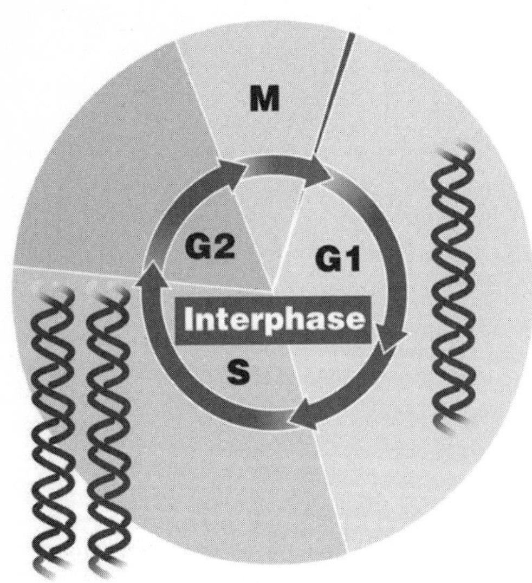

34. During the cell cycle, DNA is transmitted from one generation to the next. Which portion of the cell cycle makes it possible for identical copies of DNA to be passed on to each daughter cell?

(A) G0 or resting phase
(B) G1 or first growth phase
(C) S or synthesis phase
(D) G2 or second growth phase

35. One of the unifying features of life is that DNA is almost universally used as the primary means of carrying genetic information. Some viruses, such as HIV (which can cause AIDS), utilize RNA to carry their genetic information. Which of the below illustrates why the HIV virus is not a deviation from the idea that DNA is the universal carrier of genetic information.

(A) The HIV virus carries an enzyme called reverse transcriptase that converts the RNA of HIV to a DNA code.
(B) The membrane surrounding the HIV virus is derived from the host cell machinery.
(C) Viruses are not considered to be living organisms and utilize different molecules to carry their genetic code.
(D) Viruses are surrounded by a protein capsule that is coded for by the host cell DNA.

The illustration shows the structure for two amino acids, phenylalanine and cysteine.

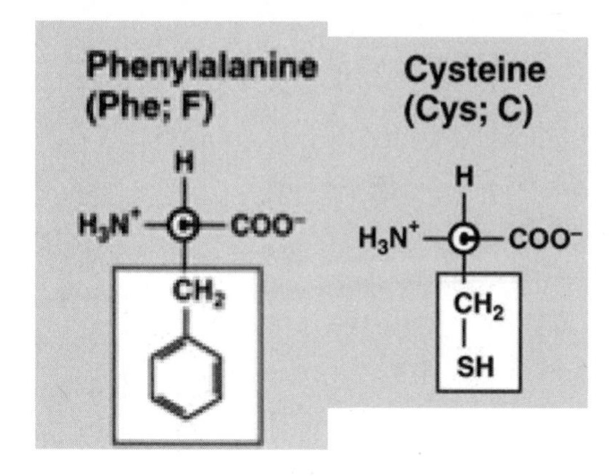

36. Which of the below explains what could happen if a substitution mutation of phenylalanine for cysteine occurs.

    (A) Both have a small "R" group, so there will be little to no change in the function of the protein.
    (B) Both phenylalanine and cysteine have very polar side groups so there will be little to no change in the protein shape.
    (C) The sulfur in cysteine forms disulfide bridges which will no longer occur causing major changes in the shape and stability of the protein.
    (D) The R group of phenylalanine will interact with an aqueous environment causing a change in the secondary structure of the protein.

37. The functions of macromolecules are an outgrowth of the structure of their sub-units. Which of the below macromolecules is correctly paired with its structural determining properties?

    (A) RNA: the shape of the molecule determines the sequence of its protein.
    (B) Protein: the shape of the molecule determines its function.
    (C) DNA: the sequence of monomers determines how much energy the molecule can carry.
    (D) Carbohydrate: the sequence of monosaccharides determines the shape of the molecule.

38. The graph below shows two typical growth rate patterns. The curve marked "B" depicts a logistic growth curve that is typical for many large mammals. Which of the below explains why the line levels out?

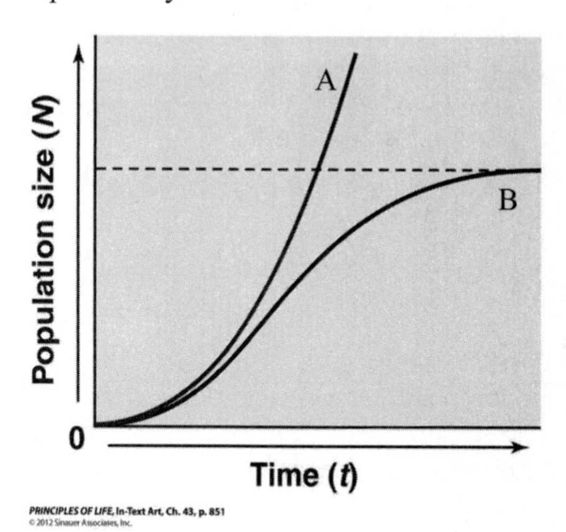

    (A) A population's density rarely exceeds the carrying capacity.
    (B) The logistic growth model is generally a result of density dependent controls.
    (C) The population reached a maximum and crashed with limited resources available.
    (D) Large mammals are limited in their population size because of their high metabolic rates.

Questions 39 – 40

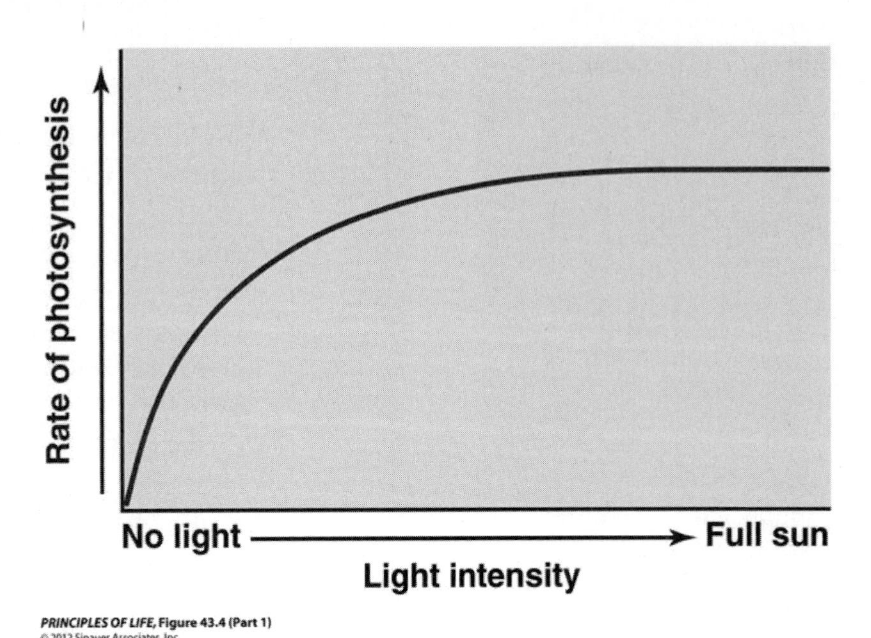

39. As light intensity increases, the rate of photosynthesis in a plant will increase until a maximum is reached. After this point, more light does not increase the plant's rate of photosynthesis. Which of the below is a correct explanation for this phenomenon?

    (A) The intense sunlight causes the enzymes for photosynthesis to denature and stop functioning.
    (B) The increased shade caused by the plant's growth slows its rate of photosynthesis.
    (C) Increased light increases the formation of hydrogen ions, and the acidity slows the plant's growth.
    (D) The light-harvesting apparatus of the plant becomes saturated and cannot make use of additional light.

40. A student decided to test the idea represented in the graph above. Which of the below would be the best method to test this idea?

    (A) Expose a group of plants to different colored lights and measure their change in height after a week.
    (B) Expose a group of different plants to sunlight throughout the day and measure their change in height after a week.
    (C) Expose three different groups of plants to low, medium, and high light-intensity, and then measure their change in height after a week.
    (D) Expose a plant to sunlight throughout the day and measure the rate of CO_2 production throughout the day.

41. Choose the most likely set of quantities related to a specific human gene.

(A) 2,378 nucleotides in DNA; 191 nucleotides in mRNA; 191 amino acids in protein
(B) 2,378 nucleotides in DNA; 573 nucleotides in mRNA; 191 amino acids in protein
(C) 2,378 nucleotides in DNA; 2,378 nucleotides in mRNA; 793 amino acids in protein
(D) 2,378 nucleotides in DNA; 793 nucleotides in mRNA; 264 amino acids in protein

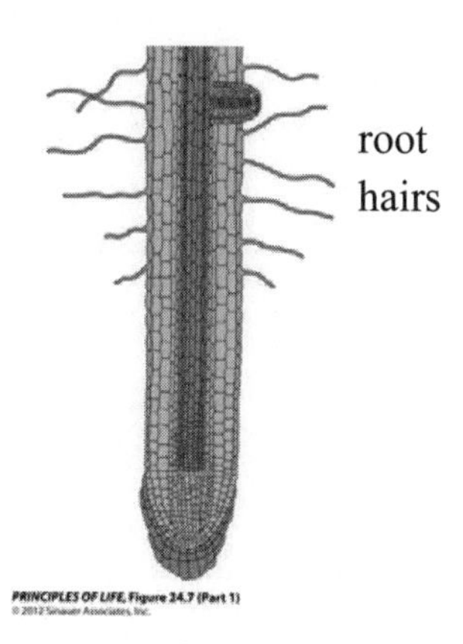

42. The roots hairs of plants are extensions of the membrane of epidermal root cells and

(A) enhance water conservation by closing the stomata on windy days much like hair cells in the cochlea move in response to ripples in the cochlear fluid.
(B) increase a plant's surface area for water absorption in much the same way as the intestinal villi increase the surface area for nutrient absorption in animals.
(C) protect the root tissue from insect attack much like nasal hairs protect the nasal cavity from insect attack.
(D) reduce water loss from the roots much like tight junctions between epithelial cells in the skin reduce water loss.

43. The origin of life on our planet is a persistent question in biology, one based on the premise that life arose from non-life. One hypothesis is that an "RNA world" preceded the development of proteins, which emerged later. This idea is partially support by observations that RNA molecules can work together in an enzyme-like manner. Since the origin of cells, however, protein synthesis takes place in ribosomes, which are made up of proteins and ribosomal RNA (rRNA). The RNA code for some of these ribosomal proteins has been recently analyzed and this code appears to be as old, evolutionarily, as the rRNA molecules in the ribosome, suggesting that

(A) Amino acids did not play a role in synthesizing the ribosomal proteins.
(B) DNA must have originated prior to the development of the ribosome, and a "DNA-RNA world" preceded the development of protein synthesis.
(C) The rRNA and the ribosomal proteins coded by RNA have worked together since the origin of the ribosome.
(D) Protein synthesis in pre-cellular world was directly controlled by DNA, without the involvement of RNA.

44. A cell that can give rise to the entire organism, including the extra-embryonic membranes, is considered to be a totipotent cell. Choose the description below that correctly represents this idea.

(A) A "theme" of science-fiction movies is that having any cell from an adult organism allows scientists to produce clones of that organism.
(B) Cells below your epidermis continually create new skin cells.
(C) Gametes are continually created by germ cells in the gonads.
(D) Cells in the nasal cavity that generate olfactory neurons.

45. The diagram to the right shows how a healthy kidney functions. Creatinine is a byproduct of muscle metabolism, and its presence in the blood increases following vigorous exercise. As long as the kidneys are working normally, creatinine levels in the blood return to low levels soon after exercise, most likely due to:

(A) low rates of glomerular filtration into the nephron
(B) secretion by the proximal tubule of the nephron
(C) reabsorption by the loop of Henle in the nephron
(D) the counter-current flows of blood and renal filtrate

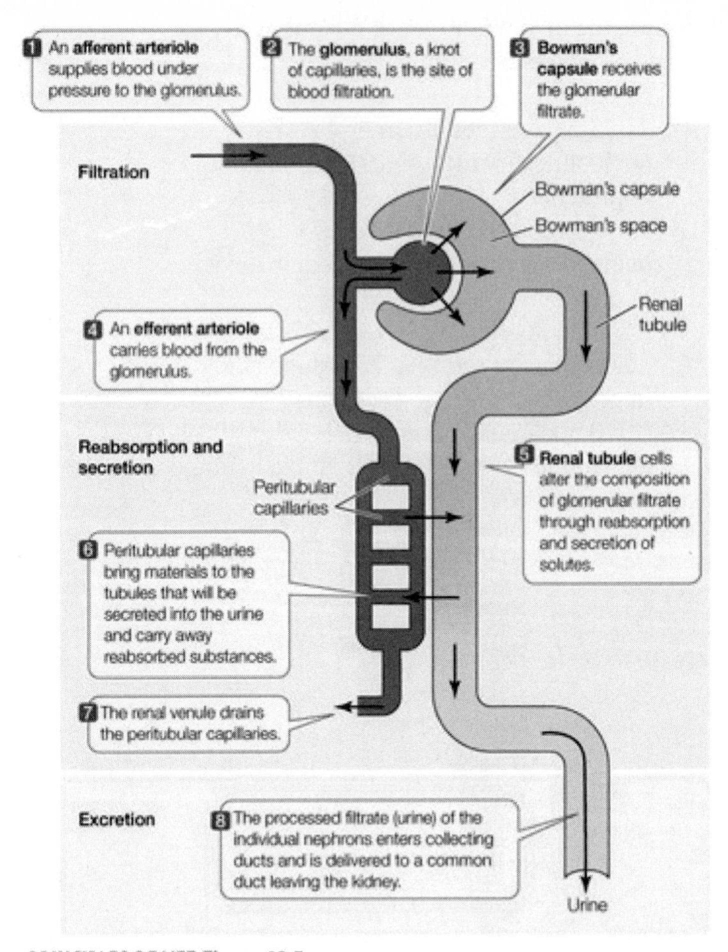

1 An afferent arteriole supplies blood under pressure to the glomerulus.
2 The glomerulus, a knot of capillaries, is the site of blood filtration.
3 Bowman's capsule receives the glomerular filtrate.
Filtration
Bowman's capsule
Bowman's space
Renal tubule
4 An efferent arteriole carries blood from the glomerulus.
Reabsorption and secretion
Peritubular capillaries
5 Renal tubule cells alter the composition of glomerular filtrate through reabsorption and secretion of solutes.
6 Peritubular capillaries bring materials to the tubules that will be secreted into the urine and carry away reabsorbed substances.
7 The renal venule drains the peritubular capillaries.
Excretion
8 The processed filtrate (urine) of the individual nephrons enters collecting ducts and is delivered to a common duct leaving the kidney.
Urine

***PRINCIPLES OF LIFE*, Figure 40.5**

46. The flow of the blood in the veins is unidirectional, i.e., one-way, back toward the heart. This is because veins have

(A) one-way valves
(B) smooth muscle that contracts in a coordinated sequence
(C) skeletal muscles that compress the veins
(D) autorhythmic contraction cycles

47. A culture of cells is poured into a high-speed blender with a mixture of oil and water. After the blender was run for several minutes, the fluid was allowed to settle. Next, the oily section was separated from the watery section. Choose the expectation that most accurately predicts results of this process.

(A) The watery component will include RNA, proteins, and enzymes; the oily component will include ions, membranes, and steroids.
(B) The watery component will include enzymes, membranes, proteins, and RNA; the oily component will include ions and steroids.
(C) The watery component will include ions, proteins, and steroids; the oily component will include enzymes, membranes, and RNA.
(D) The watery component will include enzymes, ions, proteins, and RNA; the oily component will include membranes and steroids.

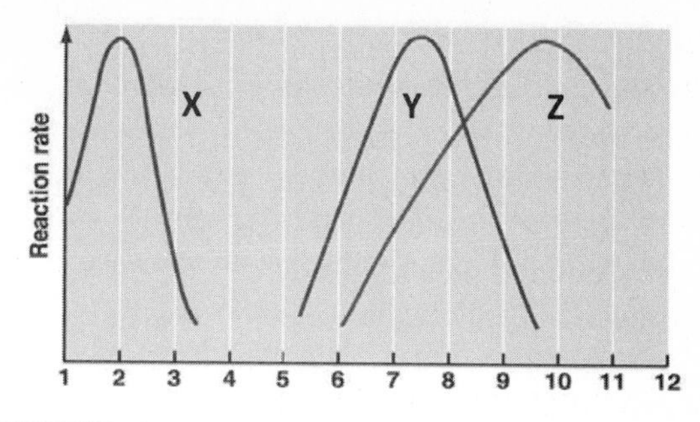

48. The graph shows reaction rates for three different (X, Y and Z) enzyme-catalyzed reactions. Assume that these enzymes are active in the digestive tract, and that their activity releases nutrients that can be absorbed. Enzymes in the stomach are activated by hydrochloric acid while a buffer from the pancreas neutralizes this acidity in the intestines. Choose the most correct statement.

(A) Enzyme X is denatured at pH 2, enzyme Y is denatured at pH 7.5, and enzyme Z is denatured at pH 9.6
(B) Enzyme X activates enzymes Y and Z
(C) Enzyme X is inhibited by feedback inhibition and enzymes Y and Z are enhanced by feed-forward activation.
(D) Enzyme X is normally active in the stomach, whereas Y and Z are active in the small intestine.

49. The hydrolysis of ATP, often called the currency of chemical energy in cells, is represented by this reaction:

ATP + H_2O → ADP + P_i + free energy

Choose the most correct set of descriptions of this reaction.

(A) The free energy released by this reaction can be used for cellular work.
(B) The hydrogen and oxygen atoms from the H_2O substrate are converted to phosphorus.
(C) The hydrolysis of ATP is an endergonic reaction.
(D) This reaction describes how light activates chlorophyll.

50. The concentration of solutes dissolved in the plasma of the blood is a particular concern when adding intravenous fluids to a patient. Suppose that a dog suffering from dehydration is brought into an emergency animal hospital. Choose the best description of the dog's RBCs at the beginning of treatment, and one-hour after treatment with an appropriate intravenous fluid.

(A) The RBCs will be larger than normal at the beginning of treatment and will become smaller after treatment.
(B) The RBCs will be normal-sized at the beginning of treatment and many will likely burst after treatment.
(C) The size of the RBCs will not be affected by dehydration or after treatment.
(D) The RBCs will be smaller than normal at the beginning of treatment and will become larger after treatment.

51. Mitosis occurs during the cell cycle, but it is not the complete cell cycle. Choose the event that is part of the cell cycle but not part of mitosis.

(A) The cell's DNA is replicated.
(B) The paired sister chromatids separate and move toward opposite poles.
(C) The chromosomes line up across the middle of the cell.
(D) The nuclear envelope forms around each set of chromosomes.

52. The relatedness of all living things on earth stems from the fact that all forms of life have

(A) the capacity to capture light energy
(B) cell motility based on flagellar contraction
(C) instructions coded in nucleotide sequences
(D) a nucleus with a membrane surrounding it

53. Mutations on the X-chromosome of a male offspring can be inherited from

(A) the father only
(B) the mother only
(C) either the mother or the father
(D) the father, but only if the mutation does not affect reproductive function

54. A group of researchers believe that they have extracted a new hormone from the cells of the pancreas. They attempt to characterize the hormone, and find it is fat soluble. Choose the most likely mechanism of action for the hormone's target cells.

(A) The hormone causes the target cells to respond rapidly, but the response is short-lived.
(B) The hormone rapidly triggers action potentials in the hypothalamus.
(C) The hormone binds to receptor proteins on the surface of the target cells.
(D) The hormone moves into a cell and causes changes in gene expression in the nucleus.

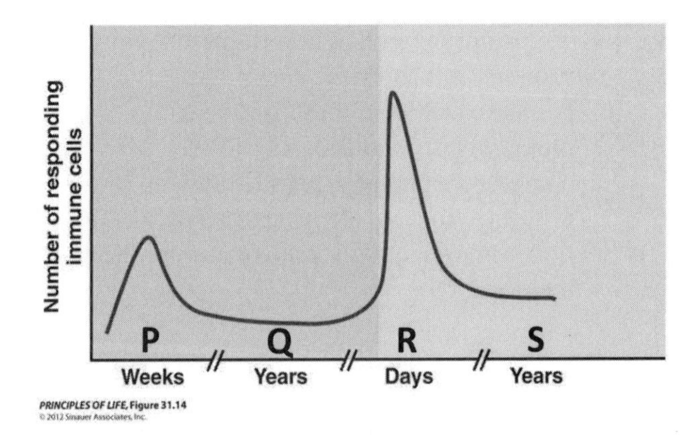

55. The cells of the immune system speed the body's recovery, following infection and reinfection. Choose the interval, as shown on the graph, when memory cells guide the production of antibodies.

(A) Q
(B) R
(C) S
(D) Q and S are both correct

56. Photosynthesis includes two sets of reactions. One set of reactions is directly light-dependent, while the other set, called the Calvin cycle, is only indirectly dependent on light. Which list has products of the Calvin cycle?

    (A) excited chlorophyll, ribulose bisphosphate carboxylase (rubisco), and electrons
    (B) carbon dioxide, hydrogen ions, and water
    (C) oxygen, ATP, and NADPH
    (D) glucose, ADP, and $NADP^+$

57. Insulin is a hormone released by the pancreas in response to eating sugary foods, and its actions help glucose become absorbed by its target cells in the body. Humulin™, a biotechnology product with actions identical to insulin, can be used to treat certain types of *diabetes mellitus.* Choose the best description of the procedure that is currently used to make biotechnology products such as Humulin™ on a scale large enough to treat many patients.

    (A) A normal copy of the gene whose product is of interest is inserted into the genome of a farm animal, and the mRNA of interest is isolated from the animal's blood, urine or milk.
    (B) Patients undergo gene-transfers to replace a defective copy of a gene with one that is not mutated.
    (C) Bacteria undergo transformation and ligation to incorporate genes whose products are of interest.
    (D) The insulin hormone is isolated from the urine of horses that are genetically bred to overproduce this hormone.

58. Although molecular nitrogen gas (N_2) makes up about 78% of the atmosphere, nitrogen is a limiting nutrient in many ecosystems. Plants cannot absorb nitrogen gas (N_2) and assimilate it into amino acids. Nitrogen atoms, in various molecular forms, move through the nitrogen cycle because there are organisms capable of using N_2 directly, which are:

    (A) bacteria
    (B) protists
    (C) insects
    (D) fish

Questions 59-63

Behavioral and genetic studies of fruit flies (*Drosophila*) have implicated similar genes and genetic networks in the behaviors of other organisms, including humans. In one such experiment, fruit flies were housed in two different feeding arrangements for several generations and then tested for their mating preferences. Assume that the feeding arrangements were either chocolate-flavored food or cinnamon-flavored food.

59. Select the prediction that indicates that the feeding arrangements could lead to the formation of new species.

(A) Male and female flies reared on cinnamon were found to be unable to detect the odor of cinnamon following several generations of rearing on cinnamon-flavored foods.
(B) The female flies reared on cinnamon-flavored foods showed little interest in mating with males, regardless of the male's food background.
(C) The male flies reared on chocolate-flavored foods showed little interest in mating with females, regardless of the female's food background.
(D) Given a choice between mating with a fly of the same or different food background, male and female "test" flies selectively mated only with flies that had been reared on the same food as the "test" flies.

UNTETHERED "TEST" SUBJECTS	TETHERED "CHOICE" PARTNERS	NUMBER OF MATINGS
Cinn Males (n=20)	Cinn Females	16
	Choc Females	4
Cinn Females (n=20)	Cinn Males	14
	Choc Males	6
Choc Males (n=20)	Cinn Females	5
	Choc Females	15
Choc Females (n=20)	Cinn Males	3
	Choc Males	17

60. Shown are hypothetical data for the experiment described in the stem. The data are observations of mating behavior when a fly (identified as *Cinn* for cinnamon rearing and *Choc* for chocolate rearing) is offered a choice between two tethered mates, one from each food background.

Choose the correct interpretation of the data presented in the table.

(A) Mating preferences were directed to tethered "choice" partner flies that were reared on foods different than that consumed by the tethered "test" subjects.
(B) Mating preferences of the flies is likely based a on single gene that is recessive.
(C) During the generations the two populations were reared on different foods, reproductive isolation appears to have developed between the two populations.
(D) Mating preferences in the "test" flies are solely dependent on the odor of the "choice" partner, and both populations of flies are able to chemically sense both flavors of food.

61. Suppose that 20 more test subjects from each background of feeding arrangements were tested in a different setting. In the second round of mating tests, assume that only one mating partner was available to the test subjects. Assuming that the results in the table above are correct, which prediction matches the previous findings?

(A) Test subjects will start mating sooner with partners of similar food background than with those of different food background, with some matings observed in each type of pairing.
(B) There would be no mating with partners that have a food background different that the test subject, but rapid and abundant mating with partners of with a food background similar to that of the test subject.
(C) Mating would occur in all combinations, due to the lack of alternatives for expressing preferences.
(D) The pacing, sequence and details of all mating interactions would be identical.

62. Consider that the isolation between the two populations in these studies continued for hundreds of generations, rather than just 10 generations. Chose the prediction that would satisfy the conclusion that the two populations had indeed become two independent species.

(A) The results would change very little, if at all, with only the beginning so differences apparent.
(B) The offspring of flies from different backgrounds would be unlikely to show any mating preference for one population or the other.
(C) After having diverged genetically for so many generations, the two populations of flies would now readily mate with partners of food background different from their own.
(D) A limited number of matings would continue to be possible between two individuals of different food backgrounds, but no viable offspring would emerge from such matings.

63. Upon closer inspection, assume that behavioral differences in food preference and in copulatory behavior were observed in comparing the two populations of fruit flies. If the two populations were mixed together in a chocolate and cinnamon enriched environment, with chocolate food resources physically separated from cinnamon food resources, the two populations of flies might continue to diverge. Given enough time, the flies would likely first differ in which of these physical characteristics?

(A) The ability to see colors.
(B) The anatomy of the reproductive system.
(C) The ability to fly.
(D) The respiratory system would change.

Part B Directions: Part B consists of six questions requiring numeric answers. Calculate the correct answer for each question.

1. The storage roots of radishes have been found to be long, round or oval. When geneticists cross long with oval radishes, the cross produces 348 long and 356 oval radishes. Crosses between round and oval produced 195 round and 189 oval radishes. Crosses between long and round produce only oval radishes. Crosses between two oval radishes produced 51 long, 112 oval and 41 round.

$$X^2 = \sum \frac{(\text{observed} - \text{expected})^2}{\text{expected}}$$

Calculate the chi-squared value using the data set that best supports the hypothesis that the oval radish is heterozygous for the radish shape gene. Give your answer to the nearest hundredth.

2. A human cell can be fused to a mouse cell in the laboratory, forming a single large cell (heterokaryon). This phenomenon was used to test whether membrane proteins can diffuse throughout the plasma membrane as a test of the fluid mosaic model.

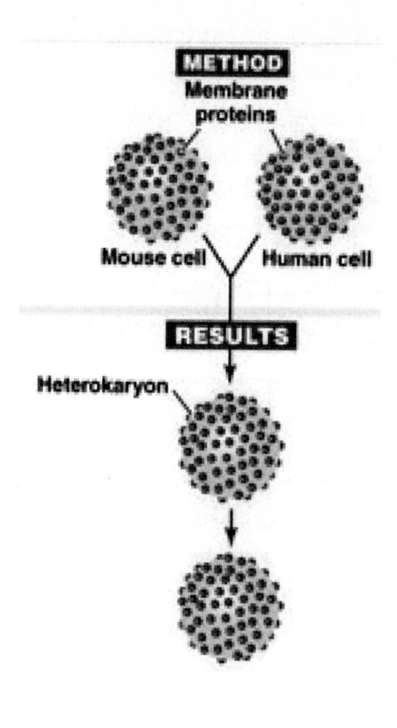

A human cell and a mouse cell were fused together to form one cell called a heterokaryon. Initially the mouse and human membrane proteins were on opposite sides of the heterokaryon. As time progressed, the proteins rapidly diffused throughout the membrane. The experiment was repeated at different temperatures with the following results.

Temperature (°C)	Cells With Mixed Proteins (%)
0	0
15	8
20	42
25	77

Calculate the rate of change from 15°C to 20°C. Give your answer to the nearest tenth per degree celsius.

3. Fiddler crabs are common residents of saltmarshes feeding on dead partially decomposed saltmarsh grass (detritus). A group of students visited a saltmarsh. Using one square meter quadrats they randomly sampled 6 different locations and counted the number of fiddler crabs in in each one square meter plot. The data are presented in the table.

Quadrat	Fiddler crabs
1	16
2	5
3	12
4	7
5	2
6	8

The size of the marsh is 20 Km^2. Calculate the average number of 1,000 fiddler crabs per square Km in the marsh. Give your answer to the nearest thousand.

4. In sheep black wool is a recessive trait. In a flock of 250 sheep, 12 sheep were found to have black wool. Assume that the population is in Hardy-Weinberg equilibrium. Calculate the percent of the population that is heterozygous for this trait. Give your answer to the nearest tenth.

5. As an action potential moves along a neuron, a sudden influx of Na^+ ions causes a brief reversal of membrane potential. Shortly after the reversal, the membrane potential is briefly more negative than was prior to the onset of action potential, a phase called the undershoot. Examine this graph to determine the minimum membrane potential during the undershoot.

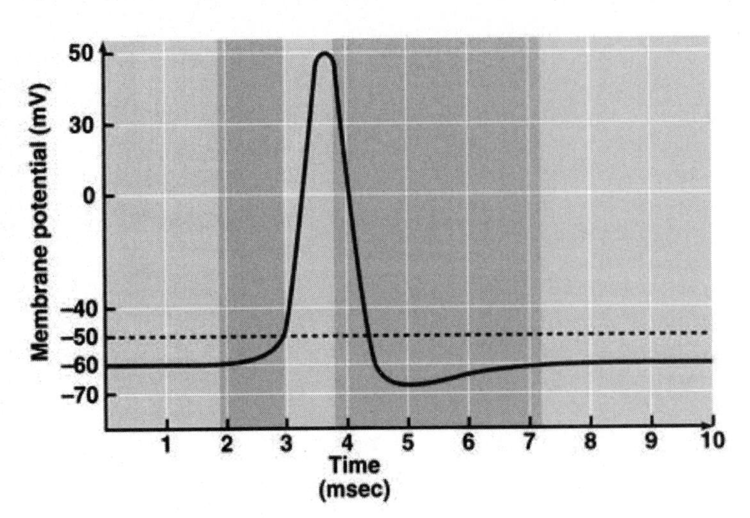

6. The axon of a giant squid can be up to 1mm in diameter and up to 2 meters in length. Calculate the SA to volume ratio for this cell – express your answer to the nearest whole number and as a fraction over 1 (as in X/1).

Some useful equations

Surface area of a circle = πr^2

Volume of a cylinder = $\pi r^2 h$

Surface area of a cylinder = $2 \pi r^2 + 2 \pi r h$

Surface area of a cube = 6 x length x width

Volume of a cube = length x width x height

END OF SECTION I

IF YOU FINISH BEFORE TIME IS CALLED, YOU MAY

CHECK YOUR WORK.

DO NOT GO ON TO SECTION II UNTIL YOU ARE TOLD TO DO SO.

BIOLOGY: Section II
8 Free-Response Questions
Time–80 Minutes

Directions: Questions 1 and 2 are long free-response questions that should require about 20 minutes each to answer. Use blank sheets of notebook paper to write your answers.

Questions 3 through 8 are short free-response questions that should require about 6 minutes each to answer. Read each question carefully and write your response in the space provided following each question. Only material written in the space will be scored. Answers must be written out. Outline form is not acceptable. It is important that you read each question completely before you begin to write.

1. The graph below shows the absorption-wavelength curves for two different organisms: *Ulva* (a green algae) and purple sulfur bacteria.

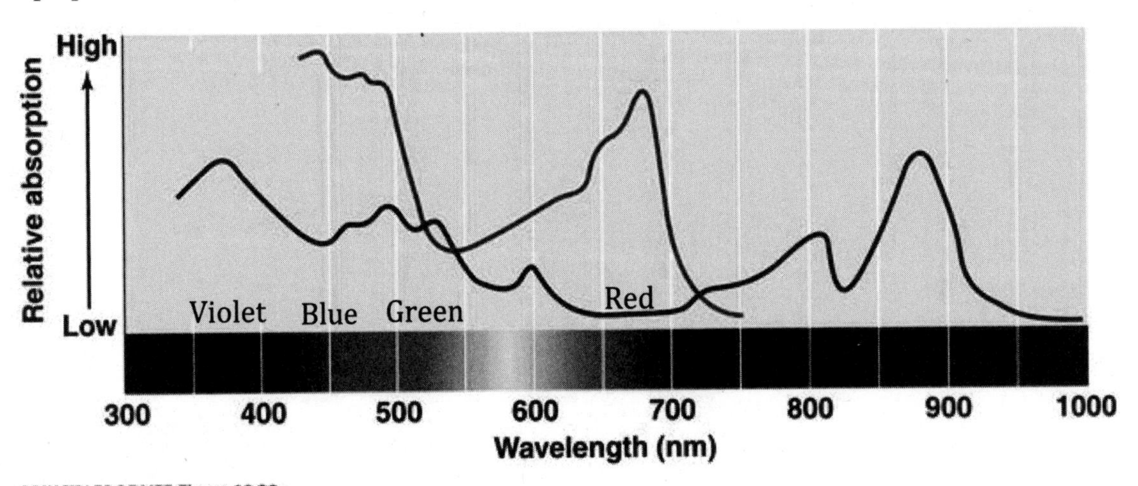

PRINCIPLES OF LIFE, **Figure 19.20**

450nm = blue
525nm = green
575nm = yellow
650nm = orange
700nm = red

(a) **Label** the two curves for *Ulva* and purple sulfur bacteria above.

(b) **Explain** how purple sulfur bacteria can carry out photosynthesis.

(c) Assume that a newly discovered species of algae that has evolved in the same region as the two shown but is not in competition with either of them for light. On the original graph, **draw** its expected absorption-wavelength curve, with close attention to the wavelengths of maximum absorption. **Explain** why your curve is shaped the way it is.

(d) **Discuss** how each of the three "light-harvesting photosynthetic pigments" in your answer above is specifically affected by the energy spectrum of light.

2. Sexual reproduction is the primary mode of reproduction among eukaryotes, including humans.

(a) **Discuss** THREE genetic and/or evolutionary benefits of sexual reproduction.

(b) **Discuss** TWO genetic and/or evolutionary costs of sexual reproduction.

(c) Among the approximately 23,000 genes in humans, about 35 are found only in one sex. **Identify** which chromosome these are found on and **propose** a function for these genes.

(d) During sexual reproduction, occasionally mutations occur that disrupt the normal development of an individual. Klinefelter's syndrome describes a condition of an individual who has a Y chromosome and two X chromosomes. **Construct** a drawing that shows how this arrangement of sex chromosome occurs during meiosis. Be sure to **explain** your drawing with written text.

3. Enzymes are critical for life. Below is a graph showing the rate of an enzyme-catalyzed biochemical reaction at $37°C$, the degradation of H_2O_2 by catalase.

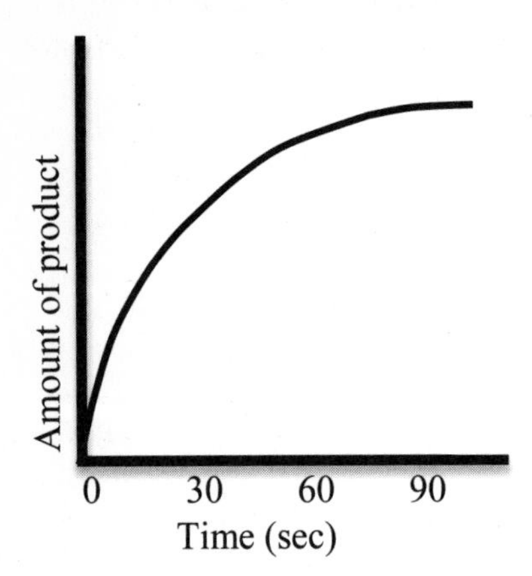

Redraw (below) the expected effects on the curve for each of the following environmental changes. **Explain** each graph in one or two sentences.

(a) The environment is cooled 20°C before repeating the experiment.

(b) After 30 seconds, a competitive inhibitor is added to the solution being tested in the experiment.

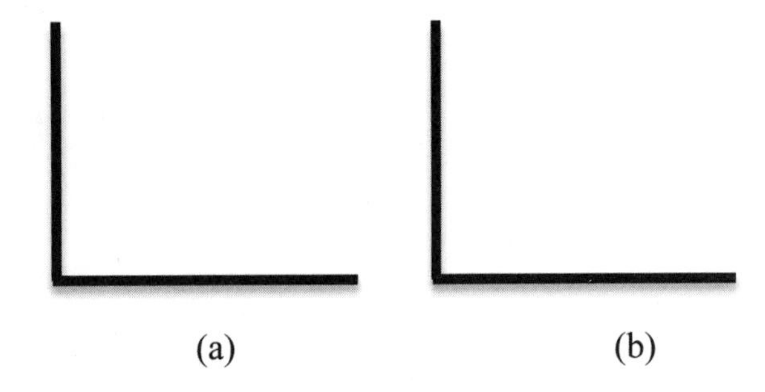

(a) (b)

4. In higher eukaryotes, the chemical ouabain can impair the activity of the "sodium pump" (Na^+/K^+-ATPase). Impaired sodium pumping can, in turn, impair cellular processes that depend on a gradient of sodium ions across the cell membrane. In lower doses, ouabain can have therapeutic effects following cardiac arrest. **Explain** why or why not the effects of ouabain are expected to be the same for nerve cells from a giant squid and from a mouse.

5. The introduction of new species often has devastating consequences on native populations.

Choose ONE of the below and **discuss** how it was introduced and the consequences of its introduction.

- Dutch elm disease
- Potato blight
- Small pox

6. The diagram to the right shows the relationship between the diversity of a grassland area and the plant biomass produced over a three-year time span.

(a) Briefly **explain** what biodiversity measures.

(b) **Summarize** the main idea shown by the graph.

(c) **Propose** an explanation as to why biomass is greater with a greater biodiversity.

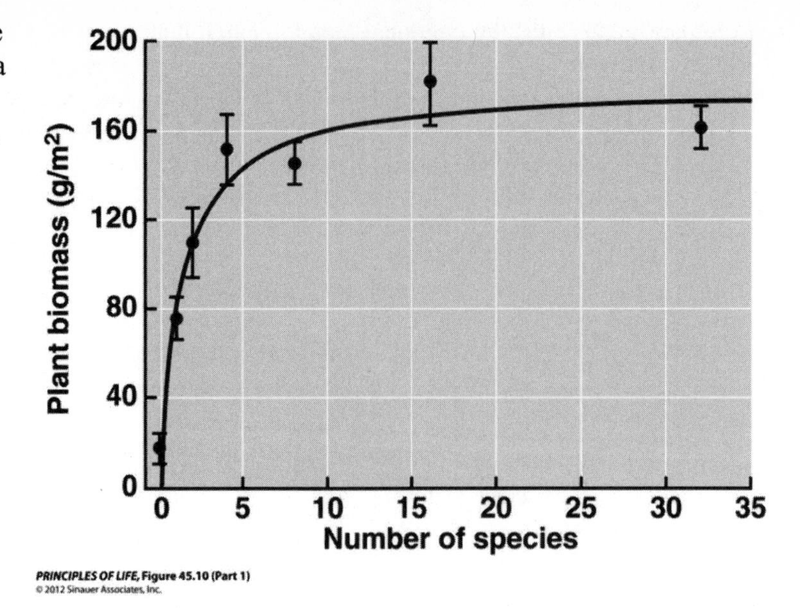

7. Plants and animals are influenced in many ways by environmental conditions, including seasonal and long-term temperature changes. Choose ONE plant and ONE animal, and, for EACH, **describe**

(a) How seasonal changes in temperature affect the phenotype.

(b) Its likely responses to global climate warming.

8. A population's ability to respond to changes in the environment is affected by its genetic diversity.

(a) **Explain** why genetic diversity is beneficial to the long-term survival of a population.

(b) Choose ONE of the examples below and **discuss** why that population is at risk for extinction.

- California condors
- Black-footed ferrets
- Prairie chickens
- Tasmanian devils and infectious cancer

STOP
END OF EXAM

Section I Answer Key

#	Answer	LO
1	C	3.24
2	B	3.34
3	A	3.15
4	C	3.45
5	D	3.4
6	D	2.16
7	A	3.25
8	A	2.21
9	B	4.5
10	B	3.36
11	D	3.12
12	C	1.4
13	B	2.24
14	A	2.28
15	C	2.23
16	B	4.19
17	D	1.9
18	C	1.2
19	A	1.10
20	B	4.8
21	C	2.5
22	D	2.8
23	C	3.11
24	D	1.18
25	A	1.14
26	B	1.17
27	B	1.20
28	C	1.20
29	A	4.13
30	D	1.17
31	B	4.4
32	A	3.6

#	Answer	LO
33	A	3.10
34	C	3.9
35	A	3.3
36	C	4.1
37	B	4.2
38	B	4.13
39	D	2.5
40	C	2.4
41	B	3.3
42	B	2.7
43	C	1.30
44	A	3.5
45	A	4.8
46	B	4.18
47	D	2.10
48	D	4.17
49	A	2.1
50	D	2.12
51	A	3.8
52	C	1.32
53	B	3.17
54	D	2.11
55	B	2.29
56	D	2.1
57	C	3.5
58	A	4.16
59	D	1.25
60	C	1.24
61	A	1.26
62	D	1.22
63	B	1.22

GRID-IN SECTION			
#	**Answer**	**Acceptable Range**	**LO**
1	**2.94**	**2.90 – 3.00**	**3.14**
2	**6.8**	**6.8**	**2.11**
3	**8333**	**8300 – 8350**	**4.19**
4	**34.2**	**33-35**	**1.1**
5	**-68**	**-66 – -69**	**2.12**
6	**2/1**	**1.8 – 2.2 / 1**	**2.6**

Sample Grid-In Calculations

1. Answer = 2.9-3.0 (actual is 2.94)

2. Answer:

$$\frac{Y2 - Y1}{X2 - X1} = \frac{DV = \%\ \text{mixed}}{IV = \text{temp}} = \frac{42-8}{20-15} = 34/5 = 6.8$$

3. Answer = 50/6 = 8.33 crabs / square meter x 1000 x 1000 = 8,333 thousand crabs / Km^2
The grid in sheets do not allow scientific notation and only 5 numbers…

4. Answer = q2 = 12/250 = .048
q = .219; p = .781
2pq = 2 (.219) (.781) = .342 or 34.2%

5. Answer = -68 (acceptable range is -66 to -69)

6. Answer = SA : volume
Hint: convert 2 meters to 2,000 millimeters

SA: Volume
2 (3.14) $(2{,}000\ mm)^2$ + 2 (3.14) (1mm) (2,000mm) : (3.14) $(2{,}000mm)^2$ (1 mm)
25120000 + 12560 : 12560000 mm^3
25132560 mm^2 : 12560000 mm^3
2:1

Sample Rubrics for Grading Section II of the Practice Exam

1. This is a 10-point question; must earn at least one point in each section to score as 10.

(a) Up to 2 points:

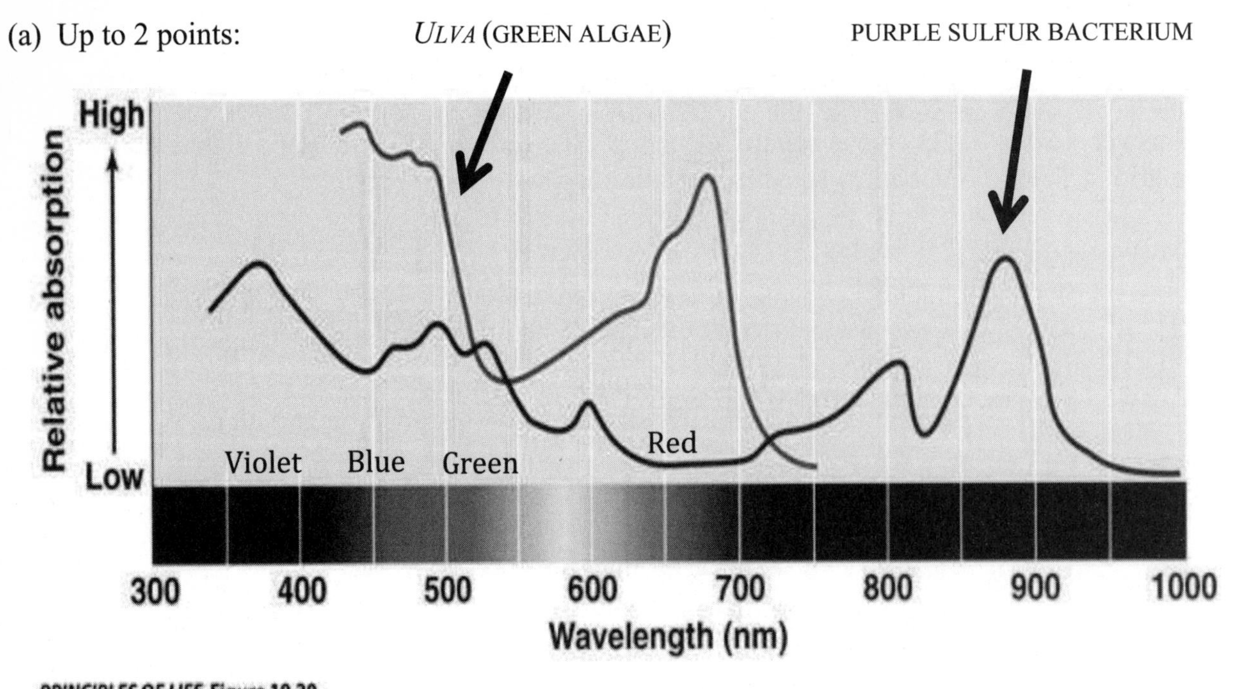

PRINCIPLES OF LIFE, Figure 19.20

(b) Up to 3 points (underlined):

- Instead of producing oxygen from water (H_2O), as do green plants,
- the purple sulfur bacteria produce sulfur from hydrogen sulfide (H_2S) as electron donor.

(c) Up to 4 points:

- Any curve added to graph that shows a pattern where relative absorption peaks at wavelengths different from the peaks of green algae and purple sulfur bacteria.
- Lack of "competition for light" by hypothetical species explains that its peaks of relative absorption are different from the peaks on curve for the *Ulva* and purple sulfur bacteria.
- Does not absorb at all wavelengths

(d) Up to 3 points:

- Green algae's pigment, chlorophyll, absorbs at about 450 and 650 nm for photosynthesis.
- Bacteriochlorophyll, in purple sulfur bacteria, is activated at long wavelength (~880 nm).
- Putative pigment of fictional bacteria activated at point indicated on student's graph.

2. This is a 10-point question; must earn at least one point in each section to score a 10.

(a) Up to 4 points; one point for elaboration:

- Offspring can vary from parents.
- Crossing over and recombination can produce diverse assemblages of genes.
- Crossing over and recombination can "repair" harmful mutations.
- Speeds up the rate of evolutionary change

(b) Up to 3 points; one point for elaboration:

- Crossing over and recombination can separate adaptive genes from each other.
- Female parents invest heavily in offspring whose genes have been "diluted" by fathers.
- Can be hard to find a mate and to accomplish gamete exchange.
- Having two parents required for each offspring slows population growth rate

(c) Up to 3 points:

- Male humans have a Y chromosome, females do not.
- The Y chromosome has about 35 genes on it, all related to testicular function.
- Female has two X chromosomes, but one is inactivated.

(d) Up to 3 points:

- Shows all ova have one X
- Shows half of sperm have X
- Shows half of sperm have Y

Klinefelter's karyotype shown. →

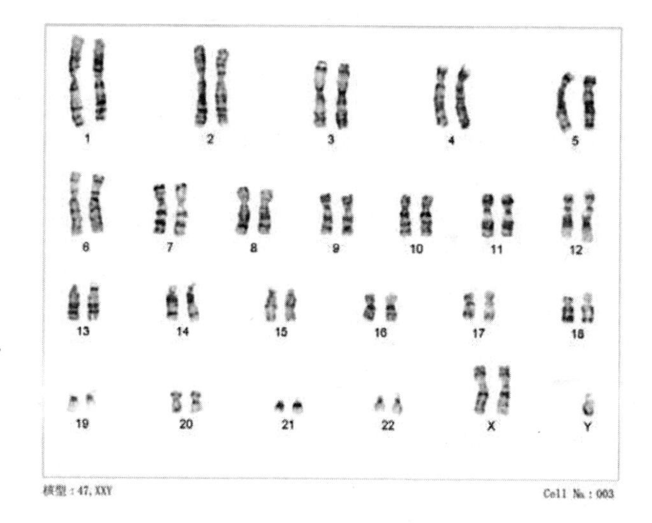

3. Up to 4 points.

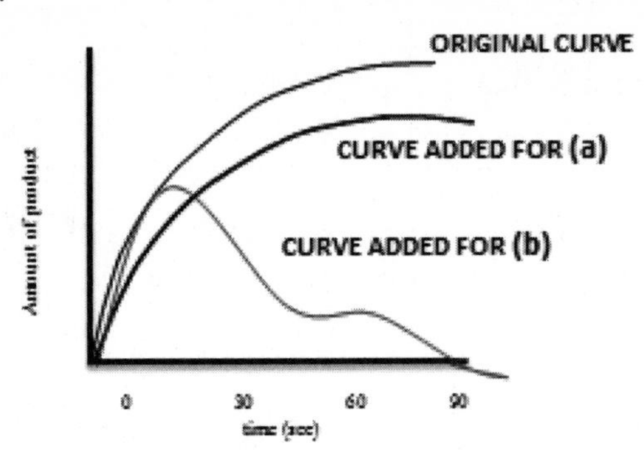

(a) the curve added to show the effects of reducing the temperature to 20°C shows a reduction in the rate of product formation due to reduced molecular interactions at cooler temperatures.
(b) the curve added to show the effects of adding a competitive inhibitor at t = 30 sec demonstrates reduced activity in the reaction due to competition for the binding site. (Note: The shape of this may vary, but it will show a marked decrease at 30 sec. possibly to zero, depending on how much competitor there is versus substrate.)
One point for each curve and one point for each explanation.

4. Up to 3 points:

Oubain inhibits Na^+/K^+-ATPase, <u>reducing</u> the sodium gradient across the membrane.
Any process utilizing the sodium gradient can be impaired, e.g.,

Calcium storage
Any type of sodium-dependent <u>secondary active transport</u>
The generation of action potentials\

The effects would be the same for mice and squid

5. Up to 4 points, only one may be answered, the first one attempted is scored:

Dutch elm disease is a fungus and is spread by the elm bark beetle. It arrived in trees accidently imported to North American from Europe (1928), but most likely had its initial origins in Asia. The disease greatly altered the types of trees found in the USA. Interestingly, elm declines have also been noted in the fossil record 6,000 and 3,000 years ago, suggesting that this is not the first time this has happened. This is another example of succession showing how forests change over time.

The potato blight, a fungus, led to the Irish Famine in the 1840s that killed a million or more people; how this fungus entered Ireland is unknown. There is evidence that it began in South America and spread via wind across North America. The earliest records of the blight in Philadelphia are 1843. A shipment of seed potatoes bound for Belgium in 1845 may well have accidently caused the blight.

Small pox is due to a virus that causes a serious, and sometimes fatal, infection of the skin. It has been part of human civilization for centuries, but ~150,000 persons in N. America died as a result of the smallpox epidemic of the late 1700s. There is also evidence that earlier travelers to N. America inadvertently introduced the virus to native populations wiping out millions of native North American Indians well before Christopher Columbus arrived. Some accounts also have European settlers deliberately giving infected blankets to native North American populations as germ warfare.

6. Up to 4 points:

(a) Biodiversity measures the diversity of species present in a given community. Biodiversity encompasses two ideas, one is the number of each species present as well as the number of different species present (species richness).

(b) As the number of plant species increases in a test plot, plant biomass also increases, up to a point, at which maximum supportable productivity is possible. The upper limit is on how much sunlight can reach the plants in the community under consideration.

(c) The primary explanation for increased biodiversity supporting greater biomass is that there are more niches for plants and animals to live in. A greater variety of species allows for different plants with different functional roles. Different plants have different traits with varied adaptations. Some perform better in cooler weather, some better in the shade, while others have varying associations with nitrogen fixing bacteria.

7. Up to 3 points:

(a) Any seasonal plant – possible answers include fruiting, leaf drop (deciduous species), responses to drought, etc.
Any animal in a seasonal environment- seasonal reproduction, seasonal changes in metabolic rate, hibernation, etc.

(b) Global warming could disrupt the seasonal patterns, extending some physiological processes too late in the season, to the extent where the individual suffers a reduction in fitness. For plants, earlier budding or flowering are trends currently seen. For animals and plants, ranges may shift as average temperatures shift.

8. Up to 3 points:

(a) Genetic diversity allows plants and animals to "hedge their bets" on what phenotype, to the extent determined by genotype, might be successful when environmental conditions shift.

(b) California condors- very few reproductive individuals persist; might not have enough genetic diversity to survive a disease outbreak.
OR
Black-footed ferrets- lack of undisturbed natural areas with prey (prairie dogs) have taken this predator to the edge of extinction.
OR
Prairie chickens- habitat loss (tall-grass prairies) has greatly curtailed its numbers; lacks adequate undisturbed areas for mating and rearing chicks.
OR
Tasmanian devils and infectious cancer – transmissible parasitic cancer, declines of up to 50-65% in areas across Australia. Density dependent. Transmitted by one devil biting another, pathology still unknown. An example of genetic drift reducing the diversity of the population.

Correlation of Free Response Questions to the Curriculum Framework

Question #	LO's Covered	Number of Points
1	2.5	10
2	3.27, 3.28, 3.12	10
3	4.17	4
4	1.16	3
5	4.21	4
6	4.27	4
7	4.23 & 4.24	3
8	4.25	3